Birkhäuser

Trends in Mathematics

Trends in Mathematics is a series devoted to the publication of volumes arising from conferences and lecture series focusing on a particular topic from any area of mathematics. Its aim is to make current developments available to the community as rapidly as possible without compromise to quality and to archive these for reference.

Proposals for volumes can be submitted using the Online Book Project Submission Form at our website www.birkhauser-science.com.

Material submitted for publication must be screened and prepared as follows:

All contributions should undergo a reviewing process similar to that carried out by journals and be checked for correct use of language which, as a rule, is English. Articles without proofs, or which do not contain any significantly new results, should be rejected. High quality survey papers, however, are welcome.

We expect the organizers to deliver manuscripts in a form that is essentially ready for direct reproduction. Any version of TeX is acceptable, but the entire collection of files must be in one particular dialect of TeX and unified according to simple instructions available from Birkhäuser.

Furthermore, in order to guarantee the timely appearance of the proceedings it is essential that the final version of the entire material be submitted no later than one year after the conference.

More information about this series at http://www.springer.com/series/4961

Willy Dörfler • Marlis Hochbruck •
Dirk Hundertmark • Wolfgang Reichel •
Andreas Rieder • Roland Schnaubelt •
Birgit Schörkhuber
Editors

Mathematics of Wave Phenomena

Birkhäuser

Editors
Willy Dörfler
Karlsruhe Institute of Technology
Baden-Württemberg
Karlsruhe, Germany

Marlis Hochbruck
Karlsruhe Institute of Technology
Baden-Württemberg
Karlsruhe, Germany

Dirk Hundertmark
Karlsruhe Institute of Technology
Baden-Württemberg
Karlsruhe, Germany

Wolfgang Reichel
Karlsruhe Institute of Technology
Baden-Württemberg
Karlsruhe, Germany

Andreas Rieder
Karlsruhe Institute of Technology
Baden-Württemberg
Karlsruhe, Germany

Roland Schnaubelt
Karlsruhe Institute of Technology
Baden-Württemberg
Karlsruhe, Germany

Birgit Schörkhuber
Karlsruhe Institute of Technology
Baden-Württemberg
Karlsruhe, Germany

ISSN 2297-0215 ISSN 2297-024X (electronic)
Trends in Mathematics
ISBN 978-3-030-47173-6 ISBN 978-3-030-47174-3 (eBook)
https://doi.org/10.1007/978-3-030-47174-3

Mathematics Subject Classification: 35-06, 35Lxx, 35Qxx, 65-06, 65Mxx, 65Nxx, 78-06

This book is published under the imprint Birkhäuser, www.birkhauser-science.com, by the registered company Springer Nature Switzerland AG.
The registered company address is: Gewerbestrasse 11, 6330 Cham, Switzerland

Preface

The mathematical modeling, simulation, and analysis of wave phenomena entail a plethora of fascinating and challenging problems both in analysis and numerical mathematics. During the past decades, these challenges have inspired a number of important approaches, developments, and results about wave-type equations in both fields of mathematics.

We organized the *Conference on Mathematics of Wave Phenomena* at Karlsruhe, July 23–27, 2018, to bring together international experts in this field with different backgrounds. The conference was a great success attracting about 250 participants, among them many leading scientists and promising young researchers working in analysis and numerics of wave-type problems and their applications. In their lectures, they presented recent high-level research in a wide range of topics which stimulated the transfer of ideas, results, and techniques within this exciting area.

This volume reflects the contents of the conference, although many of the contributions contain material that goes beyond the results presented in talks. The papers treat various types of nonlinear Schrödinger and wave-type systems, as well as water-wave problems, Helmholtz equations, and hyperbolic systems. Among the main subjects covered by the contributions are well-posedness and stability, construction of soliton solutions, dispersive estimates, invariant measures, inverse scattering, error analysis of space and time discretizations, and efficient implementation of numerical schemes.

We gratefully acknowledge the financial support by the Deutsche Forschungsgemeinschaft (DFG, German Research Foundation) through the collaborative research center "Wave Phenomena: Analysis and Numerics" (Project-ID 258734477 – SFB 1173) which enabled us to organize this successful conference.

Karlsruhe, Germany
March 2020

Willy Dörfler
Marlis Hochbruck
Dirk Hundertmark
Wolfgang Reichel
Andreas Rieder
Roland Schnaubelt
Birgit Schörkhuber

Contents

Morawetz Inequalities for Water Waves

Thomas Alazard, Mihaela Ifrim, and Daniel Tataru

Abstract Morawetz estimates capture the long time local decay properties for various linear and nonlinear dispersive flows. In these notes we provide a brief overview of recent and ongoing work concerning Morawetz estimates for water waves in two space dimensions.

1 Introduction

The water wave equations describe the motion of the free surface of the water under the action of gravity, capillarity and other physical forces. Many of these models admit a natural, conserved energy. The classical Morawetz estimates, first pioneered by Morawetz [30] in the context of the linear wave equation, describe the decay in time of the local energy, i.e. the energy restricted to a fixed spatial region. The aim of recent and ongoing work of the authors has been to explore whether Morawetz estimates, and thus local energy decay, hold for the water wave equation.

Precisely, in our work so far we consider gravity water waves in two space dimensions, with finite or infinite depth, and with or without surface tension. Assuming some uniform scale invariant Sobolev bounds for the solutions, we prove local energy decay (Morawetz) estimates globally in time. Our results are uniform in the infinite depth limit.

The article is structured as follows. In the next section we describe the water wave equations in the Eulerian formulation. After that, in Sect. 3, we provide a brief

T. Alazard
CNRS and CMLA, École Normale Supérieure de Paris-Saclay, Cachan, France
e-mail: thomas.alazard@cmla.ens-cachan.fr

M. Ifrim
University of Wisconsin, Madison, WI, USA
e-mail: ifrim@wisc.edu

D. Tataru (✉)
University of California, Berkeley, CA, USA
e-mail: tataru@berkeley.edu

W. Dörfler et al. (eds.), *Mathematics of Wave Phenomena*, Trends in Mathematics,
https://doi.org/10.1007/978-3-030-47174-3_1

introduction to Morawetz estimates. Our main results are described in Sect. 4. A key tool in our approach has been to use an alternative set of coordinates, namely holomorphic coordinates; these are described in Sect. 5, where we also outline their role in the proofs. In the last section we discuss some further questions.

2 Water Waves

We consider a fluid domain $\Omega(t)$ with a flat bottom at depth h and a free boundary $\Gamma(t)$ which represents the interface between water and air, as in the following picture:

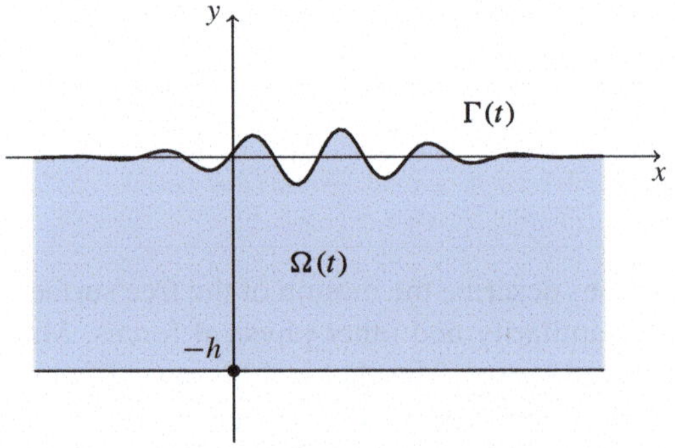

The fluid state is described by the velocity field u and the pressure p. The fluid flow is governed by the Euler equations in the moving domain $\Omega(t)$:

$$\begin{cases} u_t + u \cdot \nabla u = -\nabla p - g\mathbf{e}_2 \\ \operatorname{div} u = 0 \\ \operatorname{curl} u = 0 \\ u(0, x, y) = u_0(x, y), \end{cases} \tag{1}$$

where g represents the gravity.

To complete the description of the time evolution we need to also specify the boundary conditions. On the free boundary $\Gamma(t)$ we have

$$\begin{cases} \partial_t + u \cdot \nabla \text{ is tangent to } \bigcup \Gamma(t) & \text{(kinematic)} \\ p = 0 \text{ or } -2\kappa \mathbf{H}(\eta) \quad \text{on } \Gamma(t) & \text{(dynamic)} \end{cases}$$

where $\mathbf{H}(\eta)$ represents the curvature of the interface $\Gamma(t)$,

$$\mathbf{H}(\eta) = \partial_x \left(\frac{\eta_x}{\sqrt{1+\eta_x^2}} \right),$$

while the bottom is assumed to be impermeable,

$$u \cdot \mathbf{e}_2 = 0 \text{ on } \{y = -h\}.$$

Under the additional condition that the flow is irrotational,

$$\text{curl } u = 0 \text{ i.e. } \nabla \times u = 0,$$

the velocity field u can be described via a velocity potential ϕ,

$$u = \nabla_{x,y}\phi, \qquad \Delta_{x,y}\phi = 0 \quad \text{in } \Omega(t).$$

Then one can integrate the Euler equations to obtain the dynamic boundary condition (Bernoulli law):

$$\phi_t + \frac{1}{2}|\nabla_{x,y}\phi|^2 + p + gy = 0 \qquad \text{in } \Omega(t),$$

while the last condition satisfied by the velocity potential ϕ is

$$\partial_y \phi = 0 \quad \text{on} \quad y = -h.$$

This allows one to reduce the equations to an evolution for the free boundary in the Eulerian formulation. Here the variables are (η, ψ), where η is the elevation and $\psi(t,x) = \phi(t,x,\eta(t,x)))$. This goes back to work of Zakharov [38]; one obtains the system

$$\begin{cases} \partial_t \eta - G(\eta)\psi = 0 \\ \partial_t \psi + g\eta - \kappa \mathbf{H}(\eta) + \underbrace{\frac{1}{2}|\nabla\psi|^2 - \frac{1}{2}\frac{(\nabla\eta\nabla\psi + G(\eta)\psi)^2}{1+|\nabla\eta|^2}}_{:=N(\eta)\psi} = 0, \end{cases} \tag{2}$$

where $G(\eta)$ is the Dirichlet to Neumann map in $\Omega(t)$ relative to the free boundary:

$$G(\eta)\psi = \sqrt{1+\eta_x^2}(\nabla_{x,y}\phi \cdot N),$$

where N is the unit normal to $\Gamma(t)$.

The system (2) is Hamiltonian, with the Hamiltonian (energy) given by

$$\mathcal{H} = \frac{g}{2}\int_{\mathbb{R}} \eta^2\, dx + \kappa \int_{\mathbb{R}} \left(\sqrt{1+\eta_x^2} - 1\right) dx + \int_{\mathbb{R}} \int_{-h}^{\eta(t,x)} \left|\nabla_{x,y}\phi\right|^2 dydx.$$

It is also invariant with respect to horizontal translations, which by Noether's theorem yields the generator

$$\mathcal{M} = \int_{\mathbb{R}} \eta\psi_x dx,$$

which we refer to as the *horizontal momentum*.

For local well-posedness in Sobolev spaces one needs to work with Alihnac's *good variables* or diagonal variables

$$\eta \in H^s, \qquad \psi - T_{\phi_y|_{y=\eta}}\eta \in H^{s+\frac{1}{2}},$$

with $s \geq 2 - \epsilon$ for gravity water waves, using the standard paraproduct notation. For further references see e.g. [25].

Long time solutions are known to exist for small initial data:

$$\|(\eta, \psi - T_{\phi_y|_{y=\eta}}\eta)(0)\|_{H^2\times H^{\frac{3}{2}}} < \epsilon \quad \Rightarrow \quad T \gtrsim \epsilon^{-2}.$$

Here the norms are adapted to the gravity waves, with suitable modifications needed if capillarity is added. Results of this type were established in [1, 8, 21].

3 Morawetz Estimates

Also known as *local energy decay*, they were originally introduced in Morawetz's paper [30]. In their original form they assert that, for solutions to the linear wave equation, the local energy of the solutions is bounded, globally in time, by the initial energy. One may view this as a statement about the local decay of solutions which is invariant with respect to time translations.

Another interesting example is the Schrödinger equation. Unlike the wave equation, where one has a finite speed of propagation, in this case the group velocity increases to infinity in the high frequency limit. Because of this, the natural local energy measures a higher regularity (1/2 derivative more to be precise) than the initial data energy of the solutions; for this reason the Morawetz estimates for the Schrödinger equation are also nontrivial locally in time, and in this context they have been originally called *local smoothing*, see [16, 35].

Up to the present time, the Morawetz estimates have had a rich and complex history, which is too extensive to try to describe here. For further references we direct the reader to [28] for the wave equation, [27] for the Schrödinger equation and [33] for other models. Morawetz estimates have been proved for linear and nonlinear models, and have been used as a key ingredient in many results concerning the long time behavior of solutions in nonlinear dispersive flows. One other key development was the introduction of *interaction Morawetz estimates* in [15], which has played a major role in the study of nonlinear Schrödinger equations.

We turn our attention now to Morawetz estimates for water waves. Here additional challenges arise due to the fact that the equations are not only fully nonlinear, but also nonlocal. Another striking difference is due to the fact that in the high frequency limit the dispersive part of the group velocity goes to zero. Because of this, in the case of gravity waves we have the opposite phenomena to local smoothing, namely a loss of 1/4 derivative in the local energy. On the other hand if one adds capillarity we are back to a 1/4 derivative gain in the local energy. Combined with the nonlinear and nonlocal character of the equations, this brings substantial difficulties in the low frequency analysis.

The water wave energy corresponds to the energy space for (η, ψ)

$$E^0 = g^{-\frac{1}{2}} L^2 \cap \kappa^{-\frac{1}{2}} \dot{H}^1 \times H_h^{\frac{1}{2}}, \qquad \dot{H}_h^{\frac{1}{2}} = \dot{H}^{\frac{1}{2}} + h^{-\frac{1}{2}} \dot{H}^1,$$

whereas the momentum is associated to the space

$$E^{\frac{1}{4}} = g^{-\frac{1}{4}} H_h^{\frac{1}{4}} \times g^{\frac{1}{4}} \dot{H}_h^{\frac{3}{4}},$$

where

$$H_h^{\frac{1}{4}} := \dot{H}^{\frac{1}{4}} \cap h^{-\frac{1}{4}} L^2, \qquad \dot{H}_h^{\frac{3}{4}} = \dot{H}^{\frac{3}{4}} + h^{\frac{1}{4}} \dot{H}^1.$$

The local energy for a gravity water wave in a time interval $[0, T]$ and centered around a point x_0 is defined as

$$\|(\eta, \psi)\|_{LE_{x_0}}^2 := g \int_0^T \int_{\mathbb{R}} \chi(x - x_0)\eta^2 \, dxdt + \int_0^T \int_{\mathbb{R}} \int_{-h}^{\eta} \chi(x - x_0) \left|\nabla_{x,y}\phi\right|^2 \, dydxdt$$

where χ is a smooth compactly supported nonnegative bump function with unit support. Adding in the surface tension, the local energy centered around a point x_0 has the form

$$\|(\eta, \psi)\|_{LE_{x_0}^\sigma}^2 := \|(\eta, \psi)\|_{LE_{x_0}}^2 + \kappa \int_0^T \int_{\mathbb{R}} \chi(x - x_0)\eta_x^2 \, dxdt.$$

The choice of the center x_0 is arbitrary due to the translation invariance. Since the water wave equations are nonlocal, nonlocalized tails will occur in all estimates. Because of this, it is convenient to allow for arbitrary centers x_0 and define:

$$\|(\eta, \psi)\|_{LE}^2 := \sup_{x_0 \in \mathbb{R}} \|(\eta, \psi)\|_{LE_{x_0}}^2, \qquad \|(\eta, \psi)\|_{LE^\sigma}^2 := \sup_{x_0 \in R} \|(\eta, \psi)\|_{LE_{x_0}^\sigma}^2$$

Alternatively, one could work with localization functions χ which are not compactly supported, and instead have integrable tails at infinity.

4 Morawetz Estimates for Two Dimensional Water Waves

We are now ready to state our main results, first for pure gravity waves and then for gravity/capillary waves.

Theorem 1 (Local Energy Decay for Gravity Waves [9]) *There exist ϵ_0 and C_0 such that the following result holds. For all $g \in (0, +\infty)$, all $T \in (0, +\infty)$, all $h \in [1, +\infty)$ and all regular solutions (η, ψ) of the water wave system satisfying*

$$\sup_{[0,T]} \|\eta(t)\|_{H_h^{\frac{3}{2}}} + \|\eta_x(t)\|_{L^\infty} + g^{-\frac{1}{2}} \|\nabla\psi(t)\|_{H_h^1} \leq \epsilon_0, \tag{3}$$

the following estimate holds

$$\|(\eta, \psi)\|_{LE}^2 \leq C_0 \left(\|(\eta, \psi)(0)\|_{E^{\frac{1}{4}}}^2 + \|(\eta, \psi)(T)\|_{E^{\frac{1}{4}}}^2 \right). \tag{4}$$

We next point several key features of this result:

1. The result is uniform as $T \to \infty$ and $g > 0$; these two aspects are connected via time scaling,

$$(\eta(t, x), \psi(t, x)) \to (\eta(\lambda t, x), \lambda\psi(\lambda t, x)) \qquad (g, h) \quad \to (\lambda^2 g, h).$$

   This uniformity is the key feature of any result on local energy decay. It is also the main difficulty, as no other existing results for water waves apply independently of the time scale.
2. The window size in this result is fixed to 1 for convenience. The corresponding result for larger window sizes is also valid simply by Holder's inequality.
3. The result is uniform in infinite depth limit $h \to \infty$. This means that the infinite depth limit is tame from the perspective of local energy decay. The restriction $h \gtrsim 1$ should really be read as $depth \gtrsim window\ size$. It is there to guarantee uniformity; by scaling the result still holds for smaller h, just not uniformly as $h \to 0$.

4. The uniform apriori bound on the solutions in (3), slightly simplified from a nearby Besov structure in [9], is absolutely minimal, and corresponds exactly to the space-time scaling of the problem. For comparison, well-posedness is only known to hold at considerably higher regularity; at this regularity it would be a near impossible task, as the equations are fully nonlinear.

We continue with the similar result for gravity/capillary waves:

Theorem 2 (Local Energy Decay for Gravity–Capillary Waves [10]) *There exist* ϵ_0 *and* C_0 *such that the following result holds. For all* $T \in (0, +\infty)$*, all* $h \in [1, +\infty)$*, all* $g \in (0, +\infty)$*, all* $\kappa \in (0, +\infty)$ *with* $\kappa \lesssim g$ *and all regular solutions of the water wave system* (2) *satisfying*

$$\sup_{[0,T]} \|\eta(t)\|_{H_h^{\frac{3}{2}}} + \|\eta_x(t)\|_{L^\infty} + g^{-\frac{1}{2}}\|\nabla\psi(t)\|_{H_h^1} + \left(\frac{\kappa}{g}\right)^{\frac{1}{4}} \|\eta\|_{\dot{H}^2} \leq \epsilon_0 \tag{5}$$

the following estimate holds

$$\|(\eta, \psi)\|^2_{LE^\kappa} \leq C_0\Big(\|(\eta, \psi)(0)\|^2_{E^{\frac{1}{4}}} + \|(\eta, \psi)(T)\|^2_{E^{\frac{1}{4}}} \Big). \tag{6}$$

In addition to the comments in the gravity wave case, which still apply except for (4), we have the following remarks.

1. This result implies the pure gravity wave result in the limit $\kappa \to 0$.
2. The frequency $\xi = \left(\frac{g}{\kappa}\right)^{\frac{1}{2}}$ plays a key role here, as there we have the transition between gravity waves and capillary waves. The assumption $\kappa \lesssim g$ insures that this occurs at wavelengths below 1, and restricts our result to low bound numbers. The apriori bound (5) is identical to (3) below this frequency, and is only strengthened above.

For the remainder of this section, we provide an outline of the ideas in the proofs of these results. The starting idea is to exploit the conservation of the momentum via density-flux pairs (I, S) which by definition must satisfy

$$\mathcal{M} = \int I\, dx,$$

and also the conservation law

$$\partial_t I + \partial_x S = 0.$$

Given a Morawetz weight $m(x)$, which is a nonnegative function, increasing from 0 to 1, we multiply the last relation by $m(x)$, integrate over $[0, T] \times \mathbb{R}_x$ and then integrate by parts in both time and space to obtain

$$\int_0^T \int_{x\in\mathbb{R}} S(t, x) m_x \, dx dt = \int_{x\in\mathbb{R}} m(x) I(T, x)\, dx - \int_{x\in\mathbb{R}} m(x) I(0, x)\, dx. \tag{7}$$

Ideally, we would like $S(t,x)$ to be comparable with the energy density, in which case the proof would be essentially over. However, this does not seem to be directly possible. Adding to the difficulty, our problem is nonlocal, and that makes it less clear what should be a good density for the momentum.

Our approach will be to produce two such density functions for the momentum, whose fluxes in turn will allow us to separately capture the kinetic and the potential energy. Toward this goal, we introduce two auxiliary functions q and θ as follows:

- The function q, defined inside the fluid domain $\Omega(t)$, is the stream function, or the harmonic conjugate of ϕ, and satisfies

$$q_x = \phi_y, \qquad q_y = -\phi_x, \qquad q(t,x,-h) = 0.$$

- The function θ is also harmonic inside the fluid domain, and satisfies the Dirichlet boundary condition

$$\theta(t,x,-h) = 0, \qquad \theta(t,x,\eta(x)) = \eta(x).$$

With these notations, our two densities for the momentum are

$$I_2(t,x) = \eta(t,x)\psi_x(t,x),$$
$$I_3(t,x) = \int_{-h}^{\eta} \nabla\theta(t,x,y)\cdot\nabla q(t,x,y)\,dy.$$

Their associated fluxes are given by

$$S_2(t,x) := -\eta\psi_t - \frac{g}{2}\eta^2 + \frac{1}{2}\int_{-h}^{\eta(t,x)} (\phi_x^2 - \phi_y^2)\,dy,$$
$$S_3(t,x) := -\frac{g}{2}\eta^2 - \int_{-h}^{\eta(t,x)} \theta_y\phi_t\,dy + \int_{-h}^{\eta(t,x)} \frac{1}{2}(\phi_x^2 - \phi_y^2) + \theta_t\phi_y\,dy.$$

Recalling that at leading order we have

$$\psi_t \approx -g\eta, \quad \theta_t \approx \phi_y$$

one sees that the terms

$$-\eta\psi_t - \frac{g}{2}\eta^2 \approx \frac{g}{2}\eta^2, \qquad \int_{-h}^{\eta(t,x)} \frac{1}{2}(\phi_x^2 - \phi_y^2) + \theta_t\phi_y\,dy \approx \int_{-h}^{\eta(t,x)} \frac{1}{2}(\phi_x^2 + \phi_y^2)\,dy,$$

in the two fluxes indeed control the potential, respectively the kinetic energy. Thus starting from the multiplier identity (7), our proofs proceed as follows:

- Combine the density-flux pairs (I_2, S_2) and (I_3, S_3) with well chosen weights,
- Split the output into quadratic and higher order terms,

- Obtain the main local energy contribution form the quadratic terms,
- Estimate the cubic and higher order terms perturbatively.

Even the linear problem turns out to be quite nontrivial, due to the nonlocality of the problem, and hinges on some delicate kernel bounds for certain bilinear multipliers.

But the main difficulty is encountered in estimating the error terms. Here even extracting the cubic terms is nontrivial, let alone estimating them. To achieve this, we shift these contributions to holomorphic coordinates, which are described in the next section. This allows us to estimate all but one of the error terms perturbatively.

The remaining error term, however, turns out to be unbounded. But it does have a redeeming feature, namely that it is nonresonant. This eventually allows us to handle it using a normal form type correction to the corresponding momentum density. This correction occurs at the level of the energies and not the equations, inspired from the last two authors' *modified energy method* introduced in [21].

5 Holomorphic Coordinates

As mentioned before, these coordinates are a key tool in the proof of the Morawetz estimates for water waves. In brief, our motivation to use them is as follows:

- They appear naturally in definition of momentum density I_3.
- They diagonalize the Dirichlet to Neumann map.
- They turn nonlinear estimates into multilinear analysis.

The idea is to flatten out the surface and conformally map the fluid domain into a strip. We have a conformal (holomorphic) map defined as in the following picture:

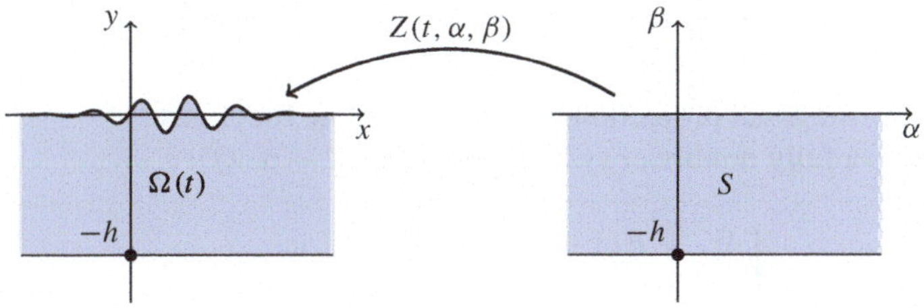

The conformal map is

$$Z : S = \mathbb{R} \times [-h, 0] \to \Omega(t)$$

and satisfies the boundary condition at infinity:

$$Z(\alpha + i\beta) \approx \alpha + i\beta.$$

The new pair of dynamical variables in the new coordinates are the functions (W, Q) defined by

$$\begin{cases} W(t,\alpha,\beta) = Z(t,\alpha,\beta) - (\alpha + i\beta) \\ Q(t,\alpha,\beta) = \phi(t,\alpha,\beta) + iq(t,\alpha,\beta). \end{cases}$$

Then η and ψ are the restriction of $\Im Z$ and $\Re Q$ on the top, whereas

$$\theta(t,\alpha,\beta) = \Im W(t,\alpha,\beta).$$

A key property here is that (W, Q) are uniquely determined by their values on the top $\beta = 0$. Furthermore, their values on the top, which will play the role of dynamical variables, replacing (η, ψ), cannot be arbitrarily chosen, instead they belong to the restricted class of functions which admit such holomorphic extensions. By a slight abuse of terminology, we will call such functions *holomorphic functions*.

These coordinates can be found in an incipient form in Ovsjannikov [32], but were further developed independently by Dyachenko et al. [17] , Wu [36]. The set-up used here is based on Hunter et al. [21], Harrop-Griffiths et al. [19].

The water wave system for the holomorphic functions (Z, Q) takes the form:

$$\begin{cases} Z_t + FZ_\alpha = 0, \\ Q_t + FQ_\alpha - ig(Z-\alpha) + P\left[\dfrac{|Q_\alpha|^2}{J}\right] = 0, \end{cases}$$

where

$$F = P\left[\frac{Q_\alpha - \bar{Q}_\alpha}{J}\right], \qquad J = |Z_\alpha|^2,$$

and P is the orthogonal projection onto holomorphic functions.

This yields a fully nonlinear system for $(W = Z - \alpha, Q)$:

$$\begin{cases} W_t + F(1 + W_\alpha) = 0 \\ Q_t + FQ_\alpha - igW + P\left[\dfrac{|Q_\alpha|^2}{J}\right] = 0. \end{cases}$$

To use the holomorphic coordinates in order to estimate the error terms in the Morawetz estimates, one simply performs the change of variables in the appropriate integrals. This yields nonlinear expressions in (W, Q), which further admit multilinear expansions beginning with cubic order. Heuristically, the cubic terms should be the problematic ones, as for quartic and higher order terms there is more room to balance the energy and local energy norms. This essentially works with one proviso, namely that one needs to also estimate certain quadrilinear terms.

Unfortunately, there is also a price to pay in order to use the holomorphic coordinates, which is related to the matter of switching strips between Eulerian and holomorphic coordinates. These are described as follows:

Eulerian strip: This is centered at x_0, with weight $m(x - x_0)$.
Holomorphic strip: This is centered at $\alpha_0(t, \beta)$, with weight $m(\alpha - \alpha_0)$.

Here, $\alpha_0 = \alpha_0(t, x_0, \beta)$ or inverting

$$x_0 = \alpha_0 + \Re W(\alpha_0, \beta).$$

Even on the top, α_0 is a function of t. This is less important, as it corresponds to a trivial one dimensional degree of freedom in the choice of holomorphic coordinates. More problematic is what happens in depth, where the holomorphic strip differs significantly from the vertical. Thus one needs to have a good estimate for the distance between strips at conformal depth β, which is given by

$$\text{dist}(strips) = \Re Z(\alpha_0, \beta) - \Re Z(\alpha_0, 0) = \Re W(\alpha_0, 0) - \Re W(\alpha_0, \beta).$$

This is done via the following

Proposition 1 *We have the relation*

$$\Re W(\alpha_0, \beta) - \Re W(\alpha_0, 0) = \beta \Im W_\alpha(\alpha_0, \beta) + \textit{better}.$$

Note: By *better* we refer to terms that can be easily controlled in the corresponding Sobolev norms.

Here $\Im W_\alpha \approx \eta_x$ on top, therefore in particular

$$|\text{dist}(strips)| \lesssim \epsilon_0 |\beta|.$$

6 Further Questions

Certainly the results presented above represent only the beginning of the study of Morawetz inequalities for water waves. In what follows we list a number of questions which should be of further interest.

First and foremost, our results only apply in the context of two dimensional water waves. One naturally expect that similar bounds should hold in higher dimension:

Conjecture *Prove that the Morawetz estimates for gravity/capillary water waves hold in all dimensions!*

Another limitation of our result is that it does not apply for pure capillary waves:

Open Question Do Morawetz estimates for capillary water waves hold in dimension $d = 2$? Also in dimension $d \geq 3$?

There is also the matter of relaxing the smallness condition in our uniform boundedness assumption.

Open Question Can the smallness condition be replaced by a qualitative assumption (e.g. nonintersecting free surface) ?

Finally, there is more than one natural way to define the localization strips.

Open Question Can the Eulerian strips be replaced by vertical holomorphic strips ? How about strips tied to a particle trajectory on the free surface ?

Acknowledgments The second author was partially supported by a Clare Boothe Luce Professorship. The third author was partially supported by the NSF grant DMS-1800294 as well as by a Simons Investigator grant from the Simons Foundation.

References

1. Ai, A.: Low regularity solutions for gravity water waves. Water Waves **1**(1), 145–215 (2017)
2. Ai, A.: Low regularity solutions for gravity water waves II: the 2D case. Annals in PDE **6**(1), (2020)
3. Alazard, T.: Stabilization of gravity water waves. J. Math. Pures Appl. **114**, 51–84 (2018)
4. Alazard, T.: Boundary observability of gravity water waves. Ann. Inst. H. Poincaré Anal. Non Linéaire **35**(3), 751–779 (2018)
5. Alazard, T., Delort, J.-M.: Sobolev estimates for two dimensional gravity water waves. Astérisque **374**, viii+241pp. (2015)
6. Alazard, T., Delort, J.-M.: Global solutions and asymptotic behavior for two dimensional gravity water waves. Ann. Sci. Norm. Supér. **48**, 1149–1238 (2015)
7. Alazard, T., Métivier, G.: Paralinearization of the Dirichlet to Neumann operator, and regularity of three-dimensional water waves. Commun. Partial Differ. Equ. **34**(10–12), 1632–1704 (2009)
8. Alazard, T., Burq, N., Zuily, C.: On the Cauchy problem for gravity water waves. Invent. Math. **198**, 71–163 (2014)
9. Alazard, T., Ifrim, M., Tataru, D.: A Morawetz inequality for water waves (2018, preprint). arXiv:1806.08443
10. Alazard, T., Ifrim, M., Tataru, D.: A Morawetz inequality for gravity-capillary water waves at low Bond number (2019, preprint). arXiv:1910.02529

11. Alazard, T., Burq, N., Zuily, C.: Strichartz estimates and the Cauchy problem for the gravity water waves equations. Mem. AMS **256**(1229), 1–120 (2018)
12. Alinhac, S.: Paracomposition et opérateurs paradifférentiels. Commun. Partial Differ. Equ. **11**(1), 87–121 (1986)
13. Alinhac, S.: Existence d'ondes de raréfaction pour des systèmes quasi-linéaires hyperboliques multidimensionnels. Commun. Partial Differ. Equ. **14**(2), 173–230 (1989)
14. Bahouri, H., Chemin, J.-Y., Danchin, R.: Fourier Analysis and Nonlinear Partial Differential Equations. In: Grundlehren der Mathematischen Wissenschaften [Fundamental Principles of Mathematical Sciences], vol. 343. Springer, Heidelberg (2011)
15. Colliander, J., Keel, M., Staffilani, G., Takaoka, H., Tao, T.-C.: Global well-posedness and scattering for the energy-critical nonlin-ear Schrödinger equation in R^3. Ann. Math.**167**(3), 767–865 (2008)
16. Constantin, P., Saut, J.-C.: Local smoothing properties of dispersive equations. J. Am. Math. Soc. **1**(2), 413–439 (1988)
17. Dyachenko, A., Kuznetsov, E., Spector, M., Zakharov, V.: Analytical description of the free surface dynamics of an ideal fluid (canonical formalism and conformal mapping). Phys. Lett. A **221**(1–2), 73–79 (1996)
18. Germain, P., Masmoudi, N., Shatah, J.: Global solutions for the gravity water waves equation in dimension 3. Ann. Math. **175**(2), 691–754 (2012)
19. Harrop-Griffiths, B., Ifrim, M., Tataru, D.: Finite depth gravity water waves in holomorphic coordinates. Ann. PDE **3**(1), 4 (2017)
20. Hörmander, L.: Lectures on linear hyperbolic differential equations. In: Mathématiques & Applications (Berlin) [Mathematics & Applications], vol. 26. Springer, Berlin (1997)
21. Hunter, J., Ifrim, M., Tataru, D.: Two dimensional water waves in holomorphic coordinates. Commun. Math. Phys. **346**(2), 483–552 (2016)
22. Ifrim, M., Tataru, D.: Two dimensional water waves in holomorphic coordinates II: global. Bull. Soc. Math. France **144**(2), 369–394 (2016)
23. Ionescu, A., Pusateri, F.: Global solutions for the gravity water waves system in $2d$. Invent. Math. **199**(3), 653–804 (2015)
24. Lannes, D.: Well-posedness of the water-waves equations. J. Am. Math. Soc. **18**(3), 605–654 (2005)
25. Lannes, D.: Water waves: mathematical analysis and asymptotics. In: Mathematical Surveys and Monographs, vol. 188. American Mathematical Society, Providence (2013)
26. Longuet-Higgins, M.S.: On integrals and invariants for inviscid, irrotational flow under gravity. J. Fluid Mech. **134**, 155–159 (1983)
27. Marzuola, J., Metcalfe, J., Tataru, D.: Strichartz estimates and local smoothing estimates for asympototically flat Schrödinger equations. J. Funct. Anal. **255**(6), 1497–1553 (2008)
28. Metcalfe, J., Tataru, D.: Decay estimates for variable coefficient wave equations in exterior domains. In: Advances in Phase Space Analysis of Partial Differential Equations. Progress in Nonlinear Differential Equations and Their Applications, vol. 78, pp. 201–216. Birkhäuser, Boston (2009)
29. Métivier, G.: Para-differential Calculus and Applications to the Cauchy Problem for Nonlinear Systems. Centro di Ricerca Matematica Ennio De Giorgi (CRM) Series, vol. 5. Edizioni della Normale, Pisa (2008)
30. Morawetz, C.S.: Time decay for the nonlinear Klein-Gordon equation. In: Proceedings of the Royal Society of London A: Mathematical, Physical and Engineering Sciences, vol. 306, pp. 291–296. London, The Royal Society (1968)
31. Nalimov, V.I.: The Cauchy-Poisson problem. Dinamika Splošn. Sredy, (Vyp. 18 Dinamika Zidkost. so Svobod. Granicami), **254**, 104–210 (1974)
32. Ovsjannikov, L.V.: To the shallow water foundation. Arch. Mech. Stos. **26**, 407–422 (1974)
33. Ozawa, T., Rogers, K.: Sharp Morawetz estimates. J. Anal. Math. **121**(1), 163–175 (2013)
34. Totz, N., Wu, S.: A rigorous justification of the modulation approximation to the 2D full water wave problem. Commun. Math. Phys. **310**, 817–883 (2012)

35. Vega, L.: Schrödinger equations: pointwise convergence to the initial data. Proc. Am. Math. Soc. **102**(4), 874–878 (1988)
36. Wu, S.: Well-posedness in Sobolev spaces of the full water wave problem in 2-D. Invent. Math. **130**(1), 39–72 (1997)
37. Wu, S.: Almost global wellposedness of the $2-D$ full water wave problem. Invent. Math. **177**, 45–135 (2009)
38. Zakharov, V.E.: Stability of periodic waves of finite amplitude on the surface of a deep fluid. J. Appl. Mech. Tech. Phys. **9**(2), 190–194 (1968)

Numerical Study of Galerkin–Collocation Approximation in Time for the Wave Equation

Mathias Anselmann and Markus Bause

Abstract The elucidation of many physical problems in science and engineering is subject to the accurate numerical modelling of complex wave propagation phenomena. Over the last decades, high-order numerical approximation for partial differential equations has become a well-established tool. Here we propose and study numerically the implicit approximation in time of wave equations by a Galerkin–collocation approach that relies on a higher order space-time finite element approach. The conceptual basis is the establishment of a direct connection between the Galerkin method for the time discretization and the classical collocation methods, with the perspective of achieving the accuracy of the former with reduced computational costs provided by the latter in terms of less complex linear algebraic systems. For the fully discrete solution, higher order regularity in time is further ensured which can be advantageous for the discretization of multi-physics systems. The accuracy and efficiency of the variational collocation approach is carefully studied by numerical experiments.

1 Introduction

The accurate and efficient numerical simulation of wave phenomena continues to remain a challenging task and attract researchers' interest. Wave phenomena are studied in various branches of natural sciences and technology. For instance, fluid-structure interaction, acoustics, poroelasticity, seismics, electro-magnetics and non-destructive material inspection represent prominent fields in that wave propagation is studied. One of our key application for wave propagation is structural health monitoring of lightweight material (for instance, carbon-fibre reinforced polymers) by ultrasonic waves in aerospace engineering. The conceptional idea of this new and intelligent approach is sketched in Fig. 1. The structure is equipped

M. Anselmann (✉) · M. Bause
Helmut Schmidt University, Hamburg, Germany
e-mail: anselmann@hsu-hh.de; bause@hsu-hh.de

W. Dörfler et al. (eds.), *Mathematics of Wave Phenomena*, Trends in Mathematics,
https://doi.org/10.1007/978-3-030-47174-3_2

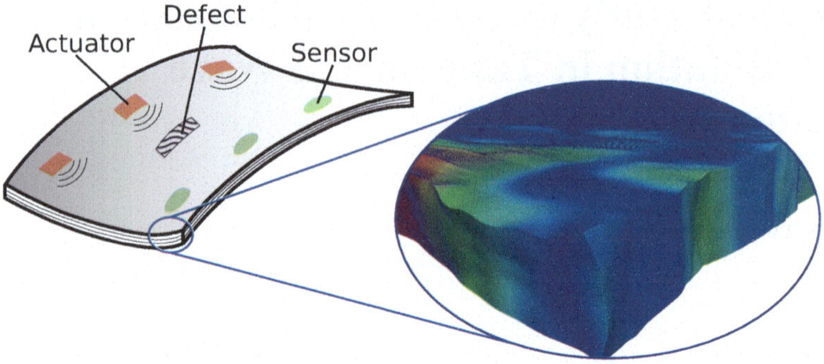

Fig. 1 Concept of structural health monitoring with finite element simulation (scaled displacement field) illustrating the expansion of elastic waves

with an integrated actuator-sensor network. The ultrasonic waves that are emitted by the actuators interact with material defects of the solid structure. By means of an inverse modelling, the signals that are recorded by the sensors monitor material failure (cf. [16]) and, as perspective for the future, may allow prognoses about the structure's residual lifetime. The design of such monitoring systems and the signal interpretation require the elucidation of wave propagation in composite material which demands for highly advanced and efficient numerical simulation techniques.

High-order numerical approximation of partial differential equations has been strongly focused and investigated in the last decades. High order methods are known to be efficient if they approximate functions with large elements of high polynomial degree in regions of high regularity. Prominent examples are *hp*- and spectral element methods in application areas such as computational fluid dynamics or computational mechanics. Their theoretical convergence analysis and the design of adaptive *hp*- and spectral element versions still experience strong development. Whereas high-order approaches have been considered for the approximation of the spatial variables, first- or second-order implicit schemes are often still used for the discretization of the time variable. We note that in this work only implicit time discretization schemes are in the scope of interest. Thus, all remarks refer to this class of methods. Our motivation for using implicit time discretization schemes comes from the overall goal to apply the proposed Galerkin–collocation techniques to mixed systems like, for instance, fluid-structure interaction for free flow modelled by the Navier–Stokes equations [24] or fully dynamic poroelasticity [22].

Driven by the tremendous increase in computing power of modern high performance computing systems and recent progress in the technology of algebraic solver, including efficient techniques of preconditioning, space-time finite element approaches have recently attracted high attention and have been brought to application maturity; cf., e.g., [7, 9, 14, 25]. Space-time finite element methods offer appreciable advantages over discretizations of mixed type based on finite difference techniques for the discretization of the time variable (e.g., by Runge–Kutta methods) and, for instance, finite element methods for the discretization of the space variables.

In particular, advantages are the natural embedding of higher order members in the various families of schemes, the applicability of functional analysis techniques in their analyses due to the uniform space-time framework and the applicability of well-known adaptive mesh refinement techniques, including goal-oriented error control [2]. In the meantime a broad variety of implementations of space-time finite element methods does exist. The families of schemes differ by the choices of the trial and test spaces. This leads to continuous or discontinuous approximations of the time variable (cf. e.g., [21, 26]). Further, the fully coupled treatment of all time steps versus time-marching approaches is discussed. In particular, the simultaneous computation of all time steps imposes high demands on the linear solver technology (cf., e.g., [7–9]).

In this work, we propose the Galerkin–collocation method for the numerical solution of wave equations. This approach combines variational approximation in time by finite element techniques with the concepts of collocation methods and follows the ideas of [5]. By imposing collocation conditions, the test space of the variational condition is downsized. The key ingredients and innovations of the approach are:

A. Higher order regularity in time of the fully discrete approximation;
B. Linear systems of reduced complexity;

Ingredient [A] is a direct consequence of the construction of the schemes. Higher order regularity is ensured by imposing collocation conditions at the discrete time nodes and endpoints of the subintervals $[t_{n-1}, t_n]$, for $n = 1, \dots, N$, of the global time interval $[0, T]$. Higher order regularity in time might offer appreciable advantages for future approximations of coupled multi-physics systems if higher order time derivatives of the discrete solution of one subproblem arise as coefficient functions in other subproblems. Ingredient [B] is ensured by the proper choice of a special basis for the discrete in time function spaces. Thereby, simple vector identities for the degrees of freedom in time are obtained at the left endpoints of the subintervals without generating computational costs. These vector identities can then be exploited to eliminate conditions from the algebraic systems and reduce its size compared to the standard continuous Galerkin–Petrov approximation in time; cf. [4]. In a further work of the authors [1], it is shown that the optimal order of convergence in time (and space) of the underlying finite element discretization is preserved by the Galerkin–collocation approach. The numerical example of Sect. 4.4 that mimics typical studies of structural health monitoring (cf. Fig. 1), demonstrates the superiority of the Galerkin–collocation approach over a standard continuous Galerkin–Petrov method admitting continuity and no differentiability in time of the discrete solution.

For the sake of brevity, standard conforming finite element methods are used for the discretization of the spatial variables in this work. This is done since we focus here on time discretization. In the literature it has been mentioned that discontinuous finite element methods in space offer appreciable advantages over continuous ones for the discretization of wave equations; cf., e.g., [6, 10]. The application of, for instance, the symmetric interior penalty discontinuous Galerkin method (cf. [16, 17] along with a Galerkin–collocation discretization in time, is straightforward.

This work is organized as follows. In Sect. 2 we introduce our prototype model. In Sect. 3 we present its discretization by two families of Galerkin–collocation methods. In Sect. 4, the discrete form of a member of theses families with C^1-regularity in time is derived. The resulting algebraic system is built and our algebraic solver is described. In Sect. 5 the discrete form of a member of the Galerkin–collocation family with C^2-regularity in time is derived. For both methods, the results of our numerical experiments are presented and evaluated.

2 Mathematical Problem and Notation

As a prototype model, we study the wave problem

$$\begin{aligned} \partial_t^2 u - c^2 \Delta u &= f\,, \text{ in } \Omega \times I\,, \\ u(0) = u_0\,, \quad \partial_t u(0) &= v_0\,, \text{ in } \Omega\,, \\ u = g^u\,, \quad \text{on } \partial\Omega_D \times I\,, \quad \partial_n u &= 0\,, \text{ on } \partial\Omega_N \times I\,. \end{aligned} \tag{1}$$

In our application of structural health monitoring (cf. Fig. 1), u denotes the scalar valued displacement field, $c \in \mathbb{R}$ with $c > 0$, is a material parameter and f an external force acting on the domain $\Omega \subset \mathbb{R}^d$, with $d = 2, 3$. Further, g^u is a prescribed trace on the Dirichlet part $\partial\Omega_D$ of the boundary $\partial\Omega = \partial\Omega_D \cup \partial\Omega_N$, with $\partial\Omega_D \cap \partial\Omega_N = \emptyset$. By ∂_n we denote the normal derivative with outer unit normal vector $\boldsymbol{n}$. Homogeneous Neumann boundary conditions on $\partial\Omega_N$ are prescribed for brevity. Finally, $I = (0, T]$ denotes the time domain. Problem (1) is well-posed and admits a unique solution $(u, \partial_t u) \in L^2(0, T; H^1(\Omega)) \times L^2(0, T; L^2(\Omega))$ under appropriate assumptions about the data; cf. [20]. By imbedding, $u \in C([0, T]; H^1(\Omega))$ and $\partial_t u \in C([0, T]; L^2(\Omega))$ is ensured; cf. [19]. Throughout, we tacitly assume that the solution admits all the (improved) regularity being necessary in the arguments.

Our notation is standard. By $H^m(\Omega)$ we denote the Sobolev space of $L^2(\Omega)$ functions with derivatives up to order m in $L^2(\Omega)$. For brevity, we let $H := L^2(\Omega)$ and $V = H^1_{0,D}(\Omega)$ be the space of all H^1-functions with vanishing trace on the Dirichlet part $\partial\Omega_D$ of $\partial\Omega$. By $\langle\cdot,\cdot\rangle_\Omega$ we denote the inner product in $L^2(\Omega)$. For the norms we use $\|\cdot\| := \|\cdot\|_{L^2(\Omega)}$ and $\|\cdot\|_m := \|\cdot\|_{H^m(\Omega)}$ for $m \in \mathbb{N}$ and $m \geq 1$. Finally, the expression $a \lesssim b$ stands for the inequality $a \leq C b$ with a generic constant C that is indepedent of the size of the space and time meshes.

By $L^2(0, T; B)$, $C([0, T]; B)$ and $C^q([0, T]; B)$, for $q \in \mathbb{N}$, we denote the standard Bochner spaces of B-valued functions for a Banach space B, equipped with their natural norms. Further, for a subinterval $J \subseteq [0, T]$, we will use the notations $L^2(J; B)$, $C^m(J; B)$ and $C^0(J; B) := C(J; B)$ for the corresponding Bochner spaces.

To derive our Galerkin–collocation approach, we first rewrite problem (1) as a first order system in time for the unknowns (u, v), with $v = \partial_t u$,

$$\partial_t u - v = 0\,, \qquad \partial_t v - c^2 \Delta u = f\,. \tag{2}$$

Further, we represent the unknowns u and v in terms of

$$u = u^0 + u^D \quad \text{and} \quad v = v^0 + v^D\,. \tag{3}$$

Here, $u^D, v^D \in C(\overline{I}; H^1(\Omega))$ are supposed to be (extended) functions with traces $u^D = g^u$ and $v^D = g^v := \partial_t g^u$ on the Dirichlet part $\partial\Omega_D$ of $\partial\Omega$.

Using (2) and (3), we then consider solving the following variational problem: *Find* $(u^0, v^0) \in L^2(0, T; H^1_{0,D}(\Omega)) \times L^2(0; T; H^1_{0,D}(\Omega))$ *such that*

$$u^0(0) = u_0 - u^D(0)\,, \qquad v^0(0) = v_0 - v^D(0)$$

and, for all $(\phi, \psi) \in (L^2(0; T; H^1_{0,D}(\Omega)))^2$,

$$\int_I \langle \partial_t u^0, \phi\rangle_\Omega - \langle v^0, \phi\rangle_\Omega \,\mathrm{d}t = 0\,, \tag{4}$$

$$\int_I \langle \partial_t v^0, \psi\rangle_\Omega + \langle c^2 \nabla u^0, \nabla\psi\rangle_\Omega \,\mathrm{d}t = \int_I \Big(\langle f, \psi\rangle_\Omega + \langle c^2 \partial_n u, \psi\rangle_{\partial\Omega_N} - \langle \partial_t v^D, \psi\rangle_\Omega - \langle c^2 \nabla u^D, \nabla\psi\rangle_\Omega\Big)\,\mathrm{d}t\,. \tag{5}$$

Remark 1

(i) We note that the correct treatment of inhomogeneous time-dependent boundary conditions is an import issue in the application of variational space-time methods. The space-time discretization that is derived below (cf. Sect. 3) and based on the variational problem (4), (5) ensures convergence rates of optimal order in space and time, also for time-dependent boundary conditions. This is confirmed by the second of the numerical experiments given in Sect. 4.3.

(ii) Our Galerkin–collocation approach is based on solving, along with some collocation conditions, the variational equations (4) and (5) in finite dimensional subspaces. In particular, the same approximation space will be used for u^0 and v^0. For this reason, the solution space for v^0 and the test space in Eq. (4) are chosen slightly stronger than usually; cf. [2]. Choosing $L^2(0; T; L^2(\Omega)))$ instead, would have been sufficient.

3 Galerkin–Collocation Schemes

In this section we introduce two families of Galerkin–collocation schemes. These families combine the concept of collocation methods, applied to the spatially discrete counterpart of the Eq. (2) with the finite element discretization of the variational equations (4) and (5). The collocation constrains then allow us to reduce the size of the discrete test spaces for the variational conditions compared to a standard Galerkin–Petrov approach; cf. [17].

First, we need some notation. For the time discretization we decompose the time interval $I = (0, T]$ into N subintervals $I_n = (t_{n-1}, t_n]$, where $n \in \{1, \dots, N\}$ and $0 = t_0 < t_1 < \cdots < t_{n-1} < t_n = T$ such that $I = \bigcup_{n=1}^{N} I_n$. We put $\tau = \max_{n=1,\dots N} \tau_n$ with $\tau_n = t_n - t_{n-1}$. Further, the set of time intervals $\mathcal{M}_\tau := \{I_1, \dots, I_n\}$ is called the time mesh. For a Banach space B and any $k \in \mathbb{N}$, we let

$$\mathbb{P}_k(I_n; B) = \left\{ w_\tau : I_n \mapsto B \;\middle|\; w_\tau(t) = \sum_{j=0}^{k} W^j t^j \,,\ \forall t \in I_n \,,\ W^j \in B \ \forall j \right\}.$$

For an integer $k \in \mathbb{N}$ we introduce the space of globally continuous functions in time

$$X_\tau^k(B) := \left\{ w_\tau \in C(\overline{I}; B) \mid w_{\tau|I_n} \in \mathbb{P}_k(I_n; B) \ \forall I_n \in \mathcal{M}_\tau \right\} ,$$

and for an integer $l \in \mathbb{N}_0$ the space of globally L^2-functions in time

$$Y_\tau^l(B) := \left\{ w_\tau \in L^2(I; B) \mid w_{\tau|I_n} \in \mathbb{P}_l(I_n; B) \ \forall I_n \in \mathcal{M}_\tau \right\} .$$

For the space discretization, let $\mathcal{T}_h = \{K\}$ be a shape-regular mesh of Ω consisting of quadrilateral or hexahedral elements with mesh size $h > 0$. Further, for some integer $p \in \mathbb{N}$ let $V_h = V_h^{(p)}$ be the finite element space that is given by

$$V_h = V_h^{(p)} = \left\{ v_h \in C(\overline{\Omega}) \mid v_{h|T} \circ T_K \in \mathbb{Q}_p \, \forall K \in \mathcal{T}_h \right\} \cap H_{0,D}^1(\Omega) \,, \tag{6}$$

where T_K is the invertible mapping from the reference cell $\hat{K}$ to the cell K of $\mathcal{T}_h$ and $\mathbb{Q}_p$ is the space of all polynomials of maximum degree p in each variable. We let $\mathcal{A}_h : H_{0,D}^1(\Omega) \mapsto V_h$ be the discrete operator that is defined by

$$\langle \mathcal{A}_h w, v_h \rangle = \langle \nabla w, \nabla v_h \rangle \quad \text{for all } v_h \in V_h \,.$$

Moreover, $(u_{0,h}, v_{0,h}) \in V_h^2$ and $(u_{\tau,h}^D, v_{\tau,h}^D) \in (C([0,T]; V_h))^2$ define suitable finite element approximations of the initial values (u_0, v_0) and the extended boundary values (u^D, v^D) in Eq. (3). Here, we use interpolation in V_h of the given data.

Now we define our classes of Galerkin–collocation schemes. We follow the lines of [1, 5]. We restrict ourselves to the schemes studied in the numerical experiments presented in Sects. 4.3, 4.4, and 5.2. The definition of classes of Galerkin–collocation schemes with even higher regularity in time is straightforward, but not done here.

Definition 1 (C^l–Regular in Time Galerkin–Collocation Schemes GCC$^l(k)$) Let $l \in \{1, 2\}$ be fixed and $k \geq 2l+1$. For $n = 1, \dots, N$ and given $(u_{\tau,h}|_{I_{n-1}}(t_{n-1}), v_{\tau,h}|_{I_{n-1}}(t_{n-1})) \in V_h^2$ for $n > 1$ and $u_{\tau,h}|_{I_0}(t_0) = u_{0,h}, v_{\tau,h}|_{I_0}(t_0) = v_{0,h}$ for $n = 1$, find $(u^0_{\tau,h}|_{I_n}, v^0_{\tau,h}|_{I_n}) \in (\mathbb{P}_k(I_n; V_h))^2$ such that, for $s_0 \in \mathbb{N}_0$, $s_1 \in \mathbb{N}$ with $s_0, s_1 \leq l$,

$$\partial_t^{s_0} w^0_{\tau,h}|_{I_n}(t_{n-1}) = \partial_t^{s_0} w^0_{\tau,h}|_{I_{n-1}}(t_{n-1}), \quad \text{for } w^0_{\tau,h} \in \left\{u^0_{\tau,h}, v^0_{\tau,h}\right\}, \tag{7}$$

$$\partial_t^{s_1} u^0_{\tau,h}|_{I_n}(t_n) - \partial_t^{s_1-1} v^0_{\tau,h}|_{I_n}(t_n) = 0, \tag{8}$$

$$\begin{aligned} \partial_t^{s_1} v^0_{\tau,h}|_{I_n}(t_n) + \mathcal{A}_h \partial_t^{s_1-1} u^0_{\tau,h}|_{I_n}(t_n) &= \partial_t^{s-1} f(t_n) \\ &\quad - \partial_t^{s_1} v^D_{\tau,h}|_{I_n}(t_n) - \mathcal{A}_h \partial_t^{s_1-1} u^D_{\tau,h}|_{I_n}(t_n), \end{aligned} \tag{9}$$

and, for all $(\varphi_{\tau,h}, \psi_{\tau,h}) \in (\mathbb{P}_0(I_n; V_h))^2$,

$$\int_{I_n} \Big(\langle \partial_t u^0_{\tau,h}, \varphi_{\tau,h}\rangle_\Omega - \langle v^0_{\tau,h}, \varphi_{\tau,h}\rangle_\Omega\Big)\, \mathrm{d}t = 0, \tag{10}$$

$$\begin{aligned} \int_{I_n} \Big(\langle \partial_t v^0_{\tau,h}, \psi_{\tau,h}\rangle_\Omega + \langle \mathcal{A} u^0_{\tau,h}, \psi_{\tau,h}\rangle_\Omega\Big)\, \mathrm{d}t &= \int_{I_n} \langle f, \psi_{\tau,h}\rangle_\Omega \, \mathrm{d}t \\ &\quad - \int_{I_n} \Big(\langle \partial_t v^D_{\tau,h}, \psi_{\tau,h}\rangle_\Omega + \langle \mathcal{A}_h u^D_{\tau,h}, \psi_{\tau,h}\rangle_\Omega\Big)\, \mathrm{d}t. \end{aligned} \tag{11}$$

Remark 2

- In Eq. (7), the discrete initial values $(\partial_t u_{\tau,h}(0), \partial_t v_{\tau,h}(0))$ arise for $s_0 = 1$. For $\partial_t u_{\tau,h}(0)$ we use a suitable finite element approximation $v_{0,h} \in V_h$ (here, an interpolation) of $v_0 \in V$. For $\partial_t v_{\tau,h}(0)$ we evaluate the wave equation in the initial time point and use a suitable finite element approximation (here, an interpolation) of $\partial_t^2 u(0) = c^2 \Delta u(0) + f(0)$. For $s_0 = 2$ in Eq. (7), the initial value $\partial_t^2 v_{\tau,h}(0)$ is computed as a suitable finite element approximation (here, an interpolation) of $\partial_t^3 u(0) = c^2 \Delta \partial_t u(0) + \partial_t f(0)$. Mathematically, this approach requires that the partial equation and its time derivative are satisfied up to the initial time point and, thereby, sufficient regularity of the continuous solution. Without such regularity assumptions, the application of higher order discretization schemes cannot be justified rigorously. Nevertheless, in practice such methods often show a superiority over lower-order ones, even for solutions without the expected high regularity (cf. Sect. 4.4).

- From Eq. (7), $(u_{\tau,h}, v_{\tau,h}) \in (C^l(\overline{I}; V_h))^2$, for fixed $l \in \{1,2\}$, is easily concluded.

An optimal order error analysis for the $\text{GCC}^1(k)$ family of schemes of Definition 1 is provided in [1]. The following theorem is proved.

Theorem 1 (Error Estimates for $(u_{\tau,h}, v_{\tau,h})$ of $\text{GCC}^1(k)$) *Let $l = 1$ and $k \geq 3$. For the error $(e^u, e^v) = (u - u_{\tau,h}, v - v_{\tau,h})$ of the fully discrete scheme $\text{GCC}^l(k)$ of Definition 1 there holds that*

$$\|e^u(t)\| + \|e^v(t)\| \lesssim \tau^{k+1} + h^{p+1}, \quad t \in \overline{I},$$
$$\|\nabla e^u(t)\| \lesssim \tau^{k+1} + h^p, \quad t \in \overline{I},$$

as well as

$$\|e^u(t)\|_{L^2(I;H)} + \|e^v(t)\|_{L^2(I;H)} \lesssim \tau^{k+1} + h^{p+1},$$
$$\|\nabla e^u(t)\|_{L^2(I;H)} \lesssim \tau^{k+1} + h^p.$$

Error estimates for the $\text{GCC}^2(k)$ family remain as a work for the future. In Sect. 5.2, the convergence of $\text{GCC}^2(5)$ is demonstrated numerically. Further, we note that a computationally cheap post-processing of improved regularity and accuracy for continuous Galerkin–Petrov methods is presented and studied in [1, 4].

In the next sections we study the schemes $\text{GCC}^1(3)$ and $\text{GCC}^2(5)$ of Definition 1 in detail. Their algebraic forms are derived and the algebraic linear solver are presented. Finally, the results of our numerical experiments with the proposed methods are presented. Here, we restrict ourselves to the lowest-order cases with $k = 3$ for $l = 1$ and $k = 5$ for $l = 2$ of Definition 1. This is sufficient to demonstrate the potential of the Galerkin–collocation approach and its superiority over the standard continuous Galerkin approach in space and time [4, 15]. An implementation of $\text{GCC}^l(k)$ for higher values of k along with efficient algebraic solvers is currenty still missing.

4 Galerkin–Collocation $\text{GCC}^1(3)$

Here, we derive the algebraic system of the $\text{GCC}^1(3)$ approach and discuss our algebraic solver for the arising block system. For brevity, the derivation is done for $k = 3$ only. The generalization to larger values of k is straightforward; cf. [1].

4.1 Fully Discrete System

To derive the discrete counterparts of the variational conditions (10), (11) and the collocation constraints (7)–(9), we let $\{\phi_j\}_{j=1}^J \subset V_h$, denote a (global) nodal Lagrangian basis of V_h. The mass matrix $\boldsymbol{M}$ and the stiffness matrix $\boldsymbol{A}$ are defined by

$$\boldsymbol{M} := \left(\langle \phi_i, \phi_j\rangle_\Omega\right)_{i,j=1}^J , \qquad \boldsymbol{A} := \left(\langle \nabla\phi_i, \nabla\phi_j\rangle_\Omega\right)_{i,j=1}^J , \tag{12}$$

On the reference time interval $\hat{I} = [0, 1]$ we define a Hermite-type basis $\{\hat{\xi}_l\}_{l=0}^3 \subset \mathbb{P}_3(\hat{I}; \mathbb{R})$ of $\mathbb{P}_3(\hat{I}; \mathbb{R})$ by the conditions

$$\begin{aligned}
&\hat{\xi}_0(0) = 1\,, && \hat{\xi}_0(1) = 0\,, && \partial_t\hat{\xi}_0(0) = 0\,, && \partial_t\hat{\xi}_0(1) = 0\,,\\
&\hat{\xi}_1(0) = 0\,, && \hat{\xi}_1(1) = 0\,, && \partial_t\hat{\xi}_1(0) = 1\,, && \partial_t\hat{\xi}_1(1) = 0\,,\\
&\hat{\xi}_2(0) = 0\,, && \hat{\xi}_2(1) = 1\,, && \partial_t\hat{\xi}_2(0) = 0\,, && \partial_t\hat{\xi}_2(1) = 0\,,\\
&\hat{\xi}_3(0) = 0\,, && \hat{\xi}_3(1) = 0\,, && \partial_t\hat{\xi}_3(0) = 0\,, && \partial_t\hat{\xi}_3(1) = 1\,.
\end{aligned} \tag{13}$$

These conditions then define the basis of $\mathbb{P}_3(\hat{I}; \mathbb{R})$ by

$$\hat{\xi}_0 = 1 - 3t^2 + 2t^3, \quad \hat{\xi}_1 = t - 2t^2 + t^3, \quad \hat{\xi}_2 = 3t^2 - 2t^3, \quad \hat{\xi}_3 = -t^2 + t^3.$$

By means of the affine transformation $T_n(\hat{t}) := t_{n-1} + \tau_n \cdot \hat{t}$, with $\hat{t} \in \hat{I}$, from the reference interval $\hat{I}$ to I_n such that $t_{n-1} \doteq T_n(0)$ and $t_n = T_n(1)$, the basis $\{\xi_l\}_{l=0}^3 \subset \mathbb{P}_3(I_n; \mathbb{R})$ is given by $\xi_l = \hat{\xi}_l \circ T_n^{-1}$ for $l = 0, \dots, 3$. In terms of basis functions, $w_{\tau,h} \in \mathbb{P}_3(I_n; V_h)$ is thus represented by

$$w_{\tau,h}(\boldsymbol{x}, t) = \sum_{l=0}^{3}\sum_{j=1}^{J} w_{n,l,j}\phi_j(\boldsymbol{x})\xi_l(t)\,, \quad (\boldsymbol{x}, t) \in \Omega \times \overline{I_n}\,. \tag{14}$$

For $\zeta_0 \equiv 1$ on $\overline{I_n}$, a test basis of $\mathbb{P}_0(I_n; V_h)$ is then given by

$$\mathcal{B} = \{\phi_1\zeta_0, \dots, \phi_J\zeta_0\}\,. \tag{15}$$

To evaluate the time integrals on the right-hand side of Eq. (11) we still apply the Hermite-type interpolation operator $I_{\tau|I_n}$, on I_n, defined by

$$I_{\tau|I_n} g(t) := \sum_{s=0}^{l} \tau_n^s \hat{\xi}_s(0)\, \underbrace{\partial_t^s g|_{I_n}(t_{n-1})}_{=:g_s} + \sum_{s=0}^{l} \tau_n^s \hat{\xi}_{s+l+1}(1)\, \underbrace{\partial_t^s g|_{I_n}(t_n)}_{=:g_{s+l+1}}\,. \tag{16}$$

Here, the values $\partial_t^s g|_{I_n}(t_{n-1})$ and $\partial_t^s g|_{I_n}(t_n)$ in (16) denote the corresponding one-sided limits the values $\partial_t^s g(t)$ from the interior of I_n.

Now, we can put the equations of the proposed GCC[1](3) approach in their algebraic forms. In the variational equations (10) and (11), we use the representation (14) for each component of $(u_{\tau,h}, v_{\tau,h}) \in (\mathbb{P}_3(I_n; V_h))^2$, choose the test functions (15) and interpolate the right-hand side of (11) by applying (16). All of the arising time integrals are evaluated analytically. Then, we can recover the variational conditions (10) and (11) on the subinterval I_n in their algebraic forms

$$\boldsymbol{M}\left(-\boldsymbol{u}_{n,0}^0 + \boldsymbol{u}_{n,2}^0\right) - \tau_n \boldsymbol{M}\left(\frac{1}{2}\boldsymbol{v}_{n,0}^0 + \frac{1}{12}\boldsymbol{v}_{n,1}^0 + \frac{1}{2}\boldsymbol{v}_{n,2}^0 - \frac{1}{12}\boldsymbol{v}_{n,3}^0\right) = \boldsymbol{0}\,, \tag{17}$$

$$\begin{aligned} \boldsymbol{M}\left(-\boldsymbol{v}_{n,0}^0 + \boldsymbol{v}_{n,2}^0\right) + \tau_n \boldsymbol{A}\left(\frac{1}{2}\boldsymbol{u}_{n,0}^0 + \frac{1}{12}\boldsymbol{u}_{n,1}^0 + \frac{1}{2}\boldsymbol{u}_{n,2}^0 - \frac{1}{12}\boldsymbol{u}_{n,3}^0\right) = \\ \tau_n \boldsymbol{M}\left(\frac{1}{2}\boldsymbol{f}_{n,0} + \frac{1}{12}\boldsymbol{f}_{n,1} + \frac{1}{2}\boldsymbol{f}_{n,2} - \frac{1}{12}\boldsymbol{f}_{n,3}\right) - \boldsymbol{M}\left(-\boldsymbol{v}_{n,0}^D + \boldsymbol{v}_{n,2}^D\right) \\ - \tau_n \boldsymbol{A}\left(\frac{1}{2}\boldsymbol{u}_{n,0}^D + \frac{1}{12}\boldsymbol{u}_{n,1}^D + \frac{1}{2}\boldsymbol{u}_2^D - \frac{1}{12}\boldsymbol{u}_{n,3}^D\right). \end{aligned} \tag{18}$$

This gives us the first two equations for the set of eight unknown solution vectors $\mathcal{L} = \{\boldsymbol{u}_{n,0}^0, \dots, \boldsymbol{u}_{n,3}^0, \boldsymbol{v}_{n,0}^0, \dots, \boldsymbol{v}_{n,3}^0\}$ on each subinterval I_n, where each of these vectors is defined by means of (14) through $\boldsymbol{w} = (w_1, \dots, w_J)^\top$ for $\boldsymbol{w} \in \mathcal{L}$.

Next, we study the algebraic forms of the collocations conditions (7)–(9). By means of the definition (13) of the basis of $\mathbb{P}(I_n; \mathbb{R})$, the constraints (7) read as

$$\boldsymbol{u}_{n,0}^0 = \boldsymbol{u}_{n-1,2}^0\,, \quad \boldsymbol{u}_{n,1}^0 = \boldsymbol{u}_{n-1,3}^0\,, \quad \boldsymbol{v}_{n,0}^0 = \boldsymbol{v}_{n-1,2}^0\,, \quad \boldsymbol{v}_{n,1}^0 = \boldsymbol{v}_{n-1,3}^0\,. \tag{19}$$

By means of (13) along with (12), the conditions (8) and (9) can be recovered as

$$\boldsymbol{M}\frac{1}{\tau_n}\boldsymbol{u}_{n,3}^0 - \boldsymbol{M}\boldsymbol{v}_{n,2}^0 = \boldsymbol{0}\,, \tag{20}$$

$$\boldsymbol{M}\frac{1}{\tau_n}\boldsymbol{v}_{n,3}^0 + \boldsymbol{A}\boldsymbol{u}_{n,2}^0 = \boldsymbol{M}\boldsymbol{f}_{n,2} - \boldsymbol{M}\frac{1}{\tau_n}\boldsymbol{v}_{n,3}^D - \boldsymbol{A}\boldsymbol{u}_{n,2}^D\,. \tag{21}$$

Putting relations (19) into the identities (17) and (18) and combining the resulting equations with (20) and (21) yields for the subinterval I_n the linear block system

$$\boldsymbol{S}\boldsymbol{x} = \boldsymbol{b} \tag{22}$$

for the vector of unknowns

$$\boldsymbol{x} = \left(\left(\boldsymbol{v}_{n,2}^0\right)^\top, \left(\boldsymbol{v}_{n,3}^0\right)^\top, \left(\boldsymbol{u}_{n,2}^0\right)^\top, \left(\boldsymbol{u}_{n,3}^0\right)^\top\right)^\top \tag{23}$$

and the system matrix $\boldsymbol{S}$ and right-hand side $\boldsymbol{b}$ given by

$$\boldsymbol{S} = \begin{pmatrix} \boldsymbol{M} & \boldsymbol{0} & \boldsymbol{0} & \frac{1}{\tau_n}\boldsymbol{M} \\ \boldsymbol{0} & \frac{1}{\tau_n}\boldsymbol{M} & \boldsymbol{A} & \boldsymbol{0} \\ -\frac{\tau_n}{2}\boldsymbol{M} & \frac{\tau_n}{12}\boldsymbol{M} & \boldsymbol{M} & \boldsymbol{0} \\ \boldsymbol{M} & \boldsymbol{0} & \frac{\tau_n}{2}\boldsymbol{A} & -\frac{\tau_n}{12}\boldsymbol{A} \end{pmatrix}, \quad \boldsymbol{b} = \begin{pmatrix} \boldsymbol{0} \\ \boldsymbol{M}\left(\boldsymbol{f}_{n,2} - \frac{1}{\tau_n}\boldsymbol{v}^D_{n,3}\right) - \boldsymbol{A}\boldsymbol{u}^D_{n,2} \\ \boldsymbol{M}\left(\boldsymbol{u}^0_{n,0} + \frac{\tau_n}{2}\boldsymbol{v}^0_{n,0} + \frac{\tau_n}{12}\boldsymbol{v}^0_{n,1}\right) \\ \boldsymbol{b}_{n,4} \end{pmatrix}, \tag{24}$$

with $\boldsymbol{b}_{n,4} = \boldsymbol{M}\big(\boldsymbol{v}^0_{n,0} + \boldsymbol{v}^D_{n,0} - \boldsymbol{v}^D_{n,2} + \frac{\tau_n}{2}(\boldsymbol{f}_{n,0} + \boldsymbol{f}_{n,2}) + \frac{\tau_n}{12}(\boldsymbol{f}_{n,1} - \boldsymbol{f}_{n,3})\big) - \boldsymbol{A}\big(\frac{\tau_n}{2}(\boldsymbol{u}^0_{n,0} + \boldsymbol{u}^D_{n,0} + \boldsymbol{u}^D_{n,2}) + \frac{\tau_n}{12}(\boldsymbol{u}^0_{n,1} + \boldsymbol{u}^D_{n,1} - \boldsymbol{u}^D_{n,3})\big)$. By means of the collocation constraints (19), the number of unknown coefficient vectors for the discrete solution $(u_{\tau,h|I_n}, v_{\tau,h|I_n}) \in (\mathbb{P}_3(I_n; V_h))^2$ is thus effectively reduced from eight to four vectors, assembled now in $\boldsymbol{x}$ by (23).

We note that the first two rows of Eq. (24) represent the collocation conditions (20) and (21). They have a sparser structure then the last two rows representing the variational conditions which can be advantageous or exploited for the construction of efficient iterative solvers for (22). Compared with a pure variational approach (cf. [12, 13, 15]), more degrees or freedom are obtained directly by computationally cheap vector identities (cf. (17)) in $\text{GCC}^l(k)$ such that they can be eliminated from the overall linear system and, thereby, used to reduce the system size.

4.2 Solver Technology

In the sequel, we present two different iterative approaches for solving the linear system (22) with the non-symmetric matrix $\boldsymbol{S}$. In Sect. 4.4, a runtime comparison between the two concepts is provided. As basic toolbox we use the deal.II finite element library [3] along with the Trilinos library [11] for parallel computations.

4.2.1 First Approach: Condensing the Linear System

The first method for solving (22) is based on the concepts developed in [16]. The key idea is to use Gaussian block elimination within the system matrix $\boldsymbol{S}$ and to end up with a linear system with matrix $\boldsymbol{S}_r$ of reduced size for one of the subvectors in $\boldsymbol{x}$ in (23) only, and to compute the remaining subvectors of (23) by computationally cheap post-processing steps afterwards. The reduced system matrix $\boldsymbol{S}_r$ should have sufficient potenial that an efficient preconditioner for the iterative solution of the reduced system can be constructed. Of course, the Gauss elimination on the block level can be done in different ways. The goal of our approach is to avoid the inversion of the stiffness matrix $\boldsymbol{A}$ in (24) in the computation of the condensed system matrix $\boldsymbol{S}_r$ such that a matrix-vector multiplication with $\boldsymbol{S}_r$ just involves calculating $\boldsymbol{M}^{-1}$. At least for discontinuous Galerkin methods in space, where $\boldsymbol{M}$ is block diagonal, this is computationally cheap; cf. [16, 17]. We note that a continuous Galerkin approach in space is used here only in order to simplify the notation and since the discretization in time by the combined Galerkin–collocation approach is in the scope of interest.

Here, we choose the subvector $\boldsymbol{u}_{n,2}^0$ of $\boldsymbol{x}$ in (23) as the essential unknown, i.e. as the unknown solution vector of the condensed system with matrix $\boldsymbol{S}_r$. By block Gaussian elemination we then end up with solving the linear system,

$$\left(\boldsymbol{M} + \frac{\tau_n^2}{12}\boldsymbol{A} + \frac{\tau_n^4}{144}\boldsymbol{A}\boldsymbol{M}^{-1}\boldsymbol{A}\right)\boldsymbol{u}_{n,2}^0 = \boldsymbol{b}_{n,r} \tag{25}$$

with right-hand side vector

$$\begin{aligned}
\boldsymbol{b}_{n,r} = {} & \boldsymbol{M}\left(\frac{1}{2}\boldsymbol{f}_{n,0}+\frac{1}{12}\boldsymbol{f}_{n,1}+\frac{1}{3}\boldsymbol{f}_{n,2}-\frac{1}{12}\boldsymbol{f}_{n,3}\right)+\boldsymbol{M}\left(2\boldsymbol{v}_{n,0}^0+\frac{1}{6}\boldsymbol{v}_{n,1}^0+\frac{2}{\tau_n}\boldsymbol{u}_{n,0}^0\right)\\
& -\boldsymbol{A}\left(\frac{2}{3}\tau_n\boldsymbol{u}_{n,0}^0+\frac{1}{12}\tau_n\boldsymbol{u}_{n,1}^0+\frac{1}{12}\tau_n^2\boldsymbol{v}_{n,2}^0+\frac{1}{72}\tau_n^2\boldsymbol{v}_{n,1}^0\right)+\boldsymbol{M}\left(2\boldsymbol{v}_{n,0}^D+\frac{1}{6}\boldsymbol{v}_{n,1}^D+\frac{2}{\tau_n}\boldsymbol{u}_{n,0}^D-\frac{2}{\tau_n}\boldsymbol{u}_{n,2}^D\right)\\
& +\boldsymbol{A}\left(\frac{1}{72}\tau_n^3\boldsymbol{f}_{n,2}-\boldsymbol{M}^{-1}\frac{1}{72}\tau_n^3\boldsymbol{u}_{n,2}^D\right)+\boldsymbol{A}\left(-\frac{2}{3}\tau_n\boldsymbol{u}_{n,0}^D-\frac{1}{12}\tau_n\boldsymbol{u}_{n,1}^D-\frac{\tau_n}{6}\boldsymbol{u}_{n,2}^D-\frac{\tau_n^2}{12}\boldsymbol{v}_{n,0}^D-\frac{\tau_n^2}{72}\boldsymbol{v}_{n,1}^D\right).
\end{aligned}$$

The product of $\boldsymbol{A}\boldsymbol{M}^{-1}\boldsymbol{A}$ in (25) mimics the discretization of a fourth order operator due to the appearance of the product of $\boldsymbol{A}$ with its "weighted" form $\boldsymbol{M}^{-1}\boldsymbol{A}$. Thereby, the condition number of the condensed system is strongly increased (cf. [16]) which is the main drawback in this concept of condensing the overall system (22) to (25) for the essential unknown $\boldsymbol{u}_{n,2}^0$. On the other hand, since $\boldsymbol{M}$ and $\boldsymbol{A}$ are symmetric and, thus, $\boldsymbol{A}\boldsymbol{M}^{-1}\boldsymbol{A} = (\boldsymbol{A}\boldsymbol{M}^{-1}\boldsymbol{A})^\top$, the condensed matrix $\boldsymbol{S}_r$ is symmetric such that the preconditioned conjugate gradient method can be applied. Solving systems of type (25) is carefully studied in [16, 17] and the references given therein.

We solve (25) by the conjugate gradient method. The left preconditioning operator

$$\boldsymbol{P} = \boldsymbol{K}_\mu \boldsymbol{M}^{-1}\boldsymbol{K}_\mu = \left(\mu\boldsymbol{M} + \frac{\tau_n^2}{4}\boldsymbol{A}\right)\boldsymbol{M}^{-1}\left(\mu\boldsymbol{M} + \frac{\tau_n^2}{4}\boldsymbol{A}\right),$$

with positive $\mu \in \mathbb{R}$, chosen such that the spectral norm of $\boldsymbol{P}^{-1}\boldsymbol{S}_r$ is minimised, is applied. For details of the choice of the parameter μ, we refer to [16, 17]. Here, we use $\mu = \sqrt{11/2}$. In order to apply the preconditioning operator $\boldsymbol{P}$ in the conjugate gradient iterations, without assembling $\boldsymbol{P}$ explicitly, i.e. to solve the auxiliary system with matrix $\boldsymbol{P}$, we have to solve linear systems for the mass matrix $\boldsymbol{M}$

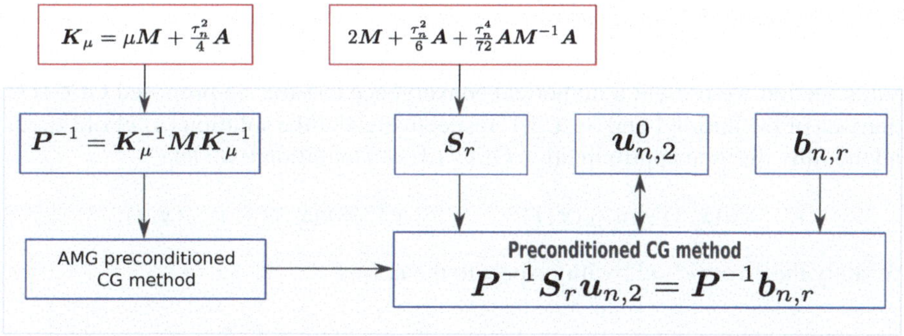

Fig. 2 Preconditioning and solver for the condensed system (25) of GCC[1](3)

and the stiffness matrix $\boldsymbol{A}$. For this, we use embedded conjugate gradient iterations combined with an algebraic multigrid preconditioner of the Trilinos library [11]. The overvall algorithm for solving (25) is sketched in Fig. 2. The advantage of this approach is that we just have to store $\boldsymbol{M}$ and $\boldsymbol{A}$ as sparse matrices in the computer memory. We never have to assemble the full matrix $\boldsymbol{S}$ from (24), nor do we have to store the reduced matrix $\boldsymbol{S}_r$ from (25). Finally, the remaining unknown subvectors $\boldsymbol{v}_{n,2}^0$, $\boldsymbol{v}_{n,3}^0$ and $\boldsymbol{u}_{n,3}^0$ in (22) are successively computed in post-processing steps.

4.2.2 Second Approach: Solving the Non-symmetric System

The second approach used to solve (22) relies on assembling the system matrix $\boldsymbol{S}$ of (24) as a sparse matrix and solving the resulting non-symmetric system. For smaller dimensions of $\boldsymbol{S}$ a parallel direct solver [18] is used. For constant time step sizes τ_n the matrix $\boldsymbol{S}$ needs to be factorized once only, which results in excellent performance properties for large sequences of time steps. For high-dimensional problems, we use the Generalized Minimal Residual (GMRES) method, an iterative Krylov subspace method, to solve (22). The drawback of this approach then comes through the necessity to provide an efficient preconditioner, i.e. an approximation to the inverse of $\boldsymbol{S}$, for the complex block matrix $\boldsymbol{S}$ of (24). Here, we use the algebraic multigrid method as preconditioning technique. We use the MueLue preconditioner [23], which is part of the Trilinos project, with non-symmetric smoothed aggregation. We use an usual V-cycle algorithm along with a symmetric successive over-relaxation (SSOR) smoother with a damping factor of 1.33. The design of efficient algebraic solvers for block systems like (22), and for higher order variational time discretizations in general, is still an active field of research. We expect further improvement in the future.

4.3 Numerical Convergence Tests

In this section we present a numerical convergence test for the proposed GCC^1(3) approach of Definition 1 and Sect. 4.1, respectively. For the solution $\{u, v\}$ of Eq. (2) and the fully discrete approximation GCC^1(3) of Definition 1 we let

$$e^u := u(\boldsymbol{x}, t) - u_{\tau,h}(\boldsymbol{x}, t), \qquad e^v := v(\boldsymbol{x}, t) - v_{\tau,h}(\boldsymbol{x}, t).$$

We study the error (e^u, e^v) with respect to the norms

$$\|e^w\|_{L^\infty(L^2)} := \max_{t\in I}\left(\int_\Omega |e^w|^2 \,\mathrm{d}x\right)^{\frac{1}{2}}, \quad \|e^w\|_{L^2(L^2)} := \left(\int_I \int_\Omega |e^w|^2 \,\underset{\cdot}{\mathrm{x}}\,\mathrm{d}t\right)^{\frac{1}{2}}, \tag{26}$$

where $w \in (u, v)$, and in the energy quantities

$$|||E|||_{L^\infty} := \max_{t\in I}\left(\int_\Omega |\nabla e^u|^2 + |e^v|^2 \,\mathrm{d}x\right)^{\frac{1}{2}}, \quad |||E|||_{L^2} := \left(\int_I \int_\Omega |\nabla e^u|^2 + |e^v|^2 \,\mathrm{d}\boldsymbol{x}\,\mathrm{d}t\right)^{\frac{1}{2}}. \tag{27}$$

All L^∞-norms in time are computed on the discrete time grid

$$I = \left\{t_n^d | t_n^d = t_{n-1} + d \cdot k_n \cdot \tau_n, \quad k_n = 0.001, d = 0, \ldots, 999, n = 1, \ldots, N\right\}.$$

For our first convergence test we prescribe the solution

$$u_1(\boldsymbol{x}, t) = \sin(4\pi t) \cdot x_1 \cdot (x_1 - 1) \cdot x_2 \cdot (x_2 - 1) \tag{28}$$

on $\Omega \times I = [0, 1]^2 \times [0, 1]$. We let $c = 1$, use a constant mesh size $h_0 = 0.25$ and start with the time step size $\tau_0 = 0.1$. We compute the errors on a sequence of successively refined time meshes by halving the step sizes in each refinement step. We choose a bicubic discretization of the space variables in V_h^3 (cf. (6)) such that the spatial part of the solution is resolved exactly by its numerical approximation. Table 1 summarizes the computed errors and experimental orders of convergence. The expected convergence rates of Theorem 1 are nicely confirmed.

In our second numerical experiment we study the space-time convergence behavior of a solution satisfying non-homogeneous Dirichlet boundary conditions,

$$u_2(\boldsymbol{x}, t) = \sin(2 \cdot \pi \cdot t + x_1) \cdot \sin(2 \cdot \pi \cdot t \cdot x_2) \tag{29}$$

on $\Omega \times I = (0, 1)^2 \times [0, 1]$. We choose a bicubic discretization in V_h^3 (cf. (6)) of the space variables. We refine the space-time mesh by halving both step sizes in each refinement step. Table 2 shows the computed errors and experimental orders of convergence for this example. In all measured norms, optimal rates in space and time (cf. Theorem 1) are confirmed. This underlines the correct treatment of the prescribed non-homogeneous Dirichlet boundary conditions.

Table 1 Calculated errors for $GCC^1(3)$ and solution (28)

τ	h	$\|e^u\|_{L^\infty(L^2)}$	EOC	$\|e^v\|_{L^\infty(L^2)}$	EOC	$\|\|\|E\|\|\|_{L^\infty}$	EOC
$\tau_0/2^0$	h_0	2.318e−04	–	1.543e−03	–	1.574e−03	–
$\tau_0/2^1$	h_0	1.541e−05	3.91	9.694e−05	3.99	1.004e−04	3.97
$\tau_0/2^2$	h_0	9.825e−07	3.97	6.260e−06	3.95	6.478e−06	3.95
$\tau_0/2^3$	h_0	6.185e−08	3.99	3.946e−07	3.99	4.082e−07	3.99
$\tau_0/2^4$	h_0	3.873e−09	4.00	2.472e−08	4.00	2.557e−08	4.00
$\tau_0/2^5$	h_0	2.422e−10	4.00	1.548e−09	4.00	1.609e−09	3.99
τ	h	$\|e^u\|_{L^2(L^2)}$	EOC	$\|e^v\|_{L^2(L^2)}$	EOC	$\|\|\|E\|\|\|_{L^2}$	EOC
$\tau_0/2^0$	h_0	1.634e−04	–	1.232e−03	–	1.441e−03	–
$\tau_0/2^1$	h_0	1.070e−05	3.93	7.864e−05	3.97	9.269e−05	3.96
$\tau_0/2^2$	h_0	6.765e−07	3.98	4.943e−06	3.99	5.836e−06	3.99
$\tau_0/2^3$	h_0	4.240e−08	4.00	3.094e−07	4.00	3.654e−07	4.00
$\tau_0/2^4$	h_0	2.652e−09	4.00	1.934e−08	4.00	2.285e−08	4.00
$\tau_0/2^5$	h_0	1.659e−10	4.00	1.212e−09	4.00	1.433e−09	3.99

Table 2 Calculated errors for $GCC^1(3)$ and solution (29)

τ	h	$\|e^u\|_{L^\infty(L^2)}$	EOC	$\|e^v\|_{L^\infty(L^2)}$	EOC	$\|\|\|E\|\|\|_{L^\infty}$	EOC
$\tau_0/2^0$	$h_0/2^0$	3.486e−03	–	3.602e−02	–	5.013e−02	–
$\tau_0/2^1$	$h_0/2^1$	2.329e−04	3.90	2.392e−03	3.90	3.338e−03	3.92
$\tau_0/2^2$	$h_0/2^2$	1.483e−05	3.97	1.527e−04	3.97	2.128e−04	3.98
$\tau_0/2^3$	$h_0/2^3$	9.320e−07	3.99	9.609e−06	3.99	1.338e−05	3.99
$\tau_0/2^4$	$h_0/2^4$	5.837e−08	4.00	6.022e−07	4.00	8.383e−08	4.00
$\tau_0/2^5$	$h_0/2^5$	3.649e−09	4.00	3.767e−08	4.00	5.243e−08	4.00
τ	h	$\|e^u\|_{L^2(L^2)}$	EOC	$\|e^v\|_{L^2(L^2)}$	EOC	$\|\|\|E\|\|\|_{L^2}$	EOC
$\tau_0/2^0$	$h_0/2^0$	2.700e−03	–	2.568e−02	–	3.458e−02	–
$\tau_0/2^1$	$h_0/2^1$	1.771e−04	3.93	1.689e−03	3.93	2.278e−03	3.92
$\tau_0/2^2$	$h_0/2^2$	1.120e−05	3.98	1.070e−04	3.98	1.444e−04	3.98
$\tau_0/2^3$	$h_0/2^3$	7.020e−07	4.00	6.713e−06	3.99	9.061e−06	3.99
$\tau_0/2^4$	$h_0/2^4$	4.391e−08	4.00	4.199e−07	4.00	5.669e−07	4.00
$\tau_0/2^5$	$h_0/2^5$	2.744e−09	4.00	2.624e−08	4.00	3.543e−08	4.00

4.4 Test Case of Structural Health Monitoring

Next, we consider a test problem that is based on [2] and related to typical problems of structural health monitoring by ultrasonic waves (cf. Fig. 1). We aim to compare the $GCC^1(3)$ approach with a standard continuous in time Galerkin–Petrov approach cGP(2) of piecewise quadratic polynomials in time; cf. [4, 12] for details. The cGP(2) scheme has superconvergence properties in the discrete time nodes t_n for $n = 1, \dots, N$ as shown in [4]. Thus, the errors $\max_{n=1,\dots,N} \|e^u(t_n)\|$ and $\max_{n=1,\dots,N} \|e^v(t_n)\|$ for the $GCC^1(3)$ and the cGP(2) scheme admit the same fourth order rate of convergence in time and, thus, are comparable with respect to accuracy.

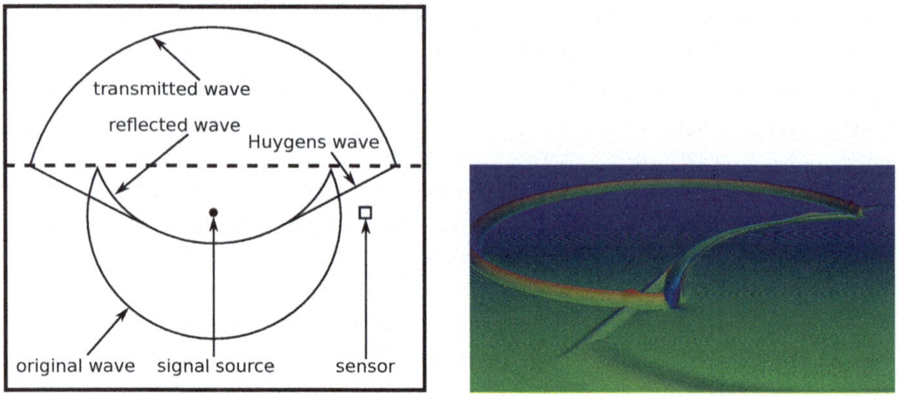

(a) (b)

Fig. 3 (**a**) Test setting (according to [1]). (**b**) Solution at $t = 0.5$ with spatial mesh

The test setting is sketched in Fig. 3a. We consider $\Omega \times I = (-1, 1)^2 \times (0, 1)$, let $f = 0$ and, for simplicity, prescribe homogeneous Dirichlet boundary conditions such that $u^D = 0$. For the initial value we prescribe a regularized Dirac impulse by

$$u_0(\boldsymbol{x}) = e^{-|\boldsymbol{x}_s|^2}\big(1 - |\boldsymbol{x}_s|^2\big)\Theta\big(1 - |\boldsymbol{x}_s|\big), \quad \boldsymbol{x}_s = 100\boldsymbol{x}\,,$$

where Θ is the Heaviside function. The coefficient function $c(\boldsymbol{x})$, mimicing a material parameter, has a jump discontinuity and is given by $c(\boldsymbol{x}) = 1$ for $x_2 < 0.2$ and $c(\boldsymbol{x}) = 9$ for $x_2 \geq 0.2$. Further we put $v_0 = 0$ for the second initial value. Finally, we define the control region $\Omega_c = (0.75 - h_c, 0.75 + h_c) \times (-h_c, h_c)$ where we calculate the signal arrival, at a sensor position for instance, in terms of

$$u_c(t) = \int_{\Omega_c} u_{\tau,h}(\boldsymbol{x}, t)\mathrm{d}\boldsymbol{x}\,. \tag{30}$$

We choose a spatial mesh of 65 536 cells and $\mathbb{Q}_7$ elements; cf. Fig. 3b. This leads to more than 3.2×10^6 degrees of freedom in space in each time step for each of the solution vectors. For each computation of the control quantity (30), with $t \in (0, 1]$, we use a constant time step size τ_n for all time steps and compare the computation with the initially chosen reference time step size of $\tau_0 = 2 \times 10^{-5}$.

Figure 4 shows the signal arrival and control quantity (30) over $t \in (0.6, 1)$ with different choices of the time step sizes for the Galerkin–collocation scheme $\mathrm{GCC}^1(3)$ and the standard Galerkin–Petrov approach cGP(2) (cf. [4, 12]) of a continuous in time approximation. For the cGP(2) approach, very small time step sizes are required to avoid over- and undershoots in the control quantity $u_c(t)$. For the $\mathrm{GCC}^1(3)$ approach with C^1 regularity in time, much larger time steps, approximately 100 times larger, can be applied without loss of accuracy compared

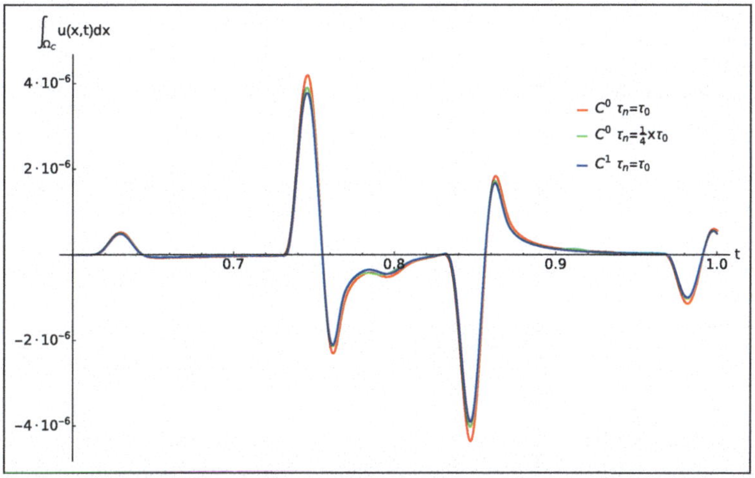

(a)

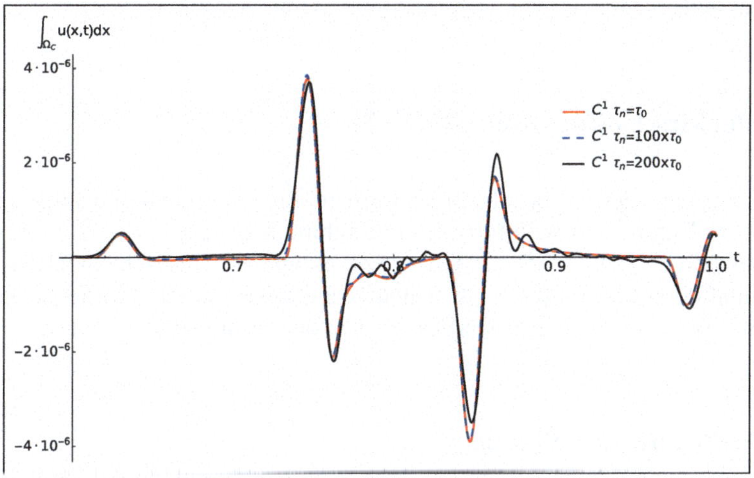

(b)

Fig. 4 Control quantity (30) for GCC^1(3) (method C^1) and cGP(2) (method C^0) for different time step sizes. (**a**) Control quantity (28) for cGP(2) (method C^0) with different time step sizes and reference solution GCC1 (3) (method C^1). (**b**) Control quantity (28) for GCC1 (3) (method C^1) with different time step sizes

to the fully converged reference solution given by GCC^1(3) with step size τ_0. This clearly shows the superiority of the Galerkin–collocation scheme GCC^1(3).

In Table 3 the computational costs are summarized, where r_1 is the runtime for solving the condensed system by the approach of Sect. 4.2.1 and r_2 is the runtime for solving the block system by the approach of Sect. 4.2.2. For the cGP(2) approach, only the first of the either iterative solver techniques was implemented. Recalling

Table 3 Runtime (wall clock time) for $GCC^1(3)$ (method C^1) and cGP(2) (method C^0) for different time step sizes and solvers of Sects. 4.2.1 (r_1) and 4.2.2 (r_2)

DoF (space)	Cores	Method	τ_n	r_1[h]	r_2[h]
3.2×10^6	224	C^0	$0.25 \times \tau_0$	219.3	–
		C^0	τ_0	40.0	–
		C^1	τ_0	46.6	25.3
		C^1	$2 \times \tau_0$	33.1	19.4
		C^1	$25 \times \tau_0$	4.5	2.3
		C^1	$35 \times \tau_0$	3.7	2.2
		C^1	$50 \times \tau_0$	2.9	1.6
		C^1	$100 \times \tau_0$	1.7	0.9
		C^1	$200 \times \tau_0$	1.1	0.7
4.2×10^6	336	C^1	$50 \times \tau_0$	3.3	1.7

from Fig. 4 that $GCC^1(3)$ with $\tau_n = 100 \times \tau_0$ leads to the fully converged solution whereas cGP(2) with $\tau_n = \tau_0$ already shows over- and undershoots, a strong superiority of $GCC^1(3)$ over cGP(2) is observed in Table 3. For both solver, a reduction in the wall clock time by a factor of about 25 is shown.

5 Galerkin–Collocation $GCC^2(5)$

Here, we briefly derive the algebraic form of the Galerkin–collocation scheme $GCC^2(k)$ of Definition 1 with fully discrete solutions $(u_{\tau,h|I_n}, v_{\tau,h|I_n}) \in (X_\tau^k(V_h))^2$ such that $(u_{\tau,h}, v_{\tau,h}) \in (C^2(\overline{I}; V_h))^2$. For brevity, we restrict ourselves to the polynomial degree in time $k = 5$ which is the lowest possible order to get C^2-regularity. The convergence properties are then demonstrated numerically.

5.1 Fully Discrete System

We follow the lines of Sect. 4.1 and use the notation introduced there. The six basis function of $\mathbb{P}_5(\hat{I}; \mathbb{R})$ on the reference interval $\hat{I}$ are defined by the conditions

$$\hat{\xi}_i^{(l)}(j) = \delta_{i-2*j-l,j} \qquad \forall i \in \{0, \cdots, 5\} \quad \wedge \quad j \in \{0, 1\}, \quad l \in \{0, 1, 2\},$$

where $\delta_{i,j}$ denotes the ususal Kronecker symbol. This gives us

$$\hat{\xi}_0 = -6t^5 + 15t^4 - 10t^3 + 1\,, \; \hat{\xi}_1 = -3t^5 + 8t^4 - 6t^3 + t\,, \; \hat{\xi}_2 = -\tfrac{1}{2}t^5 + \tfrac{3}{2}t^4 - \tfrac{3}{2}t^3 + \tfrac{1}{2}t^2\,,$$

$$\hat{\xi}_3 = 6t^5 - 15t^4 + 10t^3\,, \qquad \hat{\xi}_4 = -3t^5 + 7t^4 - 4t^3\,, \qquad \hat{\xi}_5 = \tfrac{1}{2}t^5 - t^4 + \tfrac{1}{2}t^3\,.$$

For this basis of $\mathbb{P}_5(\hat{I};\mathbb{R})$, the discrete variational conditions (10), (11) then read as

$$
\begin{aligned}
&\boldsymbol{M}\Big(-\boldsymbol{u}^0_{n,0}+\boldsymbol{u}^0_{n,3}\Big)-\tau_n\boldsymbol{M}\Big(\frac{1}{2}\boldsymbol{v}^0_{n,0}+\frac{1}{10}\boldsymbol{v}^0_{n,1}+\frac{1}{120}\boldsymbol{v}^0_{n,2}+\frac{1}{2}\boldsymbol{v}^0_{n,3}-\frac{1}{10}\boldsymbol{v}^0_{n,4}+\frac{1}{120}\boldsymbol{v}^0_{n,5}\Big)=\boldsymbol{0}\,,\\
&\boldsymbol{M}\Big(-\boldsymbol{v}^0_{n,0}+\boldsymbol{v}^0_{n,3}\Big)+\tau_n\boldsymbol{A}\Big(\frac{1}{2}\boldsymbol{u}^{n,0}_{n,0}+\frac{1}{10}\boldsymbol{u}^0_{n,1}+\frac{1}{120}\boldsymbol{u}^0_{n,2}+\frac{1}{2}\boldsymbol{u}^0_{n,3}-\frac{1}{10}\boldsymbol{u}^0_{n,4}+\frac{1}{120}\boldsymbol{u}^0_{n,5}\Big)=\\
&\tau_n\boldsymbol{M}\Big(\frac{1}{2}\boldsymbol{f}_{n,0}+\frac{1}{10}\boldsymbol{f}_{n,1}+\frac{1}{120}\boldsymbol{f}_{n,2}+\frac{1}{2}\boldsymbol{f}_{n,3}-\frac{1}{10}\boldsymbol{f}_{n,4}+\frac{1}{120}\boldsymbol{f}_{n,5}\Big)-\boldsymbol{M}\Big(-\boldsymbol{v}^D_{n,0}+\boldsymbol{v}^D_{n,3}\Big)\\
&\qquad-\tau_n\boldsymbol{A}\Big(\frac{1}{2}\boldsymbol{u}^D_{n,0}+\frac{1}{10}\boldsymbol{u}^D_{n,1}+\frac{1}{120}\boldsymbol{u}^D_{n,2}+\frac{1}{2}\boldsymbol{u}^D_{n,3}-\frac{1}{10}\boldsymbol{u}^D_{n,4}+\frac{1}{120}\boldsymbol{u}^D_{n,5}\Big).
\end{aligned}
$$

In the basis, the first collocation conditions (7) yield for $\boldsymbol{w}^0_{n,i}\in\{\boldsymbol{u}^0_{n,i},\boldsymbol{v}^0_{n,i}\}$ that

$$
\boldsymbol{w}^0_{n,0}=\boldsymbol{w}^0_{n-1,3}\,,\qquad \boldsymbol{w}^0_{n,1}=\boldsymbol{w}^0_{n-1,4}\,,\qquad \boldsymbol{w}^0_{n,2}=\boldsymbol{w}^0_{n-1,5}\,,
$$

which reduces the number of unknown solution vectors by 6 on each subinterval I_n. For the collocation conditions (8) and (9) at t_n and $s=1$ we deduce that

$$
\boldsymbol{M}\frac{1}{\tau_n}\boldsymbol{u}^0_{n,4}-\boldsymbol{M}\boldsymbol{v}^0_{n,3}=\boldsymbol{0}\,,\quad \boldsymbol{M}\frac{1}{\tau_n}\boldsymbol{v}^0_{n,4}+\boldsymbol{A}\boldsymbol{u}^0_{n,3}=\boldsymbol{M}\boldsymbol{f}_{n,3}-\boldsymbol{M}\frac{1}{\tau_n}\boldsymbol{v}^D_{n,4}-\boldsymbol{A}\boldsymbol{u}^D_{n,3}\,.
$$

Similarly, for $s=2$ the collocation conditions (8) and (9) at t_n read as

$$
\boldsymbol{M}\frac{1}{\tau_n}\boldsymbol{u}^0_{n,5}-\boldsymbol{M}\boldsymbol{v}^0_{n,4}=\boldsymbol{0}\,,\quad \boldsymbol{M}\frac{1}{\tau_n}\boldsymbol{v}^0_{n,5}+\boldsymbol{A}\boldsymbol{u}^0_{n,4}=\boldsymbol{M}\boldsymbol{f}_{n,4}-\boldsymbol{M}\frac{1}{\tau_n}\boldsymbol{v}^D_{n,5}-\boldsymbol{A}\boldsymbol{u}^D_{n,4}.
$$

Summarizing, we recover the previous conditions as the linear system $\boldsymbol{S}\boldsymbol{x}=\boldsymbol{b}$ for the vector of unknowns $\boldsymbol{x}=\Big((\boldsymbol{u}^0_{n,3})^\top,(\boldsymbol{u}^0_{n,4})^\top,(\boldsymbol{v}^0_{n,5})^\top,(\boldsymbol{u}^0_{n,5})^\top\Big)^\top$ with the system matrix $\boldsymbol{S}$ and right-hand side vector $\boldsymbol{b}$ given by

$$
\boldsymbol{S}=\begin{pmatrix}\boldsymbol{A} & \boldsymbol{0} & \boldsymbol{0} & \frac{1}{\tau_n^2}\boldsymbol{M}\\ \boldsymbol{0} & \boldsymbol{A} & \frac{1}{\tau_n}\boldsymbol{M} & \boldsymbol{0}\\ \boldsymbol{M} & -\frac{1}{2}\boldsymbol{M} & -\frac{\tau_n}{120}\boldsymbol{M} & \frac{1}{10}\boldsymbol{M}\\ \frac{\tau_n}{2}\boldsymbol{A} & \frac{1}{\tau_n}\boldsymbol{M}-\frac{\tau_n}{10}\boldsymbol{A} & \boldsymbol{0} & \frac{\tau_n}{120}\boldsymbol{A}\end{pmatrix},\quad \boldsymbol{b}=\begin{pmatrix}\boldsymbol{f}_{n,3}-\boldsymbol{A}\boldsymbol{u}^D_{n,3}-\frac{1}{\tau_n}\boldsymbol{M}\boldsymbol{v}^D_{n,4}\\ \boldsymbol{f}_{n,4}-\boldsymbol{A}\boldsymbol{u}^D_{n,4}-\frac{1}{\tau_n}\boldsymbol{M}\boldsymbol{v}^D_{n,5}\\ \boldsymbol{b}_{n,3}\\ \boldsymbol{b}_{n,4}\end{pmatrix},
\tag{31}
$$

with $\boldsymbol{b}_{n,3}=\boldsymbol{M}(\boldsymbol{u}^0_{n,0}+\boldsymbol{u}^D_{n,0}-\boldsymbol{u}^D_{n,3}+\frac{\tau_n}{2}(\boldsymbol{v}^0_{n,0}+\boldsymbol{v}^D_{n,0})+\frac{\tau_n}{10}(\boldsymbol{v}^0_{n,1}+\boldsymbol{v}^D_{n,1})+\frac{\tau_n}{120}(\boldsymbol{v}^0_{n,2}+\boldsymbol{v}^D_{n,2})+\tau_n(\frac{1}{2}\boldsymbol{v}^D_{n,3}-\frac{1}{10}\boldsymbol{v}^D_{n,4}+\frac{1}{120}\boldsymbol{v}^D_{n,5})$ and $\boldsymbol{b}_{n,4}=\boldsymbol{M}(\boldsymbol{v}^0_{n,0}+\boldsymbol{v}^D_{n,0}-\boldsymbol{v}^D_{n,3})+\tau_n(\frac{1}{2}\boldsymbol{f}_{n,3}+\frac{1}{10}\boldsymbol{f}_{n,1}+\frac{1}{120}\boldsymbol{f}_{n,2}+\frac{1}{2}\boldsymbol{f}_{n,3}-\frac{1}{10}\boldsymbol{f}_{n,4}+\frac{1}{120}\boldsymbol{f}_{n,5})-\tau_n\boldsymbol{A}(\frac{1}{2}(\boldsymbol{u}^0_{n,0}+\boldsymbol{u}^D_{n,0})2+\frac{1}{10}(\boldsymbol{u}^0_{n,1}+\boldsymbol{u}^D_{n,1})+\frac{1}{120}(\boldsymbol{u}^0_{n,2}+\boldsymbol{u}^D_{n,2})+\frac{1}{2}\boldsymbol{u}^D_{n,3}-\frac{1}{10}\boldsymbol{u}^D_{n,4}+\frac{1}{120}\boldsymbol{u}^D_{n,5})$.

5.2 Iterative Solver and Convergence Study

To solve the linear system $\boldsymbol{S}\boldsymbol{x} = \boldsymbol{b}$ with $\boldsymbol{S}$ from (31), we use block Gaussian elimination, as sketched in Sect. 4.2.1, to find a reduced system $\boldsymbol{S}_r \boldsymbol{u}_{n,4}^0 = \boldsymbol{b}_r$ for the essential unknown $\boldsymbol{u}_{n,4}^0$. All remaining unknown subvectors of $\boldsymbol{x}$ can be computed in post-processing steps. In explicit form, the condensed system reads as

$$\left(14400\boldsymbol{M} + 720\tau_n^2\boldsymbol{A} + 24\tau_n^4\boldsymbol{A}\boldsymbol{M}^{-1}\boldsymbol{A} + \tau_n^6\boldsymbol{A}\boldsymbol{M}^{-1}\boldsymbol{A}\boldsymbol{M}^{-1}\boldsymbol{A}\right)\boldsymbol{u}_{n,4}^0 = \boldsymbol{b}_r\,.$$

For brevity, we omit the exact definition of $\boldsymbol{b}_r$ that can be deduced easily from (31).

The matrix $\boldsymbol{S}_r$ is symmetric such that preconditioned conjugate gradient iterations are used for its solution. The preconditioner is constructed along the lines of Sect. 4.2.1. The remainder part $\tau_n^6\boldsymbol{A}\boldsymbol{M}^{-1}\boldsymbol{A}\boldsymbol{M}^{-1}\boldsymbol{A}$ is still ignored in the construction of the preconditioner. Even though the remainder is weighted by the small factor τ_n^6, numerical experiments indicate that this scaling is not sufficient to balance its impact on the interation process. For the construction of an efficient preconditioning technique for $\boldsymbol{S}_r$ of GCC^2(5) further improvements are still necessary.

To illustrate the convergence behavior and performance of the GCC^2(5) Galerkin–collocation approach, we present in Table 4 our numerical results for the test problem (28). The expected convergence of sixth order in time is nicely observed in all norms.

Table 4 Calculated errors for GCC^2(5) and solution (28)

τ	h	$\|e^u\|_{L^\infty(L^2)}$	EOC	$\|e^v\|_{L^\infty(L^2)}$	EOC	$\|\|\|E\|\|\|_{L^\infty}$	EOC
$\tau_0/2^0$	h_0	8.748e−06	–	4.355e−05	–	4.985e−05	–
$\tau_0/2^1$	h_0	1.370e−07	6.00	7.404e−07	5.88	8.043e−07	5.95
$\tau_0/2^2$	h_0	2.165e−09	5.98	1.202e−08	5.95	1.266e−08	5.99
$\tau_0/2^3$	h_0	3.388e−11	6.00	1.883e−10	6.00	1.980e−10	6.00
$\tau_0/2^4$	h_0	5.301e−13	6.00	2.940e−12	6.00	3.093e−12	6.00
τ	h	$\|e^u\|_{L^2(L^2)}$	EOC	$\|e^v\|_{L^2(L^2)}$	EOC	$\|\|\|E\|\|\|_{L^2}$	EOC
$\tau_0/2^0$	h_0	4.022e−06	–	2.996e−05	–	3.502e−05	–
$\tau_0/2^1$	h_0	6.353e−08	5.98	4.808e−07	5.96	5.599e−07	5.97
$\tau_0/2^2$	h_0	9.957e−10	6.00	7.565e−09	5.99	8.800e−09	5.99
$\tau_0/2^3$	h_0	1.557e−11	6.00	1.184e−10	6.00	1.377e−10	6.00
$\tau_0/2^4$	h_0	2.431e−13	6.00	1.849e−12	6.00	2.151e−12	6.00

References

1. Anselmann, M., Bause, M., Becher, S., Matthies, G.: Galerkin-collocation approximation in time for the wave equation and its post-processing. ESAIM: M2AN, 1–27 (2020). https://doi.org/10.1051/m2an/2020033
2. Bangerth, W., Geiger, M., Rannacher, R.: Adaptive Galerkin finite element methods for the wave equation. Comput. Methods Appl. Math. **10**(1), 3–48 (2010)
3. Bangerth, W., Heister, T., Kanschat, G.: `deal.II` differential equations analysis library. Technical reference (2013). http://www.dealii.org
4. Bause, M., Köcher, U., Radu, F.A., Schieweck, F.: Post-processed Galerkin approximation of improved order for wave equations. Math. Comput. 1–34 (2018); arXiv:1803.03005
5. Becher, S., Matthies, G., Wenzel, D.: Variational methods for stable time discretization of first-order differential equations. In: Georgiev, K., Todorov, M., Georgiev, M.I. (eds.) Advanced Computing in Industrial Mathematics, pp. 63–75. Springer, Berlin (2018)
6. De Basabe, J.D., Sen, M.K., Wheeler, M.F.: The interior penalty discontinuous Galerkin method for elastic wave propagation: grid dispersion. Geophys. J. Int. **175**, 83–95 (2008)
7. Dörfler, W., Findeisen, S., Wieners, C.: Space-time discontinuous Galerkin discretizations for linear first-order hyperbolic evolution systems. Comput. Methods Appl. Math. **16**, 409–428 (2016)
8. Ernesti, J.: Space-time methods for acoustic waves with applications to full waveform inversion. PhD Thesis, Department of Mathematics, Karlsruhe Institute of Technology (2017)
9. Ernesti, J., Wieners, C.: A space-time discontinuous Petrov-Galerkin method for acoustic waves CRC 1173 (2018). https://doi.org/10.5445/ir/1000085443
10. Grote, M.J., Schneebeli, A., Schötzau, D.: Discontinuous Galerkin finite element method for the wave equation. SIAM J. Numer. Anal. **44**(6), 2408–2431 (2006)
11. Heroux, M.A., et al.: An overview of the Trilinos project. ACM Trans. Math. Softw. **31**(3), 397–423 (2005)
12. Hussain, S., Schieweck, F., Turek, S.: Higher order Galerkin time discretizations and fast multigrid solvers for the heat equation. J. Numer. Math. **19**(1), 41–61 (2011)
13. Hussain, S., Schieweck, F., Turek, S.: Higher order Galerkin time discretization for nonstationary incompressible flow. In: Cangiani, A. et al. (eds.) Numerical Mathematics and Advanced Applications 2011, pp. 509–517. Springer, Berlin (2013)
14. Hussain, S., Schieweck, F., Turek, S.: Efficient Newton–multigrid solution techniques for higher order space–time Galerkin discretizations of incompressible flow. Appl. Numer. Math. **83**, 51–71 (2014)
15. Karakashian, O., Makridakis, C.: Convergence of a continuous Galerkin method with mesh modification for nonlinear wave equations. Math. Comput. **74**, 85–102 (2004)
16. Köcher, U.: Variational space-time methods for the elastic wave equation and the diffusion equation. PhD Thesis, Helmut-Schmidt-Universität (2015). http://edoc.sub.uni-hamburg.de/hsu/volltexte/2015/3112/
17. Köcher, U., Bause, M.: Variational space–time methods for the wave equation. J. Sci. Comput. **61**(2), 424–453 (2014)
18. Li, X.S., Demmel, J.W.: SuperLU_DIST: a scalable distributed-memory sparse direct solver for unsymmetric linear systems. ACM Trans. Math. Softw. **29**(2), 110–140 (2003)
19. Lions, J.L.: Optimal control of systems governed by partial differential equations. Springer, Berlin (1971)
20. Lions, J.L., Magenes, E.: Problèmes aus limites non homogènes et applications, vol. 1, 2, 3. Dunod, Paris (1968)
21. Matthies, G., Schieweck, F.: Higher order variational time discretizations for nonlinear systems of ordinary differential equations. Preprint No. 23/2011, Fakultät für Mathematik, Otto-von-Guericke-Universität, Magdeburg (2011)
22. Mikelić, A., Wheeler, M.F.: Theory of the dynamic Biot–Allard equations and their link to the quasi-static Biot system. J. Math. Phys. **53**(123702), 1–15 (2012)

23. Prokopenko, A., Hu, J.J., Wiesner, T.A., Siefert, C.M., Tuminaro, R.S.: MueLu User's Guide 1.0. Sandia National Labs. SAND2014-18874 (2014). http://trilinos.org/packages/muelu
24. Richter, T.: Fluid-structure Interactions: Models, Analysis and Finite Elements. Springer, Berlin (2017)
25. Steinbach, O.: Space–time finite element methods for parabolic problems. Comput. Methods Appl. Math. **15**, 551–566 (2015)
26. Zhao, S., Wei, W.W.: A unified discontinuous Galerkin framework for time integration. Math. Methods Appl. Sci. **37**(7):1042–1071 (2014)

Effective Numerical Simulation of the Klein–Gordon–Zakharov System in the Zakharov Limit

Simon Baumstark, Guido Schneider, and Katharina Schratz

Abstract Solving the Klein–Gordon–Zakharov (KGZ) system in the high-plasma frequency regime $c \gg 1$ is numerically severely challenging due to the highly oscillatory nature or the problem. To allow reliable approximations classical numerical schemes require severe step size restrictions depending on the small parameter c^{-2}. This leads to large errors and huge computational costs. In the singular limit $c \to \infty$ the Zakharov system appears as the regular limit system for the KGZ system. It is the purpose of this paper to use this approximation in the construction of an effective numerical scheme for the KGZ system posed on the torus in the highly oscillatory regime $c \gg 1$. The idea is to filter out the highly oscillatory phases explicitly in the solution. This allows us to play back the numerical task to solving the non-oscillatory Zakharov limit system. The latter can be solved very efficiently without any step size restrictions. The numerical approximation error is then estimated by showing that solutions of the KGZ system in this singular limit can be approximated via the solutions of the Zakharov system and by proving error estimates for the numerical approximation of the Zakharov system. We close the paper with numerical experiments which show that this method is more effective than other methods in the high-plasma frequency regime $c \gg 1$.

S. Baumstark
Institut für Angewandte und Numerische Mathematik, Karlsruher Institut für Technologie (KIT), Karlsruhe, Germany

G. Schneider (✉)
Institut für Analysis, Dynamik und Modellierung, Universität Stuttgart, Stuttgart, Germany
e-mail: guidos@mathematik.uni-stuttgart.de

K. Schratz
Laboratoire Jacques-Louis Lions, Sorbonne University, Paris, France

W. Dörfler et al. (eds.), *Mathematics of Wave Phenomena*, Trends in Mathematics,
https://doi.org/10.1007/978-3-030-47174-3_3

1 Introduction

We consider the Klein–Gordon–Zakharov (KGZ) system

$$c^{-2}\partial_t^2 u = \Delta u - c^2 u - uv, \qquad \partial_t^2 v = \Delta v + \Delta(|u|^2), \tag{1}$$

with $u(x,t)$, $v(x,t)$, $t \in \mathbb{R}$ in the limit $c \to \infty$. For pratical implementation issues we consider the system (1) posed on the d-dimensional torus $\mathbb{T}^d = (\mathbb{R}/2\pi\mathbb{Z})^d$. The subsequent approach in principle works for all space dimensions, i.e. $u(x,t)$, $x \in \mathbb{T}^d$. However, for expository reasons we restrict ourselves to $d = 1$, i.e. $\Delta = \partial_x^2$ in the following.

The KGZ system [14, Eq.(1.1)] is a model from plasma physics which is used to describe the interaction between so called Langmuir waves and ion sound waves in plasma. Here, $v(x,t)$ is proportional to the ion density fluctuation from a constant equilibrium density and $u(x,t)$ is proportional to the electric field.

We are interested in a robust numerical description of the KGZ system (1) for large values of c. Resolving the highly oscillatory behavior of the solutions in this regime is numerically very delicate, see, e.g., [1, 2, 4, 8, 11]. Severe time step restrictions need to be imposed, leading to high numerical costs. These can be avoided by passing to the regular limit system of the KGZ system (1) for $c \to \infty$. In this singular limit with the ansatz

$$\Psi_u(x,t) = \psi_u(x,t)e^{ic^2 t} + c.c., \qquad \Psi_v(x,t) = \psi_v(x,t) \tag{2}$$

for u, v the Zakharov system

$$2i\partial_t \psi_u = \Delta\psi_u - \psi_u\psi_v, \qquad \partial_t^2 \psi_v = \Delta\psi_v + 2\Delta(|\psi_u|^2) \tag{3}$$

can be derived. The numerical task to solve the KGZ equation for large c thus can be reduced to solving the corresponding non-oscillatory limit system (3). The latter can be carried out very efficiently without any additional step size restrictions. In the following we provide rigorous estimates between true solutions of the KGZ system (1) for large values of c and its numerical approximations obtained via the associated Zakharov system (3).

This *asymptotic* approach to handle highly oscillatory systems has attracted a lot of interest in the last years, cf. [2, 4, 8, 11]. In these highly oscillatory situations the approach via the regular limit system turned out to be more effective than other tools for highly oscillatory systems, such as Gautschi type approaches (see, e.g., [1]). In strong high plasma frequency limits $c \to \infty$ they are also far more effective than uniformly accurate oscillatory methods which have been recently invented for a number of Klein–Gordon type systems [4, 5, 8]. In particular, high order uniformly accurate oscillatory schemes are numerically very expensive such that the asymptotic approach of reducing the original complex system to the corresponding limit system is far more attractive from a computational point of view. Note that the

Zakharov system (3) can be solved very efficiently with high order methods (in time and in space) without any step size restrictions (see, e.g., [3, 7, 12, 13]).

A sharp estimate on the difference between the exact solution u and the limit approximation $\psi_u(t,x)e^{ic^2t} + c.c.$ is essential for the global error bound of the effective numerical scheme

$$u^n_{\text{num}} = (\psi_u)^n_{\text{num}} e^{ic^2 t_n} + c.c., \tag{4}$$

where $(\psi_u)^n_{\text{num}}$ denotes the numerical solution of the Zakharov system (3) at time $t_n = n\tau$. Such an estimate can be established with the triangle inequality. The full error can be reduced to the asymptotic error (KGZ to Zakharov) and the numerical error when solving the Zakharov system, i.e.,

$$\|u(t_n) - u_{\text{num}}(t_n)\| \le C \underbrace{\|u(t_n) - (\psi_u(t_n)e^{ic^2t_n} + c.c)\|}_{\text{asymptotic error}} + C \underbrace{\|\psi_u(t_n) - (\psi_u)^n_{\text{num}}\|}_{\text{numerical error (Zakharov)}},$$

see Sect. 3 below for the detailed error bounds.

The plan of the paper is as follows. In Sect. 2 we provide bounds for the error made by the Zakharov approximation which show that the Zakharov system (3) allows us to make correct predictions about the dynamics of the KGZ system (1) for large values of c, in particular we explain why existing error estimates for the problem posed on the real line transfer to the problem posed on the torus. In Sect. 3 we present a numerical scheme which allows us an effective simulation of the dynamics of the Zakharov system (3) and give error bounds for this numerical approximation. After that we bring together the estimates from Sects. 2 and 3 and present the error bound for this effective numerical simulation of the KGZ system via the Zakharov system (3) for large c. In Sect. 4 we close the paper with some numerical illustrations showing the strengh of the method in this highly oscillatory regime.

2 From the KGZ System to the Zakharov System

It is the purpose of this section to provide error estimates for the Zakharov approximation of the KGZ system posed on a one-dimensional torus for large values of c. This approximation question has been addressed in a number of papers. However, the results [6, 9, 16] all have been established for the KGZ system posed in $\mathbb{R}^d$. As the example of another singular limit of the KGZ system, namely the Klein–Gordon approximation, shows, such transfers can be wrong. In [10] it has been shown that for the problem posed on the torus a modified Klein–Gordon equation replaces the Klein–Gordon equation as regular limit system in this other singular limit. As we will see below also for the Zakharov approximation of the KGZ system such a transfer is non-trivial.

In [6] with the 'harmonic' ansatz

$$u(x,t) = \psi_u(x,t)e^{ic^2 t}, \qquad v(x,t) = \psi_v(x,t) \tag{5}$$

a Zakharov system with slightly different coefficients has been derived and convergence results for $c \to \infty$ have been established. Here we will concentrate on the 'real' case of the introduction.

We start with an approximation result in the spaces of 2π-spatially periodic analytic functions

$$H_{\mu,s,per} = \{u \in L^2_{\text{per}}(\mathbb{T}) : e^{\mu|\cdot|}(1+|\cdot|^2)^{\frac{s}{2}}\hat{u}(\cdot) \in \ell^2(\mathbb{Z})\},$$

equipped with the norm

$$\|u\|_{H_{\mu,s,per}} = \left(\sum_{k\in\mathbb{Z}} |\hat{u}(k)|^2 e^{2\mu|k|}(1+|k|^2)^s\right)^{\frac{1}{2}},$$

where $\mu \geq 0$ and $s \geq 0$. Functions $u \in H_{\mu,0,per}$ can be extended to functions that are analytic on the strip $\{z \in \mathbb{C}\colon |\Im(z)| < \mu\}$, cf. [15]. The KGZ system is then solved in a space

$$\mathcal{X}_{\mu_A,s} = H_{\mu_A,s+1,per} \times H_{\mu_A,s,per} \times H_{\mu_A,s,per} \times H_{\mu_A,s-1,per}.$$

We transfer [16] to the 2π-spatially periodic situation and obtain

Theorem 1 *Fix $\beta \in (0,2]$, $\mu_A > 0$, $s \geq 1$. Let $(\psi_u, \psi_v) \in C([0,T_0], H_{\mu_A,s+5,per} \times H_{\mu_A,s+4,per})$ be a solution of the Zakharov system* (3). *Then there exist $c_0 > 0$, $C_1 > 0$, $C_2 > 0$ and $T_1 \in (0,T_0]$ such that for all $c \geq c_0$ and all initial conditions $(u, \partial_t u, v, \partial_t v)(\cdot, 0)$ of the KGZ system* (1) *satisfying*

$$\|(u - \Psi_u, \varepsilon^2\partial_t(u-\Psi_u), v - \Psi_v, \partial_t(v-\Psi_v))(x,0)\|_{\mathcal{X}_{\mu_A,s}} \leq C_1 c^{-\beta},$$

there are unique solutions (u,v) of the KGZ system (1) *with*

$$\sup_{t\in[0,T_1]} \|(u - \Psi_u, \varepsilon^2\partial_t(u-\Psi_u), v - \Psi_v, \partial_t(v-\Psi_v))(x,t)\|_{\mathcal{X}_{0,s}} \leq C_2 c^{-\beta}.$$

Remark 1 The theorem is also true if the scaling ε^β of the error is replaced by a scaling function which decays to zero as $o(1)$ for $\varepsilon \to 0$.

Remark 2 The Sobolev version of this theorem for the KGZ system posed on $\mathbb{R}^d$ can be found in [9]. More precisely, the scaling of the Zakharov limit of the KGZ system is given in [9, (2.6)]. Rewriting the statements of [9] in the above form and transferring them from $x \in \mathbb{R}^d$ to $x \in \mathbb{T}^d$ yields the following theorem:

Theorem 2 *Theorem 1 remains true also in case $\mu_A = 0$, i.e. if $H_{0,s} = H^s$. Moreover we have $T_1 = T_0$.*

Remark 3 Error estimates for the Zakharov approximation are non-trivial due to some c in front of the nonlinear terms if the KGZ system is written as first order system. The problem has been solved in [9] and [16] by completely different approaches. In [9] a detailed analysis of the bilinear terms through an averaging approach has been used, whereas in [16] a Cauchy–Kowalevsjaya like approach has been chosen. Although the appraoch of [9] gives stronger results, the approach of [16] is conceptually more simple and more robust, in the sense that it applies for other systems, too, without a detailed analysis of the underlying system.

Remark 4 In order to prove Theorem 1 and Theorem 2 we have to prove that the formal error made by the Zakharov approximation is sufficiently small. The largest terms which do not cancel are $\psi_u(x,t)^2 e^{2ic^2t}$ and $\overline{\psi_u}(x,t)^2 e^{-2ic^2t}$ in the v-equation. There are two possibilities to prove that the influence of these terms on the dynamics on the given $\mathcal{O}(1)$ time interval, is less or equal order $\mathcal{O}(\varepsilon^\beta)$. These are averaging methods in the variation of constant formula or adding higher order terms to the approximation. In the following we explain for the second approach why the transfer from $x \in \mathbb{R}$ to $x \in \mathbb{T}$ is in general a non-trivial question. However, for the Zakharov approximation of the KGZ system the transfer is possible.

In order to make the residual terms to be of order $\mathcal{O}(\varepsilon^2)$ the Zakharov ansatz has to be extended to

$$\begin{aligned}
\Psi_u(x,t) &= \psi_u(x,t)e^{ic^2t} + c.c., \\
\Psi_v(x,t) &= \psi_v(x,t) + c^{-4}\psi_{v,+}(x,t)e^{2ic^2t} + c^{-4}\psi_{v,-}(x,t)e^{-2ic^2t},
\end{aligned}$$

with $-\psi_{v,+} = \Delta(\psi_u^2)$ and $-\psi_{v,-} = \Delta(\overline{\psi_u}^2)$. We have $\psi_{v,+} = \overline{\psi_{v,-}}$ and so $\Psi_v \in \mathbb{R}$. It is an easy exercise to show that the all terms down to order $\mathcal{O}(1)$ vanish. The remaining terms are

$$\begin{aligned}
\text{Res}_u &= -c^{-2}(\partial_t^2\psi_u)e^{ic^2t} + c.c. \\
&\quad -(\psi_u e^{ic^2t} + c.c.)(c^{-4}\psi_{v,+}e^{2ic^2t} + c^{-4}\psi_{v,-}e^{-2ic^2t}), \\
\text{Res}_v &= -c^{-4}(\partial_t^2\psi_{v,+})e^{2ic^2t} - c^{-4}(\partial_t^2\psi_{v,-})e^{-2ic^2t} \\
&\quad -2ic^{-2}(\partial_t\psi_{v,+})e^{2ic^2t} + 2ic^{-2}(\partial_t\psi_{v,-})e^{-2ic^2t} \\
&\quad +\Delta(c^{-4}\psi_{v,+}(x,t)e^{2ic^2t} + c^{-4}\psi_{v,-}(x,t)e^{-2ic^2t}).
\end{aligned}$$

Writing the v equation as first order system makes it necessary to estimate $\partial_x^{-1}\text{Res}_v$, cf. [16]. It can be bounded if Res_v can be written as a derivative or alternatively for $x \in \mathbb{R}^d$ with L^p–L^q estimates. In Res_v all terms have a ∂_x in front, except of the pure time derivatives $\partial_t^2\psi_{v,\pm}$ and $\partial_t\psi_{v,\pm}$. Using$-\psi_{v,+} = \Delta(\psi_u^2)$ and $-\psi_{v,-} = \Delta(\overline{\psi_u}^2)$ they can be written as $\Delta\partial_t^2(\psi_u^2)$ resp. $\Delta\partial_t(\psi_u^2)$ such that $\partial_x^{-1}\text{Res}_v$ can be estimated. See [10] for an example about what happens for $x \in \mathbb{T}$ when the residual terms cannot be written as a derivative.

Remark 5 In [17] the Zakharov approximation has been justified for the original Euler–Maxwell system. Hence the procedure of approximating the Euler–Maxwell system in the singular limit with the regular limit system is in principle possible for the original Euler–Maxwell system, too.

3 Error Bounds for the Numerical Scheme

The asymptotic approximation result given in Theorem 2 allows us to develop an efficient numerical scheme for the KGZ system (1) in the Zakharov limit. The idea is thereby the following: Let

$$(\psi_u^n)_{\text{num}}, \quad (\psi_v^n)_{\text{num}}, \quad (\psi_{v'}^n)_{\text{num}} \tag{6}$$
$$(\psi_u^0)_{\text{num}} = \frac{1}{2}(u(\cdot,0) + \frac{1}{ic^2}\partial_t u(\cdot,0)), \quad (\psi_v^0)_{\text{num}} = v(\cdot,0), \quad (\psi_{v'}^0)_{\text{num}} = \partial_t v(\cdot,0)$$

denote the numerical solution at time $t = t_n$ of the Zakahrov system (3) obtained, e.g., with the trigonometric integrator proposed in [12]. Then we choose, motivated by (5), the scheme defined through

$$u_{\text{num}}^n = (\psi_u^n)_{\text{num}} \,\mathrm{e}^{ic^2 t_n} + \overline{(\psi_u^n)_{\text{num}}}\,\mathrm{e}^{-ic^2 t_n}, \quad v_{\text{num}}^n = (\psi_v^n)_{\text{num}} \tag{7}$$

as a numerical approximation to the exact solution $(u(t_n), v(t_n), v'(t_n))$ of the KGZ system (1). The choice of initial values in (6) is thereby motivated as follows: From (2) we find for $t = 0$ that

$$u = \psi_u + \overline{\psi_u}, \qquad \partial_t u = ic^2\psi_u - ic^2\overline{\psi_u} + O(1)$$

which implies

$$\psi_u = \frac{1}{2}(u + \frac{1}{ic^2}\partial_t u)$$

when the terms indicated with $O(1)$ are ignored.

Thanks to Theorem 2, which allows us to control the asymptotic error, the scheme (7) allows for the following global error estimate.

Theorem 3 *Fix $s \geq 1$, and let $(\psi_u, \psi_v) \in C([0,T_0], H_{per}^{s+5} \times H_{per}^{s+4})$ be a solution of the Zakharov system (3). Assume that the numerical scheme (6) approximates the solution of the Zakharov system (3) with order p in H^s, i.e., there exist $C, \tau_0 > 0$ such that for all $\tau \leq \tau_0$ and $t_n \leq T$*

$$\|(\psi_u^n)_{num} - \psi_{u(t_n)}\|_{s+1} + \|(\psi_v^n)_{num} - \psi_{v(t_n)}\|_s + \|(\psi_{v'}^n)_{num} - \psi_{v'(t_n)}\|_{s-1} \leq C\tau^p. \tag{8}$$

Then the scheme (7) *converges to the solution* (u, v) *of the KGZ system* (1) *in the limit* $c \to \infty$, $\tau \to 0$. *In detail, there exist* $C, c_0, \tau_0 > 0$ *such that for all* $c > c_0$ *and* $\tau < \tau_0$ *we have*

$$\|u^n_{num} - u(t_n)\|_{s+1} + \|v^n_{num} - v(t_n)\|_s \leq C\left(\tau^p + c^{-2}\right).$$

Proof The proof follows by Theorem 2 together with the triangle inequality. Note that

$$\begin{aligned}\|u(t_n) - u^n_{\text{num}}\|_{s+1} &\leq \|u(t_n) - \Psi_u(t_n) + \Psi_u(t_n) - u^n_{\text{num}}\|_{s+1} \\ &\leq \|u(t_n) - \Psi_u(t_n)\|_{s+1} + C\|\psi_{u(t_n)} - (\psi^n_u)_{\text{num}}\|_{s+1}.\end{aligned} \tag{9}$$

Thanks to Theorem 2 we can bound the first term by Cc^{-2}. The second term in (9) is bounded by τ^p by assumption (8). This yields the assertion for the error in u. The other terms can be bounded in a similar way. □

The second-order trigonometric integrator [12, Eq. (3.9)] developed for the Zakharov system (3) together with the ansatz (7) allow us to obtain a second-order asymptotic and time convergent scheme for the KGZ system (1).

Corollary 1 (A Second-Order Scheme) *Fix* $s \geq 1$, *and let* $(\psi_u, \psi_v) \in C([0, T_0], H^{s+5}_{per} \times H^{s+4}_{per})$ *be a solution of the Zakharov system* (3). *Then the scheme [12, Eq. (3.9)] converges to the solution* (u, v) *of the KGZ system* (1) *in the limit* $c \to \infty$, $\tau \to 0$. *In detail, there exist* $C, c_0, \tau_0 > 0$ *such that for all* $c > c_0$ *and* $\tau < \tau_0$ *we have*

$$\|u^n_{num} - u(t_n)\|_{s+1} + \|v^n_{num} - v(t_n)\|_s \leq C\left(\tau^2 + c^{-2}\right).$$

4 Some Numerical Illustrations

In this section we compare various numerical schemes for the solution of the KGZ system (1). Our numerical experiments confirm that in the high plasma frequency regime $c \gg 1$ the ansatz (4), based on the Zakharov limit approximation, is more efficient than directly solving the KGZ system (1) with a uniformly accurate scheme such as [5]. Furthermore, the numerical experiments underline the second-order convergence rate (in time and in c^{-2}) established in Corollary 1.

For the practical implementation we consider $x \in \mathbb{T} = [0, 2\pi]$ and a finite time interval, i.e., $t \in [0, 1]$. For the spatial approximation we use a standard Fourier pseudospectral method with $M = 256$ Fourier modes (i.e., $\Delta x = 0.0245$) and choose the initial values

$$\begin{aligned} u(x,0) &= \frac{\sin(2x)\cos(4x)}{2 - \cos(x)\sin(2x)}, \qquad \partial_t u(x,0) = c^2(-\sin(2x)\cos(x)), \\ v(x,0) &= \frac{\sin(x)\cos(2x)}{2 - \sin(2x)^2}, \qquad \partial_t v(x,0) = \frac{\sin(x)}{2 - \cos(2x)^2}. \end{aligned} \tag{10}$$

4.1 Efficiency

In Figs. 1 and 2 we compare the error versus the computational time of different numerical methods for the KGZ system (1). The work-precision plots show the efficiency of the different methods for different values of c.

More precisely, we compare the following schemes:

- The first- and second-order schemes (4) based on the asymptotic approximation result given in Theorem 2. Thereby we use the trigonometric integration method [12] for the numerical solution of the Zakkharov system. This allows for a global error of order (cf. Corollary 1)

$$c^{-2} + \tau^p \quad \text{with } p = 1, 2.$$

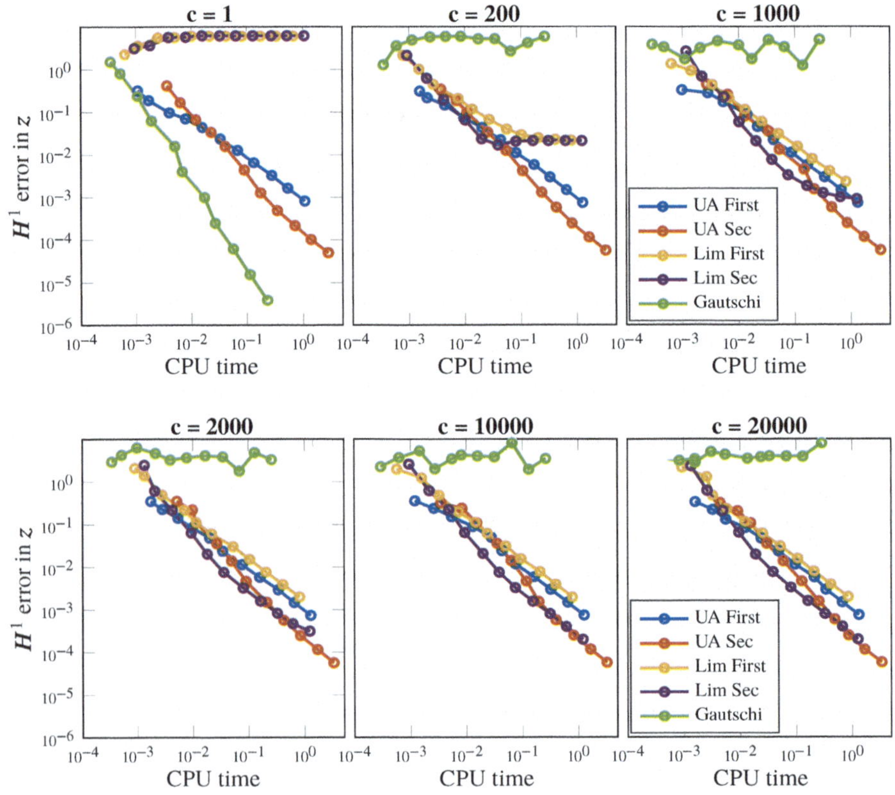

Fig. 1 Efficiency plot of z for different values of c. The blue and red lines correspond to the first- and second-order uniformly accurate method of [5]. In yellow and purple we plot the first- and second-order limit integrator based on (4) using the trigonometric integrator [12] for the numerical solution of the Zakharov limit system. The green line corresponds to the Gautschi method [2]

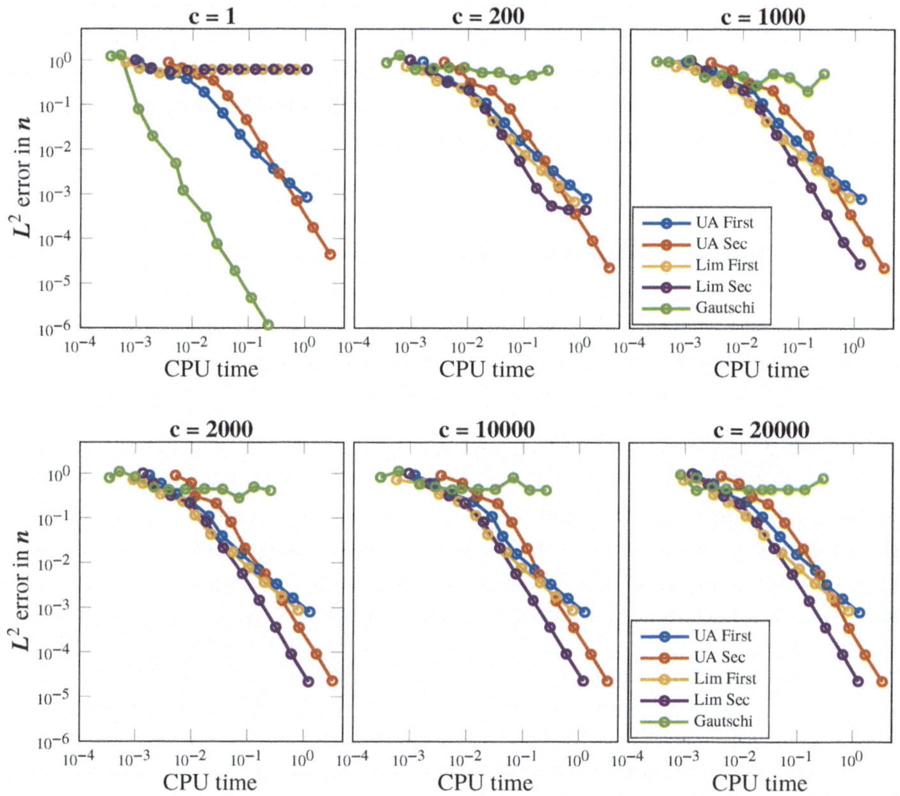

Fig. 2 Efficiency plot of n for different values of c. The blue and red lines correspond to the first- and second-order uniformly accurate method of [5]. In yellow and purple we plot the first- and second-order limit integrator based on (4) using the trigonometric integrator [12] for the numerical solution of the Zakharov limit system. The green line corresponds to the Gautschi method [2]

- The uniformly accurate methods for the KGZ system (1) developed in [5] which allow for a global error of order

$$\tau^2 \quad \text{with } p = 1, 2.$$

- A Gautschi method which was developed for the KGZ system (1) in [2]. The latter allows for a global error of order

$$c^2\tau^2.$$

We plot the corresponding error against the computation time (in seconds) of the corresponding numerical method. The reference solution is computed via the uniformly accurate method by Baumstark and Schratz [5] with a very small step size $\tau_{ref} = 1.19 \cdot 10^{-7}$.

In the numerical experiments we observe the following: Although the uniformly accurate methods allow uniform convergence (i.e., error bounds independently of c) the limit integrators are faster for very large values of c.

4.2 Asymptotic Consistency Plot

In this section we numerically underline that the Zakharov system (3) approximates the KGZ (1) with rate $O(c^{-2})$ for sufficiently smooth solutions (cf. Theorem 2). To solve the Zakharov and KGZ system we use the uniformly accurate scheme [5] and limit integrator [12], respectively.

In Figs. 3 and 4 we use the smooth initial data (10) and initial data in H^2, respectively. To test the convergence rate in c^{-2} both simulations are carried out with the constant small time step size $\tau = 1.53 \cdot 10^{-5}$ in order to not see the time discretization error in the plots. In Fig. 4 the initial values in H^2 are computed by choosing uniformly distributed random numbers in the interval [0, 1] for the real and imaginary part of the N Fourier coefficients, respectively. These coefficients are then divided by $(1+|k|)^{2+1/2}$ for $k = -\frac{N}{2}, \dots, \frac{N}{2} - 1$ and finally transformed back with the discrete Fourier transform to get the desired discrete initial data in physical space. The reference solution is computed via the uniformly accurate method [5] with $\tau_{ref} = 1.19 \cdot 10^{-7}$.

For H^2 data we numerically observe an order reduction down to order $c^{-1/2}$ in the asymptotic error of the Zakharov approximation of the KGZ system which

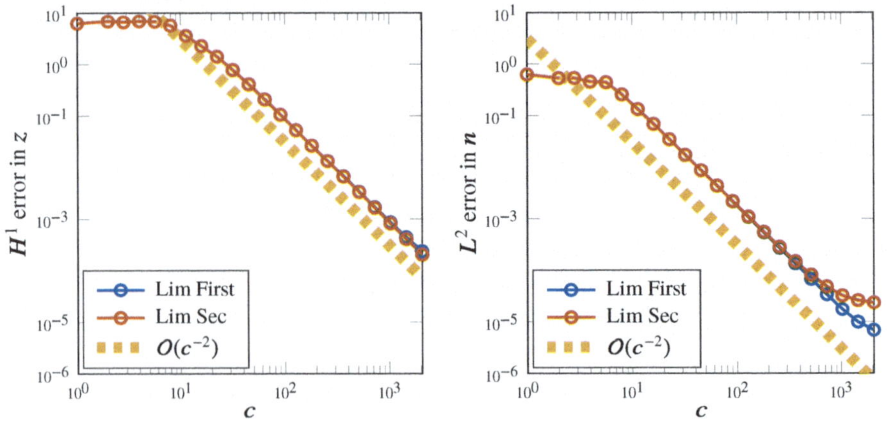

Fig. 3 Asymptotic consistency plot with initial data (10). The blue and red line correspond to the first- and second-order limit integrator for the Zaharov system of [12]. Yellow: reference line of order c^{-2}

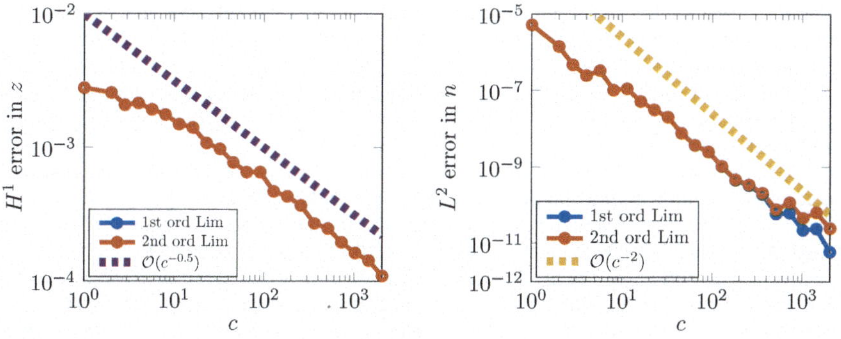

Fig. 4 Asymptotic consistency plot with initial data in H^2. The blue and red line correspond to the first- and second-order limit integrator for the Zakharov system of [12]. Reference line of order $c^{-1/2}$ and c^{-2} in purple (left) and yellow (right), respectively

underlines the necessity of sufficiently smooth solutions for the validity of the asymptotic approximation (cf. Theorem 2).

Acknowledgement The paper is partially supported by the Deutsche Forschungsgemeinschaft DFG through the SFB 1173 "Wave phenomena".

References

1. Bao, W., Dong, X.: Analysis and comparison of numerical methods for the Klein–Gordon equation in the nonrelativistic limit regime. Numer. Math. **120**(2), 189–229 (2012). https://doi.org/10.1007/s00211-011-0411-2
2. Bao, W., Dong, X., Zhao, X.: An exponential wave integrator sine pseudospectral method for the Klein–Gordon–Zakharov system. SIAM J. Sci. Comput. **35**(6), A2903–A2927 (2013). https://doi.org/10.1137/110855004
3. Bao, W., Sun, F.: Efficient and stable numerical methods for the generalized and vector Zakharov system. SIAM J. Sci. Comput. **26**(3), 1057–1088 (2005). https://doi.org/10.1137/030600941
4. Bao, W., Zhao, X.: A uniformly accurate multiscale time integrator spectral method for the Klein–Gordon–Zakharov system in the high-plasma-frequency limit regime. J. Comput. Phys. **327**, 270–293 (2016). https://doi.org/10.1016/j.jcp.2016.09.046. http://www.sciencedirect.com/science/article/pii/S0021999116304673
5. Baumstark, S., Schratz, K.: Uniformly accurate oscillatory integrators for the Klein-Gordon-Zakharov system from low- to high-plasma frequency regimes. SIAM J. Numer. Anal. (to appear)
6. Bergé, L., Bidégaray, B., Colin, T.: A perturbative analysis of the time-envelope approximation in strong Langmuir turbulence. Phys. D **95**(3–4), 351–379 (1996). https://doi.org/10.1016/0167-2789(96)00058-9
7. Chang, Q., Guo, B., Jiang, H.: Finite difference method for generalized Zakharov equations. Math. Comput. **64**(210), 537–553 (1995). http://www.jstor.org/stable/2153438
8. Chartier, P., Crouseilles, N., Lemou, M., Méhats, F.: Uniformly accurate numerical schemes for highly oscillatory Klein–Gordon and nonlinear Schrödinger equations. Numer. Math. **129**(2), 211–250 (2015). https://doi.org/10.1007/s00211-014-0638-9

9. Colin, T., Ebrard, G., Gallice, G., Texier, B.: Justification of the Zakharov model from Klein-Gordon–wave systems. Commun. Partial Differ. Equ. **29**(9–10), 1365–1401 (2004). http://dx.doi.org/10.1081/PDE-200037756
10. Daub, M., Schneider, G., Schratz, K.: From the Klein-Gordon-Zakharov system to the Klein-Gordon equation. Math. Methods Appl. Sci. **39**(18), 5371–5380 (2016). http://dx.doi.org/10.1002/mma.3922
11. Faou, E., Schratz, K.: Asymptotic preserving schemes for the Klein-Gordon equation in the non-relativistic limit regime. Numer. Math. **126**(3), 441–469 (2014)
12. Herr, S., Schratz, K.: Trigonometric time integrators for the Zakharov system. IMA J. Numer. Anal. **37**(4), 2042–2066 (2016). https://dx.doi.org/10.1093/imanum/drw059
13. Jin, S., Markowich, P.A., Zheng, C.: Numerical simulation of a generalized Zakharov system. J. Comput. Phys. **201**(1), 376–395 (2004). http://dx.doi.org/10.1016/j.jcp.2004.06.001
14. Masmoudi, N., Nakanishi, K.: From the Klein-Gordon-Zakharov system to the nonlinear Schrödinger equation. J. Hyperbolic Differ. Equ. **2**(4), 975–1008 (2005). https://doi.org/10.1142/S0219891605000683
15. Reed, M., Simon, B.: Methods of modern mathematical physics. II: Fourier analysis, self-adjointness. Academic Press, London. A subsidiary of Harcourt Brace Jovanovich, Publishers. XV, 361 p. (1975)
16. Schneider, G.: The Zakharov limit of Klein-Gordon-Zakharov like systems for analytic solutions. Tech. Rep., CRC Preprint (2019)
17. Texier, B.: Derivation of the Zakharov equations. Arch. Ration. Mech. Anal. **184**(1), 121–183 (2007). https://doi.org/10.1007/s00205-006-0034-4

Exponential Dichotomies for Elliptic PDE on Radial Domains

Margaret Beck, Graham Cox, Christopher Jones, Yuri Latushkin, and Alim Sukhtayev

Abstract It was recently shown by the authors that a semilinear elliptic equation can be represented as an infinite-dimensional dynamical system in terms of boundary data on a shrinking one-parameter family of domains. The resulting system is ill-posed, in the sense that solutions do not typically exist forward or backward in time. In this paper we consider a radial family of domains and prove that the linearized system admits an exponential dichotomy, with the unstable subspace corresponding to the boundary data of weak solutions to the linear PDE. This generalizes the spatial dynamics approach, which applies to infinite cylindrical (channel) domains, and also generalizes previous work on radial domains as we impose no symmetry assumptions on the equation or its solutions.

M. Beck
Department of Mathematics and Statistics, Boston University, Boston, MA, USA
e-mail: mabeck@bu.edu

G. Cox
Department of Mathematics and Statistics, Memorial University of Newfoundland, St. John's, NL, Canada
e-mail: gcox@mun.ca

C. Jones
Department of Mathematics, University of North Carolina at Chapel Hill, Chapel Hill, NC, USA
e-mail: ckrtj@email.unc.edu

Y. Latushkin (✉)
Department of Mathematics, University of Missouri, Columbia, MO, USA
e-mail: latushkiny@missouri.edu

A. Sukhtayev
Department of Mathematics, Miami University, Oxford, OH, USA
e-mail: sukhtaa@miamioh.edu

W. Dörfler et al. (eds.), *Mathematics of Wave Phenomena*, Trends in Mathematics,
https://doi.org/10.1007/978-3-030-47174-3_4

1 Introduction

The fundamental idea of spatial dynamics is to write a partial differential equation on a cylindrical domain $\Omega = \mathbb{R} \times \Omega' \subset \mathbb{R}^n$ as an ordinary differential equation with respect to the longitudinal variable $x \in \mathbb{R}$. For instance, $\Delta u + F(x, y, u) = 0$ becomes

$$\frac{du}{dx} = v, \quad \frac{dv}{dx} = -F(x, y, u) - \Delta_{\Omega'} u,$$

where $(x, y) \in \mathbb{R} \times \Omega'$ and $\Delta_{\Omega'}$ denotes the Laplacian on the cross-section $\Omega' \subset \mathbb{R}^{n-1}$. This idea first appeared in [11]; see also [1, 2, 8, 9, 12, 15–19] and references therein.

In [3] we extended this ODE–PDE correspondence to semi-linear elliptic equations on bounded domains. Assuming Ω is smoothly deformed through a one-parameter family Ω_t, we obtain a dynamical system satisfied by the boundary data of solutions to $\Delta u + F(x, u) = 0$ on $\partial\Omega_t$.

In the current paper we start to investigate the application of dynamical systems methodology to the resulting system of equations, which we call the Spatial Evolutionary System (SES). In particular, we construct an exponential dichotomy, and prove that the unstable subspace coincides with the space of boundary data for weak solutions to the PDE.

Our results are valid for systems of equations; functions are thus assumed to take values in $\mathbb{C}^N$ unless stated otherwise. We abbreviate $H^s(S^{n-1}; \mathbb{C}^N) = H^s(S^{n-1})$ etc.

Suppose u is a smooth solution to the linear elliptic system

$$\Delta u = V u \tag{1}$$

on $\mathbb{R}^n$, where V is an $N \times N$ matrix-valued function. Writing $u = u(r, \theta)$ in terms of generalized polar coordinates $(r, \theta) \in (0, \infty) \times S^{n-1}$, we define the functions

$$f(t) := u(t, \cdot), \quad g(t) := \frac{\partial u}{\partial r}(t, \cdot),$$

which are in $C^\infty(S^{n-1})$ for $t > 0$, and combine these to form the trace

$$\operatorname{Tr}_t u := (f(t), g(t)). \tag{2}$$

Using the fact that

$$\Delta u = \frac{\partial^2 u}{\partial r^2} + \frac{n-1}{r}\frac{\partial u}{\partial r} + \frac{1}{r^2}\Delta_{S^{n-1}} u,$$

a direct computation shows that for all $t > 0$, f and g satisfy the linear system

$$\frac{d}{dt}\begin{pmatrix} f \\ g \end{pmatrix} = \begin{pmatrix} 0 & 1 \\ V_t - t^{-2}\Delta_{S^{n-1}} & -(n-1)t^{-1} \end{pmatrix}\begin{pmatrix} f \\ g \end{pmatrix}, \tag{3}$$

where $V_t := V(t, \cdot)$ and $\Delta_{S^{n-1}}$ is the Laplace–Beltrami operator on the sphere.

In [3] it was shown that the equivalence between (1) and (3) extends to weak H^1 solutions. To state this precisely, consider the Hilbert spaces

$$\mathcal{H} = H^{1/2}(S^{n-1}) \oplus H^{-1/2}(S^{n-1}), \quad \mathcal{H}^1 = H^{3/2}(S^{n-1}) \oplus H^{1/2}(S^{n-1}).$$

The results can then be summarized as follows, where B_T denotes the open ball of radius T.

Theorem 1 *Let $u \in H^1(B_T)$ be a weak solution to* (1) *for some $T > 0$. Then $(f(t), g(t)) = \mathrm{Tr}_t\, u$ satisfies the regularity conditions*

$$(f, g) \in C^0\big((0, T), \mathcal{H}^1\big) \cap C^1\big((0, T), \mathcal{H}\big) \cap C^0\big((0, T], \mathcal{H}\big), \tag{4}$$

solves (3) *for $0 < t < T$, and has $\|f(t)\|_{H^{1/2}(S^{n-1})} + \|g(t)\|_{H^{-1/2}(S^{n-1})}$ bounded near $t = 0$.*

On the other hand, if (f, g) satisfies (4), *solves* (3) *for $0 < t < T$, and has $t^p\|f(t)\|_{H^{1/2}(S^{n-1})} + t^{n-p-1}\|g(t)\|_{H^{-1/2}(S^{n-1})}$ bounded near $t = 0$ for some $p \in (0, n/2)$, then there exists a weak solution $u \in H^1(B_T)$ to* (1) *with $\mathrm{Tr}_t\, u = (f(t), g(t))$ for all $t \in (0, T)$.*

This equivalence also extends to semilinear equations on non-radial domains; see [3] for the general statement.

The system (3) is ill-posed, in the sense that solutions do not necessarily exist forward (or backward) in time for given initial (or terminal) data at prescribed at time $t_0 > 0$. However, we will prove that $\mathcal{H}$ splits into two infinite-dimensional subspaces for which the system admits solutions forward and backward in time, respectively. This property is described using the language of exponential dichotomies. The system (3) does not admit an exponential dichotomy in the strict sense. Rather, a dichotomy exists for a suitably rescaled and reparameterized system of equations.

We let $t = e^\tau$ and then define

$$\tilde{f}(\tau) = e^{\alpha\tau} f(e^\tau), \quad \tilde{g}(\tau) = e^{(1+\alpha)\tau} g(e^\tau) \tag{5}$$

for some constant α to be determined. A direct computation shows that if (f, g) solves (3), then

$$\frac{d}{d\tau}\begin{pmatrix} \tilde{f} \\ \tilde{g} \end{pmatrix} = \begin{pmatrix} \alpha & 1 \\ e^{2\tau} V_{e^\tau} - \Delta_{S^{n-1}} & \alpha + 2 - n \end{pmatrix}\begin{pmatrix} \tilde{f} \\ \tilde{g} \end{pmatrix} \tag{6}$$

for all $\tau \in \mathbb{R}$. For convenience we let $\tilde{h} = (\tilde{f}, \tilde{g})$.

Our main result is that (6) has an exponential dichotomy on the half line $(-\infty, 0]$ for most values of α. Let

$$\Sigma(n) = \big((-\infty, 2-n] \cup [0, \infty)\big) \cap \mathbb{Z}, \tag{7}$$

so that $\Sigma(2) = \Sigma(3) = \mathbb{Z}$, $\Sigma(4) = \mathbb{Z} \setminus \{-1\}$, etc. We also define the interpolation spaces

$$\mathcal{H}^\beta = H^{1/2+\beta}(S^{n-1}) \oplus H^{-1/2+\beta}(S^{n-1}),$$

so that $\mathcal{H}^0 = \mathcal{H}$ and $\mathcal{H}^1$ agrees with the definition given above.

Theorem 2 *If $-\alpha \notin \Sigma(n)$ and $V \in C^{0,\gamma}(B_1)$ for some $\gamma \in (0, 1)$, then for each $\beta \in [0, 1)$ there exists a Hölder continuous family of projections $P^u : (-\infty, 0] \to B(\mathcal{H}^\beta)$ and constants $K, \eta^u, \eta^s > 0$ such that, for every $\tau_0 \leq 0$ and $z \in \mathcal{H}^\beta$ there exists a solution $\tilde{h}^u(\tau; \tau_0, z)$ of (6), defined for $\tau \leq \tau_0$, such that*

- $\tilde{h}^u(\tau_0; \tau_0, z) = P^u(\tau_0)z$,
- $\|\tilde{h}^u(\tau; \tau_0, z)\|_{\mathcal{H}^\beta} \leq K e^{\eta^u(\tau-\tau_0)}\|z\|_{\mathcal{H}^\beta}$ *for all* $\tau \leq \tau_0$,
- $\tilde{h}^u(\tau; \tau_0, z) \in R(P^u(\tau))$ *for all* $\tau \leq \tau_0$,

and a solution $\tilde{h}^s(\tau; \tau_0, z)$ of (6), defined for $\tau_0 \leq \tau \leq 0$, such that

- $\tilde{h}^s(\tau_0; \tau_0, z) = P^s(\tau_0)z$,
- $\|\tilde{h}^s(\tau; \tau_0, z)\|_{\mathcal{H}^\beta} \leq K e^{\eta^s(\tau_0-\tau)}\|z\|_{\mathcal{H}^\beta}$ *for all* $\tau_0 \leq \tau \leq 0$,
- $\tilde{h}^s(\tau; \tau_0, z) \in R(P^s(\tau))$ *for all* $\tau_0 \leq \tau \leq 0$,

where $P^s(\tau) = I - P^u(\tau)$.

We will see below that the exponential dichotomy on $(-\infty, 0]$ carries information about bounded solutions to the linear PDE (1) on the unit ball, B_1. By the same method we can also obtain an exponential dichotomy on $(-\infty, \log T]$ for any $T > 0$, corresponding to the PDE on the ball B_T.

The exponential dichotomy can also be described in terms of operators $\Phi^s(\tau, \tau_0)$ and $\Phi^u(\tau, \tau_0)$, defined by

$$\Phi^{s,u}(\tau, \tau_0)z = \tilde{h}^{s,u}(\tau; \tau_0, z) \tag{8}$$

for $z \in \mathcal{H}^\beta$, so that $\Phi^{s,u}(\tau_0, \tau_0) = P^{s,u}(\tau_0)$. Note that $\Phi^u(\tau, \tau_0)z$ is defined for $\tau \leq \tau_0 \leq 0$ and $\Phi^s(\tau, \tau_0)z$ is defined for $\tau_0 \leq \tau \leq 0$. From Theorem 2 we have the estimates

$$\|\Phi^u(\tau, \tau_0)z\|_{\mathcal{H}^\beta} \leq K e^{\eta^u(\tau-\tau_0)}\|z\|_{\mathcal{H}^\beta}, \quad \tau \leq \tau_0$$

and

$$\|\Phi^s(\tau, \tau_0)z\|_{\mathcal{H}^\beta} \leq K e^{\eta^s(\tau_0-\tau)}\|z\|_{\mathcal{H}^\beta}, \quad \tau_0 \leq \tau \leq 0.$$

The precise growth and decay rates depend on α. We will see below that it is convenient to choose $0 < \alpha < n - 2$ (assuming $n > 2$), in which case a dichotomy will exist for any numbers η^u and η^s satisfying $0 \leq \eta^u < \alpha$ and $0 \leq \eta^s < n-2-\alpha$.

To simplify the exposition we now assume $\beta = 0$. For any $\tau \leq 0$ we define the unstable subspace $\widetilde{E}^u(\tau) = R(P^u(\tau))$, and then let

$$E^u(t) = \left\{\left(t^{-\alpha}\tilde{f}(\log t), t^{-1-\alpha}\tilde{g}(\log t)\right) : \left(\tilde{f}(\log t), \tilde{g}(\log t)\right) \in \widetilde{E}^u(\log t)\right\} \tag{9}$$

for $0 < t \leq 1$.

As in [3], for an appropriate choice of α we have that $E^u(t)$ corresponds to the space of boundary data of weak solutions to (1) on the ball B_t. For $t > 0$ let

$$K_t = \{u \in H^1(B_t) : \Delta u = Vu \text{ on } B_t\},$$

where the equality $\Delta u = Vu$ is meant in a distributional sense. Since K_t is a subset of $\{u \in H^1(B_t) : \Delta u \in L^2(B_t)\}$, the trace map Tr_t can be applied, and we have $\mathrm{Tr}_t\, u \in H^{1/2}(S^{n-1}) \oplus H^{-1/2}(S^{n-1})$ for each $u \in K_t$. We thus define

$$\mathrm{Tr}_t(K_t) = \{\mathrm{Tr}_t\, u : u \in K_t\} \subset \mathcal{H}.$$

The following result is then an immediate consequence of Theorem 2 and [3, Theorem 3.10].

Theorem 3 *Assume, in addition to the hypotheses of Theorem 2, that V is smooth in a neighborhood of the origin. If*

$$-\eta^s < \alpha < \eta^u + \frac{n}{2} - 1, \tag{10}$$

then $E^u(t) = \mathrm{Tr}_t(K_t)$ for each $t > 0$.

To verify (10) we must understand the dependence of η^u and η^s on α. When $n > 2$ there is always an α for which (10) is satisfied.

Corollary 1 *If $n > 2$ and $0 < \alpha < n - 2$, then $E^u(t) = \mathrm{Tr}_t(K_t)$ for each $t > 0$.*

On the other hand, no such α exists when $n = 2$. This observation, which will be proved in Sect. 2.4 below, was also made in [3, Remark 2.1]. Below we provide a different (but equivalent) explanation in terms of the spectrum of the limiting (as $\tau \to -\infty$) operator in (6). For harmonic functions (i.e. when $V = 0$) it can be shown that $E^u(t) = \mathrm{Tr}_t(K_t)$ if and only if $0 < \alpha < n - 2$. This is proved in Sect. 3 for $n = 3$, and follows from a similar computation for other n.

1.1 Outline of the Paper

The remainder of the paper is organized as follows. In Sect. 2 we construct the half-line exponential dichotomy, proving Theorem 2 and Corollary 1. In Sect. 3 we illustrate our results for the case of harmonic functions in $\mathbb{R}^3$, where the dichotomy projections can be found explicitly. Finally, in Sect. 4 we use the exponential dichotomy to reformulate a nonlinear elliptic equation as a fixed point problem for an integral equation, and give a dynamical interpretation of a linear eigenvalue problem.

2 Construction of the Exponential Dichotomy

We prove Theorem 2 using the results of [16]. We start by decomposing the right-hand side of (6) as

$$\begin{pmatrix} \alpha & 1 \\ e^{2\tau} V_{e^\tau} - \Delta_{S^{n-1}} & \alpha + 2 - n \end{pmatrix} = \underbrace{\begin{pmatrix} \alpha & 1 \\ -\Delta_{S^{n-1}} & \alpha + 2 - n \end{pmatrix}}_{A} + \underbrace{\begin{pmatrix} 0 & 0 \\ e^{2\tau} V_{e^\tau} & 0 \end{pmatrix}}_{B(\tau)} \tag{11}$$

where A is an unbounded operator on $\mathcal{H} = H^{1/2}(S^{n-1}) \oplus H^{-1/2}(S^{n-1})$ with domain $\mathcal{H}^1 = H^{3/2}(S^{n-1}) \oplus H^{-1/2}(S^{n-1})$, and $B(\tau)$ is a bounded operator on $\mathcal{H}$.

Before proceeding, we remark on the definition of the fractional Sobolev spaces appearing in our analysis. Following [14], we define $H^s(S^{n-1})$ through local coordinate charts and a partition of unity. On the other hand, following [20, 21], one can also define

$$\widetilde{H}^s = \left\{ f \in L^2 : f = (I - \Delta)^{-s/2} g \text{ for some } g \in L^2 \right\}, \qquad \|f\|_{\widetilde{H}^s} = \|g\|_{L^2}$$

for $s > 0$ and

$$\widetilde{H}^s = \left\{ f \in \mathcal{D} : f = (I - \Delta)^{\ell} g \text{ for some } g \in \widetilde{H}^{2\ell+s} \text{ with } \ell \in \mathbb{N} \text{ and } 2\ell + s > 0 \right\},$$
$$\|f\|_{\widetilde{H}^s} = \|g\|_{\widetilde{H}^{2\ell+s}}$$

for $s < 0$, where $\mathcal{D}$ denotes the space of distributions. In either case we have that

$$\|f\|^2_{\widetilde{H}^s} = \sum_{k=1}^{\infty} (1 + \lambda_k)^s |c_k|^2 \tag{12}$$

where (λ_k, ϕ_k) are the eigenvalues and eigenfunctions of $-\Delta$ and $c_k = \langle f, \phi_k \rangle$. This is equivalent to the local definition (see, for instance, [10, Theorem 3.9]), so we can use the H^s and $\widetilde{H}^s$ norms interchangeably.

When $s = -\frac{1}{2}$ we choose $\ell = 1$, so that $2\ell + s = \frac{3}{2}$, and thus obtain $\|f\|_{\widetilde{H}^{-1/2}} = \|g\|_{\widetilde{H}^{3/2}}$, where $g \in \widetilde{H}^{3/2}$ solves $(I - \Delta)g = f$. In particular, this implies

$$\|g\|_{\widetilde{H}^{3/2}} \leq \|\Delta g\|_{\widetilde{H}^{-1/2}} + \|g\|_{\widetilde{H}^{-1/2}} \tag{13}$$

for any $g \in \widetilde{H}^{3/2}$.

2.1 The Limiting Operator

In this section we describe the relevant properties of A.

Lemma 1 *A is a closed operator with compact resolvent.*

Proof We first prove that the resolvent set of A is nonempty. First consider

$$A_0 := \begin{pmatrix} 0 & 1 \\ -\Delta_{S^{n-1}} & 0 \end{pmatrix}. \tag{14}$$

A direct computation shows that

$$(A_0 - i\mu)^{-1} = \begin{pmatrix} i\mu D(\mu)^{-1} & D(\mu)^{-1} \\ -\Delta_{S^{n-1}} D(\mu)^{-1} & i\mu D(\mu)^{-1} \end{pmatrix} \tag{15}$$

where $D(\mu) := -\Delta_{S^{n-1}} + \mu^2$ is invertible for any $\mu \neq 0$. In particular, this implies the spectrum of A_0 is real. Since

$$A - A_0 = \begin{pmatrix} \alpha & 0 \\ 0 & 2 + \alpha - n \end{pmatrix}$$

is a bounded operator on $\mathcal{H}$, the spectrum of A is contained in a bounded strip around the real axis, and hence the resolvent set is nonempty. The compactness of the resolvent operator now follows from the compactness of the embedding $\mathcal{H}^1 \hookrightarrow \mathcal{H}$.

We next prove that A is closed. It suffices to prove that A_0 is closed, since $A - A_0$ is bounded. To that end, let (f_k, g_k) be a sequence in $H^{3/2}(S^{n-1}) \oplus H^{1/2}(S^{n-1})$ such that $(f_k, g_k) \to (f, g)$ in $\mathcal{H}$ and $A_0(f_k, g_k) \to (F, G)$ in $\mathcal{H}$. This means $g_k \longrightarrow F$ in $H^{1/2}(S^{n-1})$ and $-\Delta_{S^{n-1}} f_k \longrightarrow G$ in $H^{-1/2}(S^{n-1})$. From (13) we have the estimate

$$\|f\|_{H^{3/2}(S^{n-1})} \leq C\left(\left\|\Delta_{S^{n-1}} f\right\|_{H^{-1/2}(S^{n-1})} + \|f\|_{H^{-1/2}(S^{n-1})}\right)$$

for all $f \in H^{3/2}(S^{n-1})$. Since $f_k \to f$ in $H^{1/2}(S^{n-1})$ and $\Delta_{S^{n-1}} f_k \to -G$ in $H^{-1/2}(S^{n-1})$, the estimate implies that $f_k \to f$ in $H^{3/2}(S^{n-1})$. Therefore, $(f_k, g_k) \to (f, g)$ in $H^{3/2}(S^{n-1}) \oplus H^{1/2}(S^{n-1})$, and so $A_0(f_k, g_k) \to A_0(f, g) = (F, G)$ in $H^{1/2}(S^{n-1}) \oplus H^{-1/2}(S^{n-1})$. This completes the proof that A_0 (and hence A) is closed. □

We now compute the spectrum of A.

Lemma 2 *The spectrum of A is $\alpha + \Sigma(n)$, where $\Sigma(n)$ is the set defined in* (7).

Proof It suffices to show that the spectrum is $\Sigma(n)$ when $\alpha = 0$. Since A has compact resolvent, the spectrum is discrete and contains only eigenvalues. For $\alpha = 0$ the eigenvalue equation is

$$\begin{pmatrix} 0 & 1 \\ -\Delta_{S^{n-1}} & 2-n \end{pmatrix} \begin{pmatrix} f \\ g \end{pmatrix} = \nu \begin{pmatrix} f \\ g \end{pmatrix}$$

hence $g = \nu f$ and $-\Delta_{S^{n-1}} f + (2-n)g = \nu g$, which we combine to obtain

$$-\Delta_{S^{n-1}} f = \nu(\nu + n - 2) f.$$

The distinct eigenvalues of $-\Delta_{S^{n-1}}$ are of the form $l(l+n-2)$ for $l \in \mathbb{N} \cup \{0\}$. Setting $\nu(\nu+n-2) = l(l+n-2)$, we obtain $\nu = l, 2-n-l$ as claimed. □

Finally, we prove a resolvent estimate for A.

Lemma 3 *For $-\alpha \notin \Sigma(n)$ there exists $C > 0$ such that*

$$\left\|(A - i\mu)^{-1}\right\|_{B(\mathcal{H})} \leq \frac{C}{1+|\mu|} \tag{16}$$

for all $\mu \in \mathbb{R}$.

Proof From Lemma 2, the hypothesis on α guarantees $A - i\mu$ is boundedly invertible for any $\mu \in \mathbb{R}$, so we just need to prove that (16) holds when $|\mu|$ is sufficiently large.

We next observe that it is enough to prove the estimate for the operator A_0 defined in (14). If the estimate holds for A_0 we can choose μ large enough that $\|(A_0 - i\mu)^{-1}(A - A_0)\|_{B(\mathcal{H})} \leq 1/2$, since $A - A_0 \in B(\mathcal{H})$. This implies $I + (A_0 - i\mu)^{-1}(A - A_0)$ is invertible, with

$$\left\|\big(I + (A_0 - i\mu)^{-1}(A - A_0)\big)^{-1}\right\|_{B(\mathcal{H})} \leq \sum_{k=0}^{\infty} \left(\frac{1}{2}\right)^k = 2.$$

Writing $A - i\mu = (A_0 - i\mu)\big(I + (A_0 - i\mu)^{-1}(A - A_0)\big)$, we thus obtain $\|(A - i\mu)^{-1}\|_{B(\mathcal{H})} \leq 2C/(1+|\mu|)$.

It remains to prove the resolvent estimate (16) for A_0 when $|\mu|$ is large. The resolvent is given by (15). Therefore it suffices to prove the estimates

$$\|D(\mu)^{-1}\|_{B(H^{-1/2}(S^{n-1}))} \leq \frac{C}{1+\mu^2}$$

$$\|D(\mu)^{-1}\|_{B(H^{1/2}(S^{n-1}))} \leq \frac{C}{1+\mu^2}$$

$$\|D(\mu)^{-1}\|_{B(H^{-1/2}(S^{n-1}),H^{1/2}(S^{n-1}))} \leq \frac{C}{1+|\mu|}$$

$$\|D(\mu)^{-1}\|_{B(H^{1/2}(S^{n-1}),H^{3/2}(S^{n-1}))} \leq \frac{C}{1+|\mu|}$$

for sufficiently large $|\mu|$.

Letting (λ_k) denote the eigenvalues of $-\Delta_{S^{n-1}}$, and (ϕ_k) the corresponding eigenfunctions, we can compute the H^s norm of f by

$$\|f\|^2_{H^s(S^{n-1})} = \sum_k (1+\lambda_k)^s |c_k|^2, \tag{17}$$

where $c_k = \langle f, \phi_k \rangle$. For smooth f we have

$$\|D(\mu)f\|^2_{H^s(S^{n-1})} = \sum_k (1+\lambda_k)^s (\lambda_k + \mu^2)^2 |c_k|^2.$$

Using the inequality $(\lambda_k + \mu^2)^2 \geq \mu^4$, we obtain

$$\|D(\mu)f\|^2_{H^s(S^{n-1})} \geq \mu^4 \sum_k (1+\lambda_k)^s |c_k|^2 = \mu^4 \|f\|^2_{H^s(S^{n-1})}.$$

Similarly, assuming without loss of generality that $|\mu| \geq 1$, we find that

$$(\lambda_k + \mu^2)^2 = \lambda_k^2 + 2\lambda_k \mu^2 + \mu^4 \geq \lambda_k \mu^2 + \mu^2 = \mu^2(1+\lambda_k) \geq \frac{1}{2}(1+|\mu|)^2(1+\lambda_k)$$

and hence

$$\|D(\mu)f\|^2_{H^s(S^{n-1})} \geq \frac{1}{2}(1+|\mu|)^2 \sum_k (1+\lambda_k)^{s+1} |c_k|^2 = \frac{1}{2}(1+|\mu|)^2 \|f\|^2_{H^{s+1}(S^{n-1})},$$

which completes the proof. □

2.2 The Perturbation

We now establish the required continuity and decay properties of the perturbation B.

Lemma 4 *$B(\cdot) \in C^{0,\gamma}\big((-\infty, 0], B(\mathcal{H}^\beta, \mathcal{H})\big)$ and $\|B(\tau)\|_{B(\mathcal{H}^\beta,\mathcal{H})} \leq Ce^{2\tau}$ for $\tau \leq 0$.*

Proof From the definition of $B(\tau)$ in (11) we obtain

$$\|B(\tau)\|_{B(\mathcal{H}^\beta,\mathcal{H})} = t^2 \|V_t\|_{B(H^{1/2+\beta}(S^{n-1}),H^{-1/2}(S^{n-1}))},$$

where V_t denotes the operator on $H^{1/2+\beta}(S^{n-1})$ that is multiplication by V_t followed by inclusion into $H^{-1/2}(S^{n-1})$. For any $f \in H^{1/2}(S^{n-1})$ we have

$$\|V_t f\|_{H^{-1/2}(S^{n-1})} = \sup_{g\neq 0} \frac{|\langle V_t f, g\rangle|}{\|g\|_{H^{1/2}(S^{n-1})}} \leq \left(\sup_{\theta\in S^{n-1}} |V(t,\theta)|\right) \|f\|_{H^{-1/2}(S^{n-1})}$$

and so

$$\|V_t\|_{B(H^{1/2+\beta}(S^{n-1}),H^{-1/2}(S^{n-1}))} = \sup_{f\neq 0} \frac{\|V_t f\|_{H^{-1/2}(S^{n-1})}}{\|f\|_{H^{1/2+\beta}(S^{n-1})}} \leq C \sup_{\theta\in S^{n-1}} |V(t,\theta)|,$$

where C depends on the norm of the embedding $H^{1/2+\beta}(S^{n-1}) \hookrightarrow H^{-1/2}(S^{n-1})$. This proves the claimed decay estimate for $B(\tau)$.

By the same argument we obtain

$$\|B(\tau_1) - B(\tau_2)\|_{B(\mathcal{H}^\beta,\mathcal{H})} \leq C \sup_{\theta\in S^{n-1}} \left|t_1^2 V(t_1,\theta) - t_2^2 V(t_2,\theta)\right|.$$

For any $0 < t_1, t_2 \leq 1$ and $\theta \in S^{n-1}$ we compute

$$\begin{aligned}\left|t_1^2 V(t_1,\theta) - t_2^2 V(t_2,\theta)\right| &\leq \left|t_1^2 - t_2^2\right| |V(t_1,\theta)| + t_2^2 |V(t_1,\theta) - V(t_2,\theta)| \\ &\leq 2\,|t_1 - t_2|\,|V(t_1,\theta)| + |V(t_1,\theta) - V(t_2,\theta)|\end{aligned}$$

and so $\|B(\tau_1) - B(\tau_2)\|_{B(\mathcal{H}^\beta,\mathcal{H})} \leq C'\,|t_1 - t_2|^\gamma$. The required estimate now follows from the fact that $|t_1 - t_2| = |e^{\tau_1} - e^{\tau_2}| \leq |\tau_1 - \tau_2|$ for all $\tau_1, \tau_2 \leq 0$. □

2.3 Unique Continuation

We next prove a unique continuation result for the rescaled system (6) and its adjoint. Given the equivalence established in Theorem 1, this is an easy consequence of the unique continuation principle for elliptic equations; see, for instance [4].

Lemma 5 *Suppose* $(\tilde{f}, \tilde{g})$ *is a solution of* (6) *on* $(-\infty, 0)$. *If* $(\tilde{f}(0), \tilde{g}(0)) = 0$, *then* $(\tilde{f}(\tau), \tilde{g}(\tau)) = 0$ *for all* $\tau \leq 0$.

Proof Let $(f(t), g(t))$ denote the corresponding solution to (3), obtained by undoing the transformation (5). Using the results of [3], we can write $(f(t), g(t)) = \mathrm{Tr}_t\, u$, where $u \in H^1(B_1 \setminus \{0\})$ is a weak solution to $\Delta u = Vu$. Then

$$u|_{\partial B_1} = f(1) = 0, \qquad \left.\frac{\partial u}{\partial \nu}\right|_{\partial B_1} = g(1) = 0$$

and so u must be identically zero. It follows that $f(t) = 0$ and $g(t) = 0$ for all $t \in (0, 1]$, hence $\tilde{f}(\tau)$ and $\tilde{g}(\tau)$ vanish for $\tau \leq 0$. □

We also need a unique continuation result for the adjoint system

$$\frac{d}{d\tau}\begin{pmatrix}\tilde{f}\\ \tilde{g}\end{pmatrix} = \begin{pmatrix}-\alpha & \Delta_{S^{n-1}} - e^{2\tau}V_{e^\tau}\\ -1 & n-2-\alpha\end{pmatrix}\begin{pmatrix}\tilde{f}\\ \tilde{g}\end{pmatrix}. \tag{18}$$

A direct calculation shows that $(f(t), g(t))$ satisfies (3) if and only if the rescaled quantity

$$\begin{pmatrix}-e^{(n-1-\alpha)\tau} g(e^\tau)\\ e^{(n-2-\alpha)\tau} f(e^\tau)\end{pmatrix} \tag{19}$$

satisfies (18). Therefore, the adjoint system (18) is also equivalent to the PDE (1), in the sense of Theorem 1, and so the argument of Lemma 5 applies.

2.4 Proof of Theorem 2 and Corollary 1

Given Lemmas 1, 2, 3, 4, and 5, Theorem 2 is an immediate consequence of [16, Theorem 1]. In fact, we are in the even better situation of [16, Corollary 2], which guarantees that $P^u(\tau)$ decays exponentially to the projection onto the unstable subspace for the autonomous operator A as $\tau \to -\infty$.

To prove Corollary 1, suppose $0 < \alpha < n-2$, so the condition $-\eta^s < \alpha$ is satisfied for any $\eta^s \geq 0$. Moreover, the smallest positive eigenvalue of A is α, so we can choose any $\eta^u \in [0, \alpha)$. Therefore it suffices to choose $\eta^u \in (\alpha + 1 - n/2, \alpha)$. This interval is nonempty because $n > 2$, and contains positive numbers because $\alpha > 0$. This completes the proof of the corollary.

Finally, we prove the claim that no such α exists when $n = 2$. To see this, let $-\alpha \notin \Sigma(2) = \mathbb{Z}$, so $\alpha + k \in (0, 1)$ for some $k \in \mathbb{Z}$. The growth and decay rates must satisfy

$$0 \le \eta^u < \alpha + k, \quad 0 \le \eta^s < 1 - \alpha - k.$$

Assuming (10) holds with $n = 2$, the condition $\alpha < \eta^u$ implies $k \ge 1$, hence $\eta^s < 1 - \alpha - k \le -\alpha$, which contradicts the other inequality in (10).

As mentioned in the introduction, the non-existence of suitable α for $n = 2$ is related to the spectrum of the asymptotic operator A. When $\alpha = 0$ the spectrum is given by the set $\Sigma(n)$ defined in (7). Note that 0 is always an eigenvalue of A, corresponding to the space of constant functions. When $n > 2$ the eigenvalue $2 - n$ corresponds to the fundamental solution r^{2-n}, which is singular at the origin. The exponential dichotomy distinguishes between these solutions provided $\alpha \in (0, n - 2)$; this is precisely the content of Corollary 1. On the other hand, when $n = 2$ the eigenvalue 0 is repeated, on account of the harmonic function $\log r$, which blows up at the origin at a slower rate than any polynomial, in the sense that $r^\alpha \log r \to 0$ as $r \to 0$ for any $\alpha > 0$.

3 Dichotomy Subspaces and Spherical Harmonics

We illustrate the results of the previous section for harmonic functions on $\mathbb{R}^3$. In this case $V = 0$, so (6) becomes

$$\frac{d}{d\tau}\begin{pmatrix} \tilde{f} \\ \tilde{g} \end{pmatrix} = \begin{pmatrix} \alpha & 1 \\ -\Delta_{S^2} & \alpha - 1 \end{pmatrix}\begin{pmatrix} \tilde{f} \\ \tilde{g} \end{pmatrix}. \tag{20}$$

In particular, $B(\tau) = 0$, so we are in the simpler case of [16, Lemma 2.1], which guarantees the existence of a dichotomy for (6) on the entire real line, with τ-independent projections P^s and P^u.

3.1 The Dichotomy Subspaces

From Lemma 2 the eigenvalues of A are

$$\nu_l^+ = \alpha + l, \quad \nu_l^- = \alpha - l - 1,$$

for $l = 0, 1, 2, \ldots$. Each $\nu_l^{\pm}$ has multiplicity $2l + 1$. The eigenfunctions can be expressed in terms of spherical harmonics Y_l^m as

$$\begin{pmatrix} \tilde{f}_{lm}^{+}(\tau) \\ \tilde{g}_{lm}^{+}(\tau) \end{pmatrix} = e^{(\alpha+l)\tau} Y_l^m \begin{pmatrix} 1 \\ l \end{pmatrix}, \quad \begin{pmatrix} \tilde{f}_{lm}^{-}(\tau) \\ \tilde{g}_{lm}^{-}(\tau) \end{pmatrix} = e^{(\alpha-l-1)\tau} Y_l^m \begin{pmatrix} 1 \\ -(l+1) \end{pmatrix}$$

for $-l \leq m \leq l$, and so

$$\begin{pmatrix} f_{lm}^{+}(t) \\ g_{lm}^{+}(t) \end{pmatrix} = t^{l} Y_l^m \begin{pmatrix} 1 \\ l/t \end{pmatrix}, \quad \begin{pmatrix} f_{lm}^{-}(t) \\ g_{lm}^{-}(t) \end{pmatrix} = t^{-l-1} Y_l^m \begin{pmatrix} 1 \\ -(l+1)/t \end{pmatrix}.$$

Note that $\big(f_{lm}^{+}(t), g_{lm}^{+}(t)\big)$ is the boundary data of the harmonic function $u(r, \theta, \phi) = r^l Y_l^m(\theta, \phi)$ on the surface $\{|x| = t\}$, and $\big(f_{lm}^{-}(t), g_{lm}^{-}(t)\big)$ is the boundary data of $u(r, \theta, \phi) = r^{-l-1} Y_l^m(\theta, \phi)$. Solutions corresponding to ν_l^{+} are bounded at the origin and blow up at infinity, whereas solutions corresponding to ν_l^{-} blow up at the origin and decay to zero at infinity.

The unstable subspace $\widetilde{E}^u(\tau)$ is spanned by the eigenfunctions for which the corresponding eigenvalue $\nu_l^{\pm}$ is positive, and similarly for the stable subspace $\widetilde{E}^s(\tau)$. For any $\alpha \in (0, 1)$ we have $\nu_0^{-} < 0 < \nu_0^{+}$, and hence $\nu_l^{-} < 0 < \nu_l^{+}$ for all l. Therefore, for any such α, $E^u(t)$ is precisely the set of boundary data of harmonic functions that are bounded at the origin, as was shown more generally in Corollary 1.

3.2 *The Dichotomy Projections*

We assume the spherical harmonics Y_l^m are normalized so that $\langle Y_l^m, Y_k^n \rangle = \delta_{mn}\delta_{lk}$, where $\langle \cdot, \cdot \rangle$ denotes the $L^2(S^2)$ inner product:

$$\langle f, g \rangle = \int_0^{2\pi} \int_0^{\pi} f(\theta, \phi)\overline{g(\theta, \phi)} \sin\theta \, d\theta d\phi.$$

Expanding $z = (z_1, z_2) \in \mathcal{H}$ as

$$z = \sum_{l=0}^{\infty} \sum_{m=-l}^{l} \left(A_{lm} \begin{pmatrix} 1 \\ l \end{pmatrix} + B_{lm} \begin{pmatrix} 1 \\ -(l+1) \end{pmatrix} \right) Y_l^m,$$

we find that

$$A_{lm} = \frac{1}{2l+1} \langle (l+1)z_1 + z_2, Y_l^m \rangle, \quad B_{lm} = \frac{1}{2l+1} \langle l z_1 - z_2, Y_l^m \rangle,$$

and so the dichotomy projections are given by

$$P^u z = \sum_{l=0}^{\infty} \sum_{m=-l}^{l} Y_l^m \frac{1}{2l+1} \left\langle (l+1)z_1 + z_2, Y_l^m \right\rangle \begin{pmatrix} 1 \\ l \end{pmatrix}, \tag{21}$$

$$P^s z = \sum_{l=0}^{\infty} \sum_{m=-l}^{l} Y_l^m \frac{1}{2l+1} \left\langle l z_1 - z_2, Y_l^m \right\rangle \begin{pmatrix} 1 \\ -(l+1) \end{pmatrix}. \tag{22}$$

3.3 The Evolution Operators

We next give explicit formulas for the operators $\Phi^{s,u}(\tau, \tau_0)$ defined in (8).

For arbitrary $z \in \mathcal{H}$, $\Phi^u(\tau, \tau_0)z$ must be of the form $\sum\sum C_{lm} e^{(\alpha+l)\tau}(Y_l^m, lY_l^m)$. Using the formula for P^u obtained above, and the fact that $\Phi^u(\tau_0, \tau_0)z = P^u z$, we find that $C_{lm} = e^{-(\alpha+l)\tau_0} \left\langle (l+1)z_1 + z_2, Y_l^m \right\rangle / (2l+1)$, and hence

$$\Phi^u(\tau, \tau_0)z = \sum_{l=0}^{\infty} \sum_{m=-l}^{l} e^{(\alpha+l)(\tau-\tau_0)} Y_l^m \frac{1}{(2l+1)} \langle (1+l)z_1 + z_2, Y_l^m \rangle \begin{pmatrix} 1 \\ l \end{pmatrix} \tag{23}$$

for $\tau \leq \tau_0$. Similarly, we obtain

$$\Phi^s(\tau, \tau_0)z = \sum_{l=0}^{\infty} \sum_{m=-l}^{l} e^{(\alpha-l-1)(\tau-\tau_0)} Y_l^m \frac{1}{(2l+1)} \langle l z_1 - z_2, Y_l^m \rangle \begin{pmatrix} 1 \\ -(l+1) \end{pmatrix} \tag{24}$$

for $\tau \geq \tau_0$.

3.4 Liouville-Type Theorems

Since (20) is autonomous, the exponential dichotomy exists on the entire real line; cf. Theorem 2 which only guarantees the existence of a half-line dichotomy. Therefore, [16, Theorem 2] says that the only bounded solution to (20) is $(\tilde{f}(\cdot), \tilde{g}(\cdot)) = (0, 0)$. Using this, we obtain the following Liouville-type result, which rules out the existence of slowly-growing harmonic functions.

Corollary 2 *Suppose u is an entire harmonic function on $\mathbb{R}^n$. If $\|u\|_{H^1(\Omega_t)} \leq Ct^r$ for some $r < n/2 - 1$, then u is identically zero.*

Proof From [3] we have the estimates

$$\|f(t)\|_{H^{1/2}(S^{n-1})} \leq Ct^{-n/2}\|u\|_{H^1(\Omega_t)},$$
$$\|g(t)\|_{H^{-1/2}(S^{n-1})} \leq Ct^{-n/2}\|u\|_{H^1(\Omega_t)},$$

and hence

$$\|\tilde{f}(\tau)\|_{H^{1/2}(S^{n-1})} \leq Ct^{\alpha-n/2}\|u\|_{H^1(\Omega_t)},$$
$$\|\tilde{g}(\tau)\|_{H^{-1/2}(S^{n-1})} \leq Ct^{1+\alpha-n/2}\|u\|_{H^1(\Omega_t)}.$$

Choose a number $0 < \alpha < (n/2 - 1) - r$ with $-\alpha \notin \Sigma(n)$, so that Theorem 2 applies. It follows from elliptic regularity that u and ∇u are uniformly bounded in a neighborhood of the origin, say $|u(x)|, |\nabla u(x)| \leq c$ for all $x \in \Omega_t$, with t sufficiently small. Then

$$\|u\|^2_{H^1(\Omega_t)} = \int_{\Omega_t} \left(|u|^2 + |\nabla u|^2\right) \leq 2c^2|\Omega_t| = 2c^2\omega_n t^n,$$

and so $\|u\|_{H^1(\Omega_t)} \leq Ct^{n/2}$ for small t. Since $\alpha > 0$, both $\|\tilde{f}(\tau)\|_{H^{1/2}(S^{n-1})}$ and $\|\tilde{g}(\tau)\|_{H^{-1/2}(S^{n-1})}$ are thus bounded as $\tau \to -\infty$. On the other hand, the hypothesis $\|u\|_{H^1(\Omega_t)} \leq Ct^r$ implies

$$\|\tilde{g}(\tau)\|_{H^{-1/2}(S^{n-1})} \leq Ct^{1+\alpha-n/2+r}$$

is bounded as $\tau \to \infty$, since $1+\alpha-n/2+r < 0$, and similarly for $\|\tilde{f}(\tau)\|_{H^{1/2}(S^{n-1})}$. □

4 Applications

The previous sections gave a dynamical interpretation of the linear elliptic equation (1), expanding on the results in [3] in the radial case. We conclude by presenting some applications of these ideas to linear and nonlinear PDE. In particular, we show that the presence (or absence) of unstable eigenvalues is encoded in the dichotomy subspaces, and demonstrate how the exponential dichotomy can be used to construct solutions to nonlinear equations on bounded and unbounded domains.

4.1 Eigenvalue Problems

Here we use Corollary 1 to give a dynamical interpretation of the eigenvalue problem

$$-\Delta u + Vu = \lambda u \tag{25}$$

with Dirichlet boundary conditions. To do so we let $E^u(t)$ denote the unstable subspace corresponding to (25), with α chosen to satisfy the hypotheses of Theorem 3, and define the *Dirichlet subspace*

$$\mathcal{D} = \{(0, g) : g \in H^{-1/2}(S^{n-1})\} \subset \mathcal{H}. \tag{26}$$

Theorem 4 *λ is an eigenvalue of the Dirichlet problem* (25) *on B_t if and only if the unstable subspace $E^u(t)$ intersects the Dirichlet subspace $\mathcal{D}$ nontrivially. Moreover, the multiplicity of λ equals* $\dim\left(E^u(t) \cap \mathcal{D}\right)$.

Other boundary conditions (Neumann, Robin, etc.) can be characterized in a similar way by replacing $\mathcal{D}$ accordingly; see [5, 6] for details.

Therefore we have given a dynamical perspective on elliptic eigenvalue problems, similar to the Evans function [17], which counts intersections between stable and unstable subspaces. This is also closely related to the Maslov index, a symplectic winding number that counts intersections of Lagrangian subspaces in a symplectic Hilbert space; see [5–7, 13].

4.2 Reformulation of Two Nonlinear Problems

In this section we illustrate how to reformulate equations of the form

$$\Delta u - V(x)u = F(x, u),$$

where F is smooth with $F(x, 0) = D_u F(x, 0) = 0$, using the dichotomy constructed above.

We emphasize that this approach allows for the construction of solutions that are not radially symmetric, even though spherical subdomains $\Omega_t = \{|x| < t\}$ are used in constructing the dichotomy.

4.2.1 A Nonlinear Boundary Value Problem

First, we consider the case where $x \in B_T = \{x \in \mathbb{R}^n : |x| < T\}$, with some appropriate boundary condition:

$$\begin{aligned} \Delta u - V(x)u &= F(x,u), \quad x \in B_T \\ \left(u|_{\partial B_T}, \partial_\nu u|_{\partial B_T}\right) &\in \mathcal{B}, \end{aligned} \tag{27}$$

for some subspace $\mathcal{B} \subset \mathcal{H}$. Using the framework introduced above, we write this as the equivalent spatial evolutionary system

$$\frac{d}{dt}\begin{pmatrix} f \\ g \end{pmatrix} = \begin{pmatrix} 0 & 1 \\ V(t,\theta) - t^{-2}\Delta_{S^{n-1}} & -t^{-1}(n-1) \end{pmatrix}\begin{pmatrix} f \\ g \end{pmatrix} + \begin{pmatrix} 0 \\ F(t,\theta,f) \end{pmatrix}. \tag{28}$$

Applying the change of variables used above, $t = e^\tau$, $\tilde{f}(\tau) = e^{\alpha\tau} f(e^\tau)$, $\tilde{g}(\tau) = e^{(\alpha+1)\tau} g(e^\tau)$, we find

$$\frac{d}{d\tau}\begin{pmatrix} \tilde{f} \\ \tilde{g} \end{pmatrix} = \begin{pmatrix} \alpha & 1 \\ e^{2\tau}V(e^\tau,\theta) - \Delta_{S^{n-1}} & \alpha + 2 - n \end{pmatrix}\begin{pmatrix} \tilde{f} \\ \tilde{g} \end{pmatrix} + \begin{pmatrix} 0 \\ e^{(\alpha+2)\tau} F(e^\tau,\theta,e^{-\alpha\tau}\tilde{f}) \end{pmatrix}. \tag{29}$$

It was shown above that for any $\beta \in [0,1)$ an exponential dichotomy exists in $\mathcal{H}^\beta$ on the interval $(-\infty, \log T]$, for the linear evolution associated with the above system, as long as $-\alpha \notin \Sigma(n)$ and $V \in C^{0,\gamma}(\Omega)$, which we assume in this section. For notational convenience, write the above system as

$$\frac{d}{d\tau}\tilde{h} = \mathcal{A}(\tau)\tilde{h} + \mathcal{F}(\tau,\tilde{h}), \qquad \tilde{h} = \begin{pmatrix} \tilde{f} \\ \tilde{g} \end{pmatrix}, \tag{30}$$

where

$$\mathcal{A}(\tau) = \begin{pmatrix} \alpha & 1 \\ e^{2\tau}V(e^\tau,\theta) - \Delta_{S^{n-1}} & \alpha + 2 - n \end{pmatrix}, \quad \mathcal{F}(\tau,\tilde{h}) = \begin{pmatrix} 0 \\ e^{(\alpha+2)\tau} F(e^\tau,\theta,e^{-\alpha\tau}\tilde{f}) \end{pmatrix},$$

and we have notationally suppressed any θ-dependence. With a suitable assumption on the nonlinearity F, any solution to (30) that is bounded as $\tau \to -\infty$ can be written in terms of the operators $\Phi^{s,u}$ defined in (8) as

$$\tilde{h}(\tau) = \Phi^u(\tau, \log T)\tilde{h}_* + \int_{-\infty}^{\tau} \Phi^s(\tau,\rho)\mathcal{F}(\rho,\tilde{h}(\rho))d\rho + \int_{\log T}^{\tau} \Phi^u(\tau,\rho)\mathcal{F}(\rho,\tilde{h}(\rho))d\rho \tag{31}$$

for some $\tilde{h}_* \in \mathcal{H}^\beta$. For instance, it is sufficient to have $\mathcal{F} \in C^{1,1}\big((-\infty, \log T] \times \mathcal{H}^\beta, \mathcal{H}\big)$, which is equivalent to requiring that the map $(\tau, \tilde{f}) \mapsto e^{(\alpha+2)\tau} F(e^\tau, \theta, e^{-\alpha\tau}\tilde{f})$ is in $C^{1,1}\big((-\infty, \log T] \times H^{1/2+\beta}(S^{n-1}), H^{-1/2}(S^{n-1})\big)$; see [16, p. 294].

Using the fact that

$$\frac{d}{d\tau}\Phi^{s,u}(\tau,\rho) = \mathcal{A}(\tau)\Phi^{s,u}(\tau,\rho), \qquad \Phi^s(\tau,\tau) + \Phi^u(\tau,\tau) = \mathrm{Id},$$

once can directly check that $\tilde{h}(\tau)$ given in (31) is indeed a solution of (30). The exponential bounds for $\Phi^{s,u}(\tau,\rho)$ ensure that it is well-behaved as $\tau \to -\infty$. At the moment, $\tilde{h}_* \in \mathcal{H}^\beta$ is arbitrary. However, we have not yet made reference to the boundary condition. We need

$$\tilde{h}(\log T) = P^u(\log T)\tilde{h}_* + \int_{-\infty}^{\log T} \Phi^s(\log T, \rho)\mathcal{F}(\rho, \tilde{h}(\rho))d\rho \in \mathcal{B}. \tag{32}$$

The idea is thus to choose $\tilde{h}_* \in \mathcal{H}^\beta$ so that (32) holds. Note that, since $\tilde{h}$ is defined implicitly via (31), the integral term in (32) depends on the choice of $\tilde{h}_*$ through $\tilde{h}$. The best way to understand (32) would depend on the details of the dichotomy and the boundary conditions.

4.2.2 A Nonlinear Problem on $\mathbb{R}^n$

Next consider

$$\Delta u - V(x)u = F(x,u), \qquad x \in \mathbb{R}^n. \tag{33}$$

If we reformulate this as the evolutionary system (29), then the linear part admits an exponential dichotomy on the negative half line $(-\infty, 0]$, by Theorem 2. We denote this by $\Phi^{s,u}_-$. Moreover, if $|x|^2V(x) \to 0$ as $|x| \to \infty$, the proof of Theorem 2 also yields a dichotomy on the positive half line $[0,\infty)$, which we denote by $\Phi^{s,u}_+$. (When $V = 0$ we have a dichotomy on the whole line, so $\Phi^u_\pm$ and $\Phi^s_\pm$ are given explicitly by (23) and (24) for $n = 3$, and can be expressed similarly for $n > 3$.)

As in the previous section, with a suitable assumption on the nonlinearity F, bounded solutions on $(-\infty, 0]$ are given by

$$\tilde{h}^-(\tau) = \Phi^u_-(\tau,0)\tilde{h}_1 + \int_{-\infty}^{\tau} \Phi^s_-(\tau,\rho)\mathcal{F}(\rho, \tilde{h}^-(\rho))d\rho + \int_0^{\tau} \Phi^u_-(\tau,\rho)\mathcal{F}(\rho, \tilde{h}^-(\rho))d\rho \tag{34}$$

and bounded solutions on $[0, \infty)$ are given by

$$\tilde{h}^+(\tau) = \Phi^s_+(\tau, 0)\tilde{h}_2 + \int_{+\infty}^{\tau} \Phi^u_+(\tau, \rho)\mathcal{F}(\rho, \tilde{h}^+(\rho))d\rho + \int_0^{\tau} \Phi^s_+(\tau, \rho)\mathcal{F}(\rho, \tilde{h}^+(\rho))d\rho, \tag{35}$$

where $\tilde{h}_{1,2} \in \mathcal{H}^\beta$ are, for the moment, arbitrary.

To find a solution to (29) that is bounded for all $\tau \in \mathbb{R}$, we must match (34) and (35) at $\tau = 0$. This leads to the matching condition

$$\begin{aligned} 0 &= \tilde{h}^+(0) - \tilde{h}^-(0) \\ &= P^u_-(0)\tilde{h}_1 + \int_{-\infty}^{0} \Phi^s_-(0, \rho)\mathcal{F}(\rho, \tilde{h}^-(\rho))d\rho \\ &\quad - P^s_+(0)\tilde{h}_2 - \int_{+\infty}^{0} \Phi^u_+(0, \rho)\mathcal{F}(\rho, \tilde{h}^+(\rho))d\rho. \end{aligned}$$

Similar to the previous example, the best way to understand this matching condition depends on the details of the nonlinearity. In the $V = 0$ case one has the advantage of having an explicit formula for the dichotomy and the projection operators.

Acknowledgments The authors would like to acknowledge the support of the American Institute of Mathematics and the Banff International Research Station, where much of this work was carried out. M.B. acknowledges the support of NSF grant DMS-1411460 and of an AMS Birman Fellowship. G.C. acknowledges the support of NSERC grant RGPIN-2017-04259. C.J. was supported by ONR grant N00014-18-1-2204. Y.L. was supported by the NSF grant DMS-1710989, by the Research Board and Research Council of the University of Missouri, and by the Simons Foundation. A.S. was supported by NSF grant DMS-1910820.

References

1. Amick, C.J.: Semilinear elliptic eigenvalue problems on an infinite strip with an application to stratified fluids. Ann. Sc. Norm. Super. Pisa Cl. Sci. (4) **11**(3), 441–499 (1984). http://www.numdam.org/item/ASNSP_1984_4_11_3_441_0
2. Beck, M., Sandstede, B., Zumbrun, K.: Nonlinear stability of time-periodic viscous shocks. Arch. Ration. Mech. Anal. **196**(3), 1011–1076 (2010). https://doi.org/10.1007/s00205-009-0274-1
3. Beck, M., Cox, G., Jones, C., Latushkin, Y., Sukhtayev, A.: A dynamical approach to semilinear elliptic equations (2019). Preprint. arXiv:1907.09986
4. Behrndt, J., Rohleder, J.: An inverse problem of Calderón type with partial data. Commun. Partial Differ. Equ. **37**(6), 1141–1159 (2012). https://doi.org/10.1080/03605302.2011.632464. http://dx.doi.org.libproxy.lib.unc.edu/10.1080/03605302.2011.632464
5. Cox, G., Jones, C.K.R.T., Marzuola, J.L.: A Morse index theorem for elliptic operators on bounded domains. Commun. Partial Differ. Equ. **40**(8), 1467–1497 (2015). https://doi.org/10.1080/03605302.2015.1025979. http://dx.doi.org.libproxy.lib.unc.edu/10.1080/03605302.2015.1025979

6. Cox, G., Jones, C.K.R.T., Latushkin, Y., Sukhtayev, A.: The Morse and Maslov indices for multidimensional Schrödinger operators with matrix-valued potentials. Trans. Am. Math. Soc. **368**(11), 8145–8207 (2016). https://doi.org/10.1090/tran/6801
7. Deng, J., Jones, C.: Multi-dimensional Morse index theorems and a symplectic view of elliptic boundary value problems. Trans. Am. Math. Soc. **363**(3), 1487–1508 (2011). https://doi.org/10.1090/S0002-9947-2010-05129-3
8. Doelman, A., Sandstede, B., Scheel, A., Schneider, G.: The dynamics of modulated wave trains. Mem. Am. Math. Soc. **199**(934), viii+105 (2009). https://doi.org/10.1090/memo/0934
9. Gardner, R.: Existence of multidimensional travelling wave solutions of an initial-boundary value problem. J. Differ. Equ. **61**(3), 335–379 (1986). https://doi.org/10.1016/0022-0396(86)90111-7
10. Große, N., Schneider, C.: Sobolev spaces on Riemannian manifolds with bounded geometry: general coordinates and traces. Math. Nachr. **286**(16), 1586–1613 (2013). https://doi.org/10.1002/mana.201300007
11. Kirchgässner, K.: Wave-solutions of reversible systems and applications. J. Differ. Equ. **45**(1), 113–127 (1982). https://doi.org/10.1016/0022-0396(82)90058-4
12. Latushkin, Y., Pogan, A.: The dichotomy theorem for evolution bi-families. J. Differ. Equ. **245**(8), 2267–2306 (2008). https://doi.org/10.1016/j.jde.2008.01.023
13. Latushkin, Y., Sukhtaiev, S.: The Maslov index and the spectra of second order elliptic operators. Adv. Math. **329**, 422–486 (2018). https://doi.org/10.1016/j.aim.2018.02.027
14. McLean, W.: Strongly Elliptic Systems and Boundary Integral Equations. Cambridge University Press, Cambridge (2000)
15. Mielke, A.: A reduction principle for nonautonomous systems in infinite-dimensional spaces. J. Differ. Equ. **65**(1), 68–88 (1986). https://doi.org/10.1016/0022-0396(86)90042-2
16. Peterhof, D., Sandstede, B., Scheel, A.: Exponential dichotomies for solitary-wave solutions of semilinear elliptic equations on infinite cylinders. J. Differ. Equ. **140**(2), 266–308 (1997). https://doi.org/10.1006/jdeq.1997.3303
17. Sandstede, B.: Stability of travelling waves. In: Handbook of Dynamical Systems, vol. 2, pp. 983–1055. North-Holland, Amsterdam (2002). https://doi.org/10.1016/S1874-575X(02)80039-X
18. Sandstede, B., Scheel, A.: On the structure of spectra of modulated travelling waves. Math. Nachr. **232**, 39–93 (2001). https://doi.org/10.1002/1522-2616(200112)232:1<39::AID-MANA39>3.3.CO;2-X
19. Scheel, A.: Radially symmetric patterns of reaction-diffusion systems. Mem. Am. Math. Soc. **165**(786), viii+86 (2003). https://doi.org/10.1090/memo/0786
20. Strichartz, R.S.: Analysis of the Laplacian on the complete Riemannian manifold. J. Funct. Anal. **52**(1), 48–79 (1983). https://doi.org/10.1016/0022-1236(83)90090-3
21. Triebel, H.: Theory of Function Spaces. II. Monographs in Mathematics, vol. 84. Birkhäuser Verlag, Basel (1992). https://doi.org/10.1007/978-3-0346-0419-2

Stability of Slow Blow-Up Solutions for the Critical Focussing Nonlinear Wave Equation on $\mathbb{R}^{3+1}$

Stefano Burzio

Abstract In this brief survey we outline the recent advances on the stability issues of certain finite time type II blow-up solutions for the energy critical focusing wave equation $\Box u = -u^5$ in $\mathbb{R}^{3+1}$. Hereafter we use the convention $\Box = -\partial_t^2 + \Delta$. The objective of this article is twofold: firstly we describe the construction of singular solutions contained in Krieger et al. (Duke Math J 147(1):1–53, 2009) and Krieger and Schlag (J Math Pures Appl 101(6):873–900, 2014), and secondly we undertake a detailed analysis of their stability properties enclosed in Krieger (On the stability of type II blow up for the critical NLW on $\mathbb{R}^{3+1}$. Mem Am Math Soc, 2018) and Burzio and Krieger (Mem Am. Math. Soc. arXiv preprint math/1709.06408, 2017)

1 Introduction

Despite its naive appearance, the semilinear wave equation

$$\Box u = -u^5, \quad u : \mathbb{R}^{1+3}_{t,x} \to \mathbb{R} \tag{1}$$

is an excellent simplistic model since its main features are shared with multiple geometric and physical equations such as critical Wave-Maps and Yang-Mills equations. However, as we shall see, the price to pay to avoid many technical issues is the ubiquity of type I blow-up solutions which constitute the generic blow-up scenario.

Local well-posedness up to the optimal regularity class $H^1(\mathbb{R}^3) \times L^2(\mathbb{R}^3)$ of the Cauchy problem for Eq. (1) coupled with initial data was proved by Lindblad and Sogge [20] and it relies on the celebrated Strichartz estimates, see also [24] for a

S. Burzio (✉)
EPFL, Bâtiment des Mathématiques, Lausanne, Switzerland
e-mail: stefano.burzio@eplf.ch

W. Dörfler et al. (eds.), *Mathematics of Wave Phenomena*, Trends in Mathematics,
https://doi.org/10.1007/978-3-030-47174-3_5

detailed description. Moreover, as a typical trademark for focusing equations, the conserved energy

$$E(u)(t) = \int_{\mathbb{R}^3} \frac{1}{2}|\nabla_{t,x}u|^2 - \frac{1}{6}|u|^6 dx$$

is not positive definite, making the extensions of local solutions to global one a highly non-trivial question.

In fact, several obstructions to long-time existence of solution of (1) have been uncovered. For instance, Levine [19] demonstrated via a convexity argument that break down in finite time occur for initial data with negative energy. Nonetheless, Levine's argument is indirect, and it does not provide much information about the exact nature of the blow-up. More primitive blow-up solutions can be explicitly constructed by the ODE technique: let $\phi \in C_0^\infty(\mathbb{R}^3)$ such that $\phi(x) = 1$ if $|x| \leq 2T$, set the initial data $u_0(0, x) = (\frac{3}{4})^{1/4} T^{-1/2}\phi(x)$, and $u_t(0, x) = (\frac{3}{64})^{1/4} T^{-3/2}\phi(x)$. Then the solution of (1) behaves like the so called *fundamental self-similar solution*

$$u(t, x) = (\frac{3}{4})^{1/4}(T - t)^{-1/2} \tag{2}$$

for $0 < t < T$ and $|x| < T - t$. As this example shows, singularities can arise in finite time even for smooth compactly supported initial data. Observe that for these solutions the critical Sobolev norm diverges as time approaches the maximum time of existence:

$$\limsup_{t \to T} \|\nabla_{t,x}u(t, \cdot)\|_{L^2(\mathbb{R}^3)} \to +\infty. \tag{3}$$

Motivated by such blow-up mechanism it is common to define a blow-up solution u with maximum forward time of existence $T < +\infty$ of *type I* if (3) holds, and of *type II* otherwise, that is if $\|\nabla_{t,x}u(t, \cdot)\|_{L^2}$ remains bounded up to the break down time. The dichotomy between type I and type II blow-up solutions is well understood at this point in time.

Another explicit solution of (1) is the Aubin-Talenti function

$$W(x) = \left(1 + \frac{|x|^2}{3}\right)^{-1/2}$$

which is the unique (up to symmetries) positive solution to the associated elliptic equation $\triangle W = -W^5$ and it is the minimizer of the Sobolev embedding $\dot{H}^1(\mathbb{R}^3) \hookrightarrow L^6(\mathbb{R}^3)$, see [1] and [25]. Through a remarkable series of works Duyckaerts et al. [6–9] provided a complete abstract classification of all possible type II blow-up solutions in finite time in terms of a finite number of rescaled W plus a small radiation term.

Theorem 1 ([9]) *Let u be a radial type II solution of* (1) *which breaks down in finite time T. Then there exist finitely many continuous functions $\lambda_j(t)$, $j = 1, \ldots, k$, with $\lim_{t\to T}(T-t)\lambda_j(t) = +\infty$, and*

$$\lim_{t\to T}\left|\log\left(\frac{\lambda_i(t)}{\lambda_j(t)}\right)\right| = +\infty, \qquad \textit{for } i \neq j,$$

such that

$$u(t,x) = \sum_{j=1}^{k} \pm W_{\lambda_j(t)}(x) + \eta(t,x)$$

and where $(\eta, \partial_t\eta) \in C([0,T], \dot{H}^1 \times L^2)$ and $W_\lambda(x) = \lambda^{1/2}W(\lambda x)$.

The extent of the previous result is essential in the progress of understanding type II blow-up solutions. However, due to the nature of the arguments, namely the famous concentration compactness method, the Duyckaerts, Kenig, and Merle program does not demonstrate the existence of all such possible blow-up dynamics. In fact, at the best of author's knowledge it emerges that only finite time blow-up solutions with one bulk term W are known to exist. Moreover, the precise blow-up dynamics is unknown and it does not appear to give any information on the stability of such solutions.

Complementary, an explicit finite time type II blow-up was constructed by Krieger et al. [16]. The breakthrough [16] consists in establishing the existence of a family of rough blow-up solutions displaying a continuum of blow-up rates slower then the one provided by the self-similar blow-up. In addition, all previously known blow-up solutions become singular along a hypersurface, vice-versa the ones furnished in [16] exhibit a one-point blow-up. In a subsequent work [13], the first two authors extended the range of allowed blow-up speeds up to reach arbitrary close the self-similar blow-up speed.

Other concrete realizations for finite-time type II dynamics where established: Hillairet and Raphaël [10] constructed type II smooth solution for the energy critical semilinear wave equation $\Box u = -u^3$ in $\mathbb{R}^{4+1}$, with the fixed scaling law

$$\lambda(t) = t^{-1}e^{\sqrt{|\log t|}}, \qquad \text{as } t \to 0.$$

The set of initial data leading to such type II blow-up is given by a co-dimension one Lipschitz manifold. Another constructive approach was given in Krieger et al. [17], where the authors provided a finite time blow-up solution of type II with oscillating scaling law, that is of the form $u(t,x) = W_{\lambda(t)}(x) + \eta(t,x)$ where $\lambda(t) = t^{-\nu(t)}$ and $\nu(t) = \nu + \epsilon_0\frac{sin(\log t)}{\log t}$, with $\nu > 3$ and $|\epsilon_0| \ll 1$ be arbitrary and η a small error.

A deeper study of the stability of such blow-up scenarios has been the subject of a number of recent works. Persuaded by numerical evidence provided by Bizon et al. [2], which suggested that finite time blow-up for (1) are generically of type I, in a sequence of pioneering works Donninger and Schörkhuber [5] and Donninger [4] settled the asymptotic stability of the ODE blow-up solution (2) in the energy norm. On the other hand, Krieger et al. [18] elucidated that type II solutions are unstable in the energy norm in the following precise sense.

Theorem 2 ([18]) *Let $\lambda(t) \to +\infty$ as $t \to T$, and*

$$u(t,x) = W_{\lambda(t)}(x) + \eta(t,x)$$

be a type II blow up solution on $I \times \mathbb{R}^3$ for (1), such that

$$\sup_{t\in I} \left\| \nabla_{t,x}\eta(t,\cdot) \right\|_{L^2_x} \le \delta \ll 1$$

for some sufficiently small $\delta > 0$, where $I = [0,T]$ denotes the maximal life span of the Shatah-Struwe solution u. Also, assume that $t_0 \in I$. Then there exists a co-dimension one Lipschitz manifold Σ in a small neighborhood of the data $\big(u(t_0,\cdot), u_t(t_0,\cdot)\big) \in \Sigma$ in the energy topology $\dot{H}^1(\mathbb{R}^3) \times L^2(\mathbb{R}^3)$, such that initial data $\big(f,g\big) \in \Sigma$ result in a type II solution, while initial data

$$\big(f,g\big) \in B_\delta \backslash \Sigma,$$

where $B_\delta \subset \dot{H}^1(\mathbb{R}^3) \times L^2(\mathbb{R}^3)$ is a sufficiently small ball centered at $\big(u(t_0,\cdot), u_t(t_0,\cdot)\big)$, either lead to blow-up in finite time, or solutions scattering to zero, depending on the 'side of Σ' these data are chosen from.

In spite of the universality of type I blow-up for Eq. (1), with the purpose of study more sophisticated equations at the critical regime where only type II dynamic is present, it is fundamental to investigate further type II blow-up solutions and its stability properties. The stability of solutions constructed in [13] and [16] was analyzed by Krieger [11] where a conditional result requiring two extra co-dimensions was obtained for solutions which blow-up at a rate sufficiently close to the self-similar one. The optimal stability result was achieved by the author and Krieger in [3]. In the second part of this article we outline the proof of the latter results.

To place these results in a proper context, some more discussion on similar results for different equations is in order. As a matter of fact, the work [16] is an occurrence in a triplets of works [14–16], dedicated to the explicit construction of rough type II singular solutions respectively for semilinear wave equations, for the co-rotational critical wave maps from $\mathbb{R}^{2+1} \to S^2$, and for the critical Yangs-Mills equations in 4 + 1 dimensions under the spherically symmetric ansatz. A parallel construction of a smooth finite time type II singular solution with fixed blow-up speed was carried out by Raphaël and Rodnianski [22] for the co-rotational critical wave maps in 2+1 dimensions with S^2 target, and for the critical $SO(4)$ Yangs-Mills equations in 4 + 1 dimensions. Concerning the stability issue, the method employed by Raphaël and Rodnianski implies that their solutions are stable. Furthermore, in a recent breakthrough Krieger and Miao [12] were able to show that the solutions constructed in [14] for the co-rotational critical wave maps are stable in a suitable topology. The corresponding result for the Yang-Mills problem is still open.

2 The Construction of Slow Blow-Up Solutions

In this section we describe the construction of explicit finite time type II blow-up solutions contained in the works [13] and [16]. We shall be interested exclusively in the case of radial solutions, thus the energy critical focusing semilinear wave equation under radial symmetry can be written as:

$$- u_{tt} + u_{rr} + \frac{2}{r}u_r = -u^5. \tag{4}$$

The goal is to construct a solution $u \in C((0, t_0], H^{1+}) \times C^1((0, t_0], H^+)$ of (4) which blows-up at the space-time origin, and the blow-up is of type II, hence its space-time gradient remains bounded on the interval of existence: $\sup_{t\in[0,t_0]} \|\nabla_{t,x} u(t, \cdot)\|_{L^2(\mathbb{R}^3)} < \infty$. Notice that, due to the time reversibility of the wave equation, we start evolving the dynamics from initial data at time $t_0 > 0$ and solve the wave equation backwards in time until the blow-up time $t = 0$. Here t_0 is a small positive constant that will be defined later.

We state here the results of [16] and [13]. The main difference between them is the lower bound for ν. In [16] the restriction $\nu > 1/2$ was imposed, and in [13] the result was extended to include $\nu > 0$.

Theorem 3 ([13, 16]) *Let $\nu > 0$ and $\lambda(t) = t^{-1-\nu}$ the scaling parameter. There exists a class of solutions to Eq. (4) of the form*

$$u_\nu(t, r) = W_{\lambda(t)}(r) + \eta(t, r) =: u_0(t, r) + \eta(t, r)$$

inside the truncated light cone $K = \{(t, r) \in (0, t_0) \times \mathbb{R}^+ : t > r\}$. The term u_0 is called bulk term or non-oscillatory elliptic term and it is given by the rescaling of W. The second term η is called oscillatory radiation part and it is composed by two distinct functions: $\eta = \eta^e + \varepsilon$. Here η^e is an non-oscillatory term satisfying $\eta^e \in C^\infty(K)$ and $\mathcal{E}_{loc}(\eta^e)(t) \lesssim (t\lambda(t))^{-2}|\log t|^2$ as $t \to 0$, hence its local energy vanishes as time $t \to 0$. The local energy relative to the origin is defined as

$$\mathcal{E}_{loc}(u)(t) = \int_{|x|<t} \frac{1}{2}|\nabla_{t,x} u|^2 - \frac{1}{6}|u|^6 dx.$$

On the other hand ε is rougher, that is $(\varepsilon(t, \cdot), \varepsilon_t(t, \cdot) \in (H^{\frac{\nu}{2}+1-}(\mathbb{R}^3) \times H^{\frac{\nu}{2}-}(\mathbb{R}^3))$ and $\mathcal{E}_{loc}(\varepsilon)(t) \to 0$ as $t \to 0$. Moreover, outside the light cone we have the bound

$$\int_{|x|\geq t} \frac{1}{2}|\nabla_{t,x} u_\nu|^2 - \frac{1}{6}|u_\nu|^6 dx \leq C < \infty.$$

Notice that the bulk term $u_0 \in C^\infty(\mathbb{R}^+ \times \mathbb{R}^+)$ blows-up at the space-time origin. However $u_0(t,\cdot) \notin L^2(\mathbb{R}^3)$, since it does not decay sufficiently fast at space infinity. To obtain a solution in H^{1+} it suffices to multiply u_0 by a bump function $\chi \in C_0^\infty(\mathbb{R})$ which equals one on the ball of radius t_0. In this way, on every fixed positive time slice we clearly have $u_0(t,\cdot)\chi(t) \in H^{1+}(\mathbb{R}^3)$ and $\partial_t(u_0(t,\cdot)\chi(t)) \in H^+(\mathbb{R}^3)$. Notice that $u_0(t,\cdot)\chi(t,\cdot)$ is in $C_0^\infty(\mathbb{R}^3)$ therefore clearly is type II. The rougher part of the solution which gives the overall regularity $C((0,t_0], H^{\frac{\nu}{2}+1-}) \times C^1((0,t_0], H^{\frac{\nu}{2}-})$ is the term η^e. In fact, although it is smooth inside the light cone, namely $\eta^e \in C^\infty(K)$, it reveals a cusp singularity along the boundary ∂K of the light cone, implying that $\eta^e(t,\cdot) \in H_{loc}^{1+\frac{\nu}{2}-}(\mathbb{R}^3)$ and $\partial_t\eta^e(t,\cdot) \in H_{loc}^{\frac{\nu}{2}-}(\mathbb{R}^3)$.

Before outlining the proof of Theorem 3, a final remark on the blow-up speed of u_ν is in order. Clearly, the $L^\infty(\mathbb{R}^3)$ norm of the ODE blow-up solution (2) concentrates, as time approaches the break-down time, at a rate which is proportional to $t^{-1/2}$. Diversely, the blow-up of speed of u_ν solutions is proportional to $t^{-1/2-\nu/2}$. Hence type II solutions blow-up faster that type I and, as ν approaches 0, the different blow-up speeds become comparables. Moreover, by varying the parameter $\nu > 0$, which it is not a priori fixed, we obtain a blow-up solution with prescribed blow-up speed, i.e. a *continuum of blow-up speeds*.

The proof of Theorem 3 is based on a two steps procedure mimicking the strategy of others constructions of type II dynamics contained in [10, 21, 22]. Firstly, one constructs a sequence of approximate solutions which solve (4) up to a small error. This approximation method will not lead to an exact solution by passing to the limit due to the divergence of the coefficients. Hence one needs to terminate the process after finitely many steps. Secondly, one completes the approximate solution to an exact solution via a fixed point argument. In regard to the second step, the argument used to prove Theorem 3 differs drastically from the strategy employed in [10, 21, 22]. In the latter pioneering works the remaining error is controlled via Morawetz and viral type identities, whereas the present proof hinges on a constructive parametrix approach.

2.1 The Renormalization Step

The aim of this step is to iteratively construct a very accurate approximate solution near the singularity depending on two parameters k and ν which has the form

$$u_k(t,r) = W_{\lambda(t)}(r) + \eta_k^e(t,r)$$

where the k-th non-oscillatory term $\eta_k^e(t,r) = \sum_{j=1}^k v_j(t,r)$ is a sum of small corrections and $\lambda(t) = t^{-1-\nu}$. The bulk term $u_0(t,r) = W_{\lambda(t)}(r)$ is very far from being an approximate solution of (4), indeed it produces an error $e_0 = \Box u_0 + (u_0)^5$

which blows up like t^{-2} as $t \to 0$. In [16] the authors adopt the strategy of adding successive corrections functions v_j so that the error $e_k = \Box u_k + u_k^5$ generated by the approximate solution u_k can be made arbitrary small in a suitable sense by picking k suitable large. More precisely, the corrections v_j are chosen in order to force e_k to go to zero like t^N as $t \to 0$ in the energy norm restricted to a light cone, where N can be made arbitrarily large by taking k large.

The construction consists in a delicate bookkeeping procedure to iteratively reduce the size of the generated error by alternating between amelioration near the spatial origin and improvements near the light cone. The finite sequence of approximate solution u_k is defined recursively. Set $u_0 = W_{\lambda(t)}(r)$, then for $k \geq 1$ the k-th approximation u_k is given in terms of the previous one via the following algorithm: let u_{k-1} be the approximate solution which generates the error $e_{k-1} = \Box u_{k-1} + u_{k-1}^5$, then one updates u_{k-1} by adding a correction, i.e. $u_k = u_{k-1} + v_k = u_0 + v_1 + \cdots + v_k$, thereby the error e_k produced by the improved approximation u_k is smaller than e_{k-1} in a suitable sense. To define the appropriate correction v_k we distinguish between k even or k odd. The odd corrections are the solutions of the following inhomogeneous second order ODEs:

$$\begin{cases} \left(\partial_r^2 + \frac{2}{r}\partial_r + 5u_0^4(t,r)\right)v_{2k-1}(t,r) = e_{2k-2}(t,r) & \text{in } \mathbb{R}_r^+, \\ v_{2k-1}(t,0) = \partial_r v_{2k-1}(t,0) = 0. \end{cases} \tag{5}$$

The heuristic which leads to such formulation is that when $r \ll t$ we expect the term involving the time derivative in (4) to be negligible. On the other hand, for even corrections we improve the approximate solution near the light cone $r \approx t$, thus we can roughly estimate u_0 by zero and we are led to the $1+1$-inhomogeneous hyperbolic equation

$$\begin{cases} \left(-\partial_t^2 + \partial_r^2 + \frac{2}{r}\partial_r\right)v_{2k}(t,r) = e_{2k-1}(t,r) & \text{in } \mathbb{R}_t^+ \times \mathbb{R}_r^+, \\ v_{2k}(t,0) = \partial_r v_{2k}(t,0) = 0. \end{cases} \tag{6}$$

The Cauchy problem (5) is a standard Sturm-Liouville problem and it is solved via the variation of parameter method. Whereas the hyperbolic character of (6) is controlled by using self-similar coordinate $a = r/t$ and a brilliant ansatz on the form of the solution.

2.2 Completion to an Exact Solution

The main point of this second part of the argument is to perturb around the approximate solution constructed in the previous step, and thus to look for an exact solution of (4) of the form $u_\nu = u_{2k-1} + \varepsilon$. Notice that we stop the approximation algorithm after an odd number of cycles. By imposing that u_ν to be an exact solution we force an equation for ε:

$$\big(\partial_t^2 - \partial_r^2 - \frac{2}{r}\partial_r\big)\varepsilon - 5\lambda^2(t)W^4\big(\lambda(t)r\big)\varepsilon = e_{2k-1} + \tilde{N}(u_{2k-1}, \varepsilon) \tag{7}$$

where $\tilde{N}(u, \varepsilon) = (\varepsilon + u)^5 - u^5 - 5\varepsilon u_0^4$. To avoid treating a nonlinear hyperbolic equation with time-dependent potential one removes the time dependency of the potential by introducing new coordinates $(t, r) \to (\tau, R)$, where

$$\tau(t) = \int_t^{t_0} \lambda(s)ds + \nu^{-1}t_0^{-\nu} = \nu^{-1}t^{-\nu}, \qquad R(t, r) = \lambda(t)r.$$

The price to pay is that the time derivative ∂_t is transformed into the operator $\lambda(\tau)\partial_\tau + \frac{\lambda'(\tau)}{\lambda(\tau)}R\partial_R$. Let us set $v(\tau, R) = \varepsilon(t(\tau), \lambda^{-1}(\tau)R)$ and $\beta(\tau) = \lambda'(\tau)/\lambda(\tau)$, then Eq. (7) is transformed into

$$\Big[\Big(\partial_\tau + \beta(\tau)R\partial_R\Big)^2 - \beta(\tau)\Big(\partial_\tau + \beta(\tau)R\partial_R\Big) - \partial_R^2 - \frac{2}{R}\partial_R\Big]v(\tau, R) - 5W^4(R)v(\tau, R) =$$
$$\lambda^{-2}(\tau)e_{2k-1}(\tau, R) + \lambda^{-2}(\tau)\tilde{N}(u_{2k-1}, v)(\tau, R).$$

Subsequently, in order to get rid of the first derivative in the R variable, we consider the function $\tilde{\varepsilon}(\tau, R) = Rv(\tau, R)$, this new function satisfies the equation

$$(\mathcal{D}^2 + \beta(\tau)\mathcal{D} + \mathcal{L})\tilde{\varepsilon}(\tau, R) = f[\tilde{\varepsilon}](\tau, R), \qquad \text{in } \mathbb{R}_\tau^+ \times \mathbb{R}_R^+ \tag{8}$$

where $\mathcal{D} = \partial_\tau + \beta(\tau)(R\partial_R - 1)$, $\mathcal{L} = -\partial_R^2 - 5W^4(R)$, and

$$f[\tilde{\varepsilon}](\tau, R) = \lambda^{-2}(\tau)\Big(Re_{2k-1} + N(u_{2k-1}, \tilde{\varepsilon})\Big)$$

and

$$N(u_{2k-1}, \tilde{\varepsilon})(\tau, R) = 5\tilde{\varepsilon}(u_{2k-1}^4 - u_0^4) + R\Big(\frac{\tilde{\varepsilon}}{R} + u_{2k-1}\Big)^5 - Ru_{2k-1}^5 - 5u_{2k-1}^4\tilde{\varepsilon}.$$

To look for a solution of Eq. (8) a prototypical Fourier transform, namely the *distorted Fourier transform associated to the operator* $\mathcal{L}$, is applied imitating the procedure to convert to the frequencies sides the free wave equation. The spectral properties of the operator $\mathcal{L}$ play a pivotal role and are analyzed in details in [16]. This operator, when restricted to functions on $[0, \infty)$ with Dirichlet condition at $R = 0$, has a simple negative eigenvalue $\xi_d < 0$ (the subscript d referring to *discrete spectrum*), and a corresponding L^2-normalized positive ground state $\phi_d \in L^2(0, \infty) \cap C^\infty([0, \infty))$ decaying exponentially and vanishing at the origin $R = 0$. This mode will cause exponential growth for the linearized evolution $e^{it\sqrt{\mathcal{L}}}$. However, in [16] and [13] the authors avoid this problem by imposing vanishing initial data at $t = 0$ for the function ε, which is equivalent to impose zero data at $\tau = \infty$ for the function $\tilde{\varepsilon}$. In the subsequent works [11] and [3], where no such freedom of imposing zero initial data is acceptable, only a co-dimension one condition will ensure that the forward flow will remains bounded.

Let us present below the pivotal result which summarize the main properties of the distorted Fourier transform.

Proposition 1 ([16]) *There exists a generalized Fourier basis $\phi(R, \xi)$, $\xi \geq 0$, a eigenstate $\phi_d(R)$, and a spectral measure $\rho(\xi) \in C^\infty\big((0, \infty)\big)$ with the asymptotic behaviors*

$$\rho(\xi) \sim \begin{cases} \xi^{-\frac{1}{2}}, & \text{if } 0 < \xi \ll 1, \\ \xi^{\frac{1}{2}}, & \text{if } \xi \gg 1, \end{cases}$$

as well as symbol behaviour with respect to differentiation, and such that by defining

$$\mathcal{F}(f)(\xi) := \hat{f}(\xi) := \lim_{b \to +\infty} \int_0^b \phi(R, \xi) f(R)\, dR,$$

$$\hat{f}(\xi_d) = \int_0^\infty \phi_d(R) f(R)\, dR,$$

the map $f \longrightarrow \hat{f}$ is an isometry from L^2_{dR} to $L^2(\{\xi_d\} \cup \mathbb{R}^+, \rho)$, and we have

$$f(R) = \hat{f}(\xi_d)\phi_d(R) + \lim_{\mu \to \infty} \int_0^\mu \phi(R, \xi) \hat{f}(\xi) \rho(\xi)\, d\xi,$$

the limits being in the suitable L^2-sense.

The mayor issue in applying the distorted Fourier transform to Eq. (8) is the term involving $R\partial_R$ contained in the $\mathcal{D}$ operator since $\mathcal{F}(R\partial_R) \neq \xi \partial_\xi \mathcal{F}$. Therefore one defines the *error operator* $\mathcal{K}$ via the equation

$$\mathcal{F}[(R\partial_R - 1)u(\tau, R)](\xi) = \mathcal{A}\hat{u}(\tau, \xi) + \mathcal{K}\hat{u}(\tau, \xi)$$

where

$$\mathcal{A} = \begin{pmatrix} 0 & 0 \\ 0 & \mathcal{A}_c \end{pmatrix}, \qquad \mathcal{K} = \begin{pmatrix} \mathcal{K}_{dd} & \mathcal{K}_{dc} \\ \mathcal{K}_{cd} & \mathcal{K}_{cc} \end{pmatrix}$$

and $\mathcal{A}_c = -2\xi\partial_\xi - \left(\frac{5}{2} + \frac{\rho'(\xi)\xi}{\rho(\xi)}\right)$. We add the second term in $\mathcal{A}_c$ because later on we shall need the relation $(SM)^{-1}\partial_\tau SM = \mathcal{D}_\tau$ where $\mathcal{D}_\tau = \partial_\tau + \beta(\tau)\mathcal{A}_c$. In other words $\mathcal{K}$ is defined as the solution to the system

$$\begin{cases} \langle (R\partial_R - 1)u(\tau, R), \phi_d \rangle_{L^2_{dR}} = \mathcal{K}_{dd} \langle u(\tau, R), \phi_d \rangle_{L^2_{dR}} + \mathcal{K}_{dc}\hat{u} \\ \mathcal{F}[(R\partial_R - 1)u(\tau, R)] = \mathcal{A}_c\hat{u} + \mathcal{K}_{cd} \langle u(\tau, R), \phi_d \rangle_{L^2_{dR}} + \mathcal{K}_{cc}\hat{u} \end{cases}$$

and we have

$$\mathcal{K}_{dd} = -\frac{1}{2}, \qquad \mathcal{K}_{cd}(\xi) = k_d(\xi),$$

$$\mathcal{K}_{dc}(f) = -\int_0^\infty f(\xi)k_d(\xi)\rho(\xi)d\xi, \qquad \mathcal{K}_{cc}[f](\xi) = \int_0^\infty \frac{F(\xi, \eta)}{\xi - \eta} f(\eta)\rho(\eta)d\eta.$$

where $k_d(\xi)$ is a smooth and rapidly decaying function at $\xi = +\infty$ and the function F is of regularity at least C^2 on $(0, \infty) \times (0, \infty)$, and satisfies further smoothness and decay properties listed in Theorem 5.1 in [16].

We now proceed to transpose Eq. (8) to the Fourier side. Notice that the time variable remain invariant since we are dealing with Fourier transform in space only. Let us denote the distorted Fourier transform of the unknown function in (8) by $\big(x_d(\tau), x(\tau, \xi)\big) = \mathcal{F}(\tilde{\varepsilon})(\tau, \xi)$, that is:

$$x(\tau, \xi) = \int_0^\infty \phi(R, \xi)\tilde{\varepsilon}(\tau, R)\, dR, \qquad x_d(\tau) = \int_0^\infty \phi_d(R)\tilde{\varepsilon}(\tau, R)\, dR.$$

Notice that once the Fourier representation $\big(x_d(\tau), x(\tau, \xi)\big)$ is known one can easily recover the original function $\tilde{\varepsilon}$ via

$$\tilde{\varepsilon}(\tau, R) = x_d(\tau)\phi_d(R) + \int_0^\infty x(\tau, \xi)\phi(R, \xi)\rho(\xi)d\xi. \tag{9}$$

Applying the distorted Fourier transform to Eq. (8) yields to the following system involving one equation for the discrete spectral part and a second equation for the continuous spectral part:

$$(\mathcal{D}_\tau^2 + \beta(\tau)\mathcal{D}_\tau + \xi)x(\tau, \xi) = \mathcal{R}x(\tau, \xi) + f(\tau, \xi) \tag{10}$$

where $(\tau, \xi) \in \mathbb{R}^+ \times \mathbb{R}^+$, $x(\tau, \xi) = \big(x_d(\tau), x(\tau, \xi)\big)^T$, and

$$(\mathcal{D}_\tau^2 + \beta(\tau)\mathcal{D}_\tau + \xi) = \begin{pmatrix} \partial_\tau^2 + \beta(\tau)\partial_\tau + \xi_d & 0 \\ 0 & \mathcal{D}_\tau^2 + \beta(\tau)\mathcal{D}_\tau + \xi \end{pmatrix}.$$

The inhomogeneous terms on the right-hand-side of (10) are composed of a linear source term

$$\mathcal{R} = \begin{pmatrix} \mathcal{R}_{dd} & \mathcal{R}_{dc} \\ \mathcal{R}_{cd} & \mathcal{R}_{cc} \end{pmatrix}$$

where

$$\begin{aligned}
\mathcal{R}_{dd} &= -2\beta(\tau)\mathcal{K}_{dd}\partial_\tau - \beta^2(\tau)\Big(\mathcal{K}_{dd}^2 + \mathcal{K}_{dc}\mathcal{K}_{cd} + \mathcal{K}_{dd} + \frac{\beta'(\tau)}{\beta^2(\tau)}\mathcal{K}_{dd}\Big),\\
\mathcal{R}_{dc} &= -2\beta(\tau)\mathcal{K}_{dc}\mathcal{D}_\tau - \beta^2(\tau)\Big(\mathcal{K}_{dd}\mathcal{K}_{dc} + \mathcal{K}_{dc}\mathcal{K}_{cc} - \mathcal{K}_{dc}\mathcal{A}_c + \mathcal{K}_{dc} + \frac{\beta'(\tau)}{\beta^2(\tau)}\mathcal{K}_{dc}\Big),\\
\mathcal{R}_{cd} &= -2\beta(\tau)\mathcal{K}_{cd}\partial_\tau - \beta^2(\tau)\Big(\mathcal{K}_{cd}\mathcal{K}_{dd} + \mathcal{K}_{dc}\mathcal{K}_{cd} + \mathcal{A}_c\mathcal{K}_{cd} + \mathcal{K}_{cd} + \frac{\beta'(\tau)}{\beta^2(\tau)}\mathcal{K}_{cd}\Big),\\
\mathcal{R}_{cc} &= -2\beta(\tau)\mathcal{K}_{cc}\mathcal{D}_\tau - \beta^2(\tau)\Big(\mathcal{K}_{cd}\mathcal{K}_{dc} + \mathcal{K}_{cc}^2 + [\mathcal{A}_c, \mathcal{K}_{cc}] + \mathcal{K}_{cc} + \frac{\beta'(\tau)}{\beta^2(\tau)}\mathcal{K}_{cc}\Big),
\end{aligned} \tag{11}$$

plus the nonlinear term (observe that $\tilde{\varepsilon}$ depends on the unknown functions $x_d(\tau), x(\tau,\xi)$ via (9)):

$$f(\tau,\xi) = \begin{pmatrix} f_d(\tau) \\ f(\tau,\xi) \end{pmatrix} = \begin{pmatrix} \lambda^{-2}(\tau)\langle \phi_d, Re_{2k-1} + N(u_{2k-1},\tilde{\varepsilon})\rangle_{L^2_{dR}} \\ \lambda^{-2}(\tau)\mathcal{F}\Big(Re_{2k-1} + N(u_{2k-1},\tilde{\varepsilon})\Big)(\tau,\xi) \end{pmatrix}.$$

We coupled system (10) with initial conditions $\lim_{\tau\to\infty} x_d(\tau) = \partial_\tau x_d(\tau) = 0$, and $\lim_{\tau\to\infty} x(\tau,\xi) = \mathcal{D}_\tau x(\tau,\xi) = 0$.

The advantage of system (10) is the crucial observation that it can be solved completely explicitly. In fact, define $(Sf)(\tau,\xi) = f(\tau, \lambda^{-2}(\tau)\xi)$ and $(Mf)(\tau,\xi) = \lambda^{-5/2}(\tau)\rho^{1/2}(\xi) f(\tau,\xi)$, then we have the essential identity

$$(\mathcal{D}_\tau^2 + \beta(\tau)\mathcal{D}_\tau + \xi) = (SM)^{-1}[(\partial_\tau^2 + \beta(\tau)\partial_\tau + \lambda^{-2}(\tau)\xi)]SM$$

which provides the following parametrix

$$\begin{aligned}
x_d(\tau) &= \int_{\tau_0}^{\infty} -\frac{1}{2|\xi_d|^{1/2}} e^{-|\xi_d|^{1/2}|\tau-\sigma|}\big(g_d(\sigma) - \beta(\sigma)\partial_\sigma x_d(\sigma)\big)d\sigma,\\
x(\tau,\xi) &= \int_\tau^\infty \frac{\lambda^{3/2}(\tau)}{\lambda^{3/2}(\sigma)} \frac{\rho^{1/2}\big(\frac{\lambda^2(\tau)}{\lambda^2(\sigma)}\xi\big)}{\rho^{1/2}(\xi)} \frac{\sin\Big[\lambda(\tau)\xi^{1/2}\int_\tau^\sigma \lambda^{-1}(u)du\Big]}{\xi^{1/2}} g\Big(\sigma, \frac{\lambda^2(\tau)}{\lambda^2(\sigma)}\xi\Big)d\sigma,
\end{aligned} \tag{12}$$

where $g = (g_d, g)^T$ represent respectively the right-hand-side of (10).

A contraction argument allow us to conclude the proof by finding an appropriate solution to (10). The fix point iteration is carried out in a weighted Sobolev type spaces defined by means of the following norms. Let $\alpha \in \mathbb{R}^+$, and a function $u(\xi) = \left(u_d, u(\xi)\right)^T$, then define the norm

$$\|u\|^2_{L^{2,\alpha}_{d\rho}} = |u_d|^2 + \|u\|^2_{L^{2,\alpha}_{d\rho}} := |u_d|^2 + \int_0^\infty |u(\xi)|^2 |\langle\xi\rangle^{2\alpha} \rho(\xi) d\xi.$$

Notice that $\mathcal{F}$ is an isometry from $H^{2\alpha}_{dR}(\mathbb{R}^+)$ to $L^{2,\alpha}_{d\rho}(\mathbb{R}^+)$. Moreover for every τ-dependent function $f(\tau, \xi) = \left(f_d(\tau), f(\tau, \xi)\right)^T$ let us define the norm $\|f\|_{L^{2,\alpha,N}_\rho} = \sup_{\tau > \tau_0} \tau^N \|f(\tau, \cdot)\|_{L^{2,\alpha}_{d\rho}}$. Defining x via the explicit formulas (12), we obtain the linear estimate

$$\|(x, \mathcal{D}_\tau x)\|_{L^{2,\alpha+\frac{1}{2},N-2}_{d\rho} \times L^{2,\alpha,N-1}_{d\rho}} \lesssim \frac{1}{N} \|g\|_{L^{2,\alpha,N}_{d\rho}}.$$

The small factor N^{-1} is crucial for the fixed point argument to work. A similar estimate holds for the inhomogeneous terms on the right-hand-side of (10). More precisely the map g satisfies the bound

$$\|g\|_{L^{2,\alpha,N}_{d\rho}} \lesssim \|(x, \mathcal{D}_\tau x)\|_{L^{2,\alpha+\frac{1}{2},N-2}_{d\rho} \times L^{2,\alpha,N-1}_{d\rho}} \tag{13}$$

and it is locally Lipschitz as a map from $L^{2,\alpha+1/2,N-2}_{d\rho}$ to $L^{2,\alpha,N}_{d\rho}$. Here the smallness of the constant is a consequences of the smallness of the error generated by the approximate solution built in the first part of the argument and the smallness of the time interval $(0, t_0]$ where the construction holds. The lack of smoothness of the approximate solution limits the decay in frequencies, hence the nonlinear estimate (13) holds only for $\nu/4 > \alpha$.

In [16], to control the nonlinear factors enclosed in the $f(\tau, \xi)$ term, precisely to obtain the quintilinear bound

$$H^{2\alpha+1}(\mathbb{R}^3) \cdot H^{2\alpha+1}(\mathbb{R}^3) \cdot H^{2\alpha+1}(\mathbb{R}^3) \cdot H^{2\alpha+1}(\mathbb{R}^3) \cdot H^{2\alpha+1}(\mathbb{R}^3) \subset H^{2\alpha}(\mathbb{R}^3),$$

the authors relies on a standard application of the Leibniz rule and Sobolev embedding, which holds for $\alpha \geq 1/8$, leading to the lower bound on the blow-up speed: $\nu > 1/2$. The latter restriction was removed in [13] by a more detailed analysis of the first iterate of the exact solution, thus yielding to the full expected range $\nu > 0$.

3 The Stability of Slow Blow-Up Solutions

In what follows we outline the stability results of type II blow-up solutions u_ν constructed in [16] and [13]. The continuum of blow-up rates proper to u_ν and their limited regularity seem to indicate that these solutions are less stable than their smooth analogs built in [10]. Moreover, taking into consideration the parallel results in the parabolic setting [23, 26] it was commonly assumed that imposing a stability condition will single out a quantized set of allowed blow-up speeds. Although these observations had solid foundations, they were disproved in [11] and [3], making the stability of a family of rough solutions with varying concentration rates a unique feature of hyperbolic equations. In fact, the results [11] and [3] demonstrate that rough solutions u_ν are stable along a co-dimension one Lipschitz manifold of data perturbations in a suitable topology, provided that the blow-up speed is sufficiently close to the self-similar ones, i.e. $\nu > 0$ is sufficiently small. The result is optimal in view of [18], since any type II solution with data close enough to the ground state W can be at best stable for perturbations of the data along a co-dimension one hypersurface in energy space.

The main improvement of [3] over [11] is essentially in the number of co-dimensions imposed on the perturbations. In [11] Krieger showed that type II solutions u_ν are stable under an appropriate co-dimension three condition. Precisely, there exists a co-dimension three Lipschitz hypersurface $\Sigma_0 \subset H^{3/2+}_{rad,loc}(\mathbb{R}^3) \times H^{1/2+}_{rad,loc}(\mathbb{R}^3)$ such that if we take the perturbation of the initial data $(\varepsilon_0, \varepsilon_1) \in \Sigma_0$ small enough, then the solution of the perturbed problem

$$\begin{cases} \Box u = -u^5 & \text{in } (0, t_0] \times \mathbb{R}^3 \\ u[t_0] = u_\nu[t_0] + (\varepsilon_0, \varepsilon_1) \end{cases} \tag{14}$$

is a type II blow-up solution of exactly of same type as u_ν. In the subsequent work [3] the extra co-dimensions two condition was removed yielding to the optimal result. The precise statement is given below.

To properly enunciate the co-dimension conditions imposed in [11] we have to closely analyse the initial value problem on the Fourier side. We shall seek to construct a solution of (14) by perturbing around the exact solution u_ν, thus we make the following ansatz:

$$u(t, r) = u_\nu(t, r) + \varepsilon(t, r)$$

where $(\varepsilon, \partial_t \varepsilon)$ matches the initial data at time $t = t_0$: $(\varepsilon, \partial_t \varepsilon)\big|_{t=t_0} = (\varepsilon_0, \varepsilon_1)$. In analogy with the argument of the previous section we introduce the renormalized coordinates $(\tau, R) = (\nu^{-1} t^{-\nu}, \lambda(t) r)$, we set $\tilde{\varepsilon} = R\varepsilon$, and we apply the distorted Fourier transform to the equation satisfied by $\tilde{\varepsilon}$. Thus we obtain the following equation in terms of the Fourier variable $x(\tau, \xi) = \mathcal{F}(\tilde{\varepsilon})(\tau, \xi)$:

$$(\mathcal{D}_\tau^2 + \beta(\tau)\mathcal{D}_\tau + \xi)x(\tau, \xi) = \mathcal{R}x(\tau, \xi) + f[\tilde{\varepsilon}](\tau, \xi) \tag{15}$$

where $(\tau, R) \in [\tau_0, \infty) \times \mathbb{R}^+$ and the linear source terms $\mathcal{R}$ are as in (11) and the nonlinear terms are defined by

$$f[\tilde{\varepsilon}](\tau,\xi) = \begin{pmatrix} f_d[\tilde{\varepsilon}](\tau) \\ f[\tilde{\varepsilon}](\tau,\xi) \end{pmatrix} = \begin{pmatrix} \lambda^{-2}(\tau)\langle \phi_d, N(u_\nu, \tilde{\varepsilon})\rangle_{L^2_{dR}} \\ \lambda^{-2}(\tau)\mathcal{F}\Big(N(u_\nu, \tilde{\varepsilon})\Big)(\tau,\xi) \end{pmatrix}.$$

Instead of coupling the system (15) with vanishing initial data at $\tau = +\infty$ we shall impose initial data at the corresponding initial time $\tau = \tau_0$:

$$\begin{aligned} x_d(\tau_0) &= x_{0d}, & \partial_\tau x_d(\tau_0) &= x_{1d}, \\ x(\tau_0,\xi) &= x_0(\xi), & \mathcal{D}_\tau x(\tau_0,\xi) &= x_1(\xi). \end{aligned} \tag{16}$$

One can compute the initial data on the physical side $(\tilde{\varepsilon}_0, \tilde{\varepsilon}_1)$ in terms of the initial data on the Fourier side (x_0, x_1), and vice-versa, via the formulas:

$$\begin{aligned} \mathcal{F}(\tilde{\varepsilon}_0) &= x_0, & -\mathcal{F}\Big(\frac{\tilde{\varepsilon}_1}{\lambda}\Big) &= x_1 + \beta_\nu(\tau_0)\mathcal{K}_{cc}x_0 + \beta_\nu(\tau_0)\mathcal{K}_{cd}x_{0d}, \\ \langle \phi_d, \tilde{\varepsilon}_0\rangle_{L^2_{dR}} &= x_{0d}, & -\langle \phi_d, \frac{\tilde{\varepsilon}_1}{\lambda}\rangle_{L^2_{dR}} &= x_{1d} + \beta_\nu(\tau_0)\mathcal{K}_{dd}x_{0d} + \beta_\nu(\tau_0)\mathcal{K}_{dc}x_0. \end{aligned} \tag{17}$$

We now present the Theorem contained in [3] which states that the blow-up phenomenon described in Theorem 3 is stable under a suitable co-dimension one class of data perturbations.

Theorem 4 ([3]) *Assume $0 < \nu \ll 1$, and assume $t_0 = t_0(\nu) > 0$ is sufficiently small, so that the solutions u_ν constructed in [13] and [16] exist on $(0, t_0] \times \mathbb{R}^3$. Let $\delta_1 = \delta_1(\nu) > 0$ be small enough, and let $\mathcal{B}_{\delta_1} \subset \tilde{S} \times \mathbb{R}$ be the δ_1-vicinity of $\big((0,0),0\big) \in \tilde{S} \times \mathbb{R}$, where $\tilde{S}$ is the Banach space defined as the completion of $C_0^\infty(0,\infty) \times C_0^\infty(0,\infty)$ with respect to the norm*

$$\big\|(x_0,x_1)\big\|_{\tilde{S}} = \big\|\langle\xi\rangle^{\frac{1}{2}+2\delta_0}\min\{\tau_{0,0}\xi^{\frac{1}{2}},1\}^{-1}\xi^{\frac{1}{2}-\delta_0}x_0\big\|_{L^2_{d\xi}} + \big\|\langle\xi\rangle^{\frac{1}{2}+2\delta_0}\xi^{-\delta_0}x_1\big\|_{L^2_{d\xi}}.$$

Then there is a Lipschitz function $\gamma_1 : \mathcal{B}_{\delta_1} \to \mathbb{R}$, such that for any triple $\big((x_0,x_1),x_{0d}\big) \in \mathcal{B}_{\delta_1}$, the quadruple

$$\big((x_0,x_1),(x_{0d},x_{1d})\big),\ x_{1d} = \gamma_1(x_0,x_1,x_{0d})$$

determines a data perturbation pair $(\varepsilon_0,\varepsilon_1) \in H^{\frac{3}{2}+2\delta_0}_{rad,loc}(\mathbb{R}^3) \times H^{\frac{1}{2}+2\delta_0}_{rad,loc}(\mathbb{R}^3)$ via (17), and such that the perturbed initial data

$$u_\nu[t_0] + (\varepsilon_0,\varepsilon_1) \tag{18}$$

lead to a solution $\tilde{u}(t,x)$ on $(0,t_0]\times\mathbb{R}^3$ admitting the description

$$\tilde{u}(t,x) = W_{\tilde{\lambda}(t)}(x) + \epsilon(t,x),\ \big(\varepsilon(t,\cdot),\varepsilon_t(t,\cdot)\big) \in H^{1+\frac{\nu}{2}-}_{rad,loc} \times H^{\frac{\nu}{2}-}_{rad,loc}$$

where the parameter $\tilde{\lambda}(t)$ equals $\lambda(t)$ asymptotically

$$\lim_{t\to 0} \frac{\tilde{\lambda}(t)}{\lambda(t)} = 1.$$

The proof of Theorem 4 builds on the previous work [11] thence let us describe below the main ingredients contained in the latter breakthrough.

3.1 Conditional Stability Result

The strategy of [11] consists in solving system (15) coupled with (16) iteratively: define the following sequence $x^{(j)}(\tau,\xi) = x^{(0)}(\tau,\xi) + \sum_{k=1}^{j} \Delta x^{(k)}(\tau,\xi)$, where the *zero-th iterate* solves the homogeneous system:

$$\begin{cases} \big(\mathcal{D}_\tau^2 + \beta(\tau)\mathcal{D}_\tau + \xi\big)x^{(0)}(\tau,\xi) = 0 \\ (x^{(0)}, \mathcal{D}_\tau x^{(0)})\big|_{\tau=\tau_0} = (x_0, x_1) \end{cases}$$

and the *k-th increment* $\Delta x^{(k)}$ satisfies the inhomogeneous equation:

$$\begin{cases} \big(\mathcal{D}_\tau^2 + \beta(\tau)\mathcal{D}_\tau + \xi\big)\Delta x^{(k)}(\tau,\xi) = \mathcal{R}\Delta x^{(k-1)}(\tau,\xi) + \Delta f^{(k-1)}(\tau,\xi) \\ (\Delta x^{(k)}, \mathcal{D}_\tau \Delta x^{(k)})\big|_{\tau=\tau_0} = (\Delta\tilde{\tilde{x}}_0^{(k)}, \Delta\tilde{\tilde{x}}_1^{(k)}) \end{cases}$$

where $\Delta f^{(0)} = f[\tilde{\varepsilon}^{(0)}]$ and $\Delta f^{(k-1)} = f[\tilde{\varepsilon}^{(k-1)}] - f[\tilde{\varepsilon}^{(k-2)}]$ for $j \geq 2$.

As expected from the presence of a resonance of the operator $\mathcal{L}$, an accurate analysis of the zero-th iterate reveals that this term is fast growing toward $\tau = +\infty$. The growth of its discrete spectral part is easily controlled by imposing a vanishing condition on x_{0d} and x_{1d}. However, the growth of the continuous spectral part is more fundamental and it can be investigated via the explicit homogeneous parametrix:

$$\begin{aligned} x^{(0)}(\tau,\xi) &= S[x_0,x_1](\tau,\xi) \\ &:= \frac{\lambda^{5/2}(\tau)}{\lambda^{5/2}(\tau_0)} \frac{\rho^{1/2}\big(\frac{\lambda^2(\tau)}{\lambda^2(\tau_0)}\xi\big)}{\rho^{1/2}(\xi)} \cos\Big[\lambda(\tau)\xi^{1/2}\int_{\tau_0}^{\tau} \lambda^{-1}(u)du\Big] x_0\Big(\frac{\lambda^2(\tau)}{\lambda^2(\tau_0)}\xi\Big) \\ &+ \frac{\lambda^{3/2}(\tau)}{\lambda^{3/2}(\tau_0)} \frac{\rho^{1/2}\big(\frac{\lambda^2(\tau)}{\lambda^2(\tau_0)}\xi\big)}{\rho^{1/2}(\xi)} \frac{\sin\Big[\lambda(\tau)\xi^{1/2}\int_{\tau_0}^{\tau}\lambda^{-1}(u)du\Big]}{\xi^{1/2}} x_1\Big(\frac{\lambda^2(\tau)}{\lambda^2(\tau_0)}\xi\Big). \end{aligned}$$

Since $\lambda(\tau) \approx \tau^{1+\nu^{-1}}$, hence $x^{(0)}$ grows polynomially in τ. To control such a growth, the following natural co-dimensions two condition on the initial data (x_0, x_1) is imposed:

$$\begin{aligned} &\int_0^\infty \frac{\rho^{\frac{1}{2}}(\xi)x_0(\xi)}{\xi^{\frac{1}{4}}} \cos[\lambda(\tau_0)\xi^{\frac{1}{2}} \int_{\tau_0}^\infty \lambda^{-1}(s)\,ds]\,d\xi = 0, \\ &\int_0^\infty \frac{\rho^{\frac{1}{2}}(\xi)x_1(\xi)}{\xi^{\frac{3}{4}}} \sin[\lambda(\tau_0)\xi^{\frac{1}{2}} \int_{\tau_0}^\infty \lambda^{-1}(s)\,ds]\,d\xi = 0. \end{aligned} \tag{19}$$

Albeit such *vanishing relations* do not eliminate completely the growth of $x^{(0)}$ at infinity but only reduces it to linear growth, it is sufficient to run the iteration scheme. In fact, by choosing $\nu \leq 1/3$ and thanks to the decaying factor $\lambda^{-2}(\tau)$ appearing in the nonlinear terms $\triangle f^{(k-1)}$ one can control them in a relatively straightforward way.

Let us briefly discuss the role of the corrections $(\Delta\tilde{\tilde{x}}_0^{(k)}, \Delta\tilde{\tilde{x}}_1^{(k)})$, which a priori should be both set to zero. At each iterative step, the continuous spectral part of the k-th increments are computed via the two explicit parametrices:

$$\triangle x^{(k)} = I[\mathcal{R}\triangle x^{(k-1)} + \triangle f^{(k-1)}] + S[\Delta\tilde{\tilde{x}}_0^{(k)}, \Delta\tilde{\tilde{x}}_1^{(k)}]$$

where $I[g]$ is the Duhamel parametrix for the inhomogeneous problem with source g and vanishing initial data at $\tau = \tau_0$:

$$I[g] = \int_{\tau_0}^{\tau} \frac{\lambda^{3/2}(\tau)}{\lambda^{3/2}(\sigma)} \frac{\rho^{1/2}(\frac{\lambda^2(\tau)}{\lambda^2(\sigma)}\xi)}{\rho^{1/2}(\xi)} \frac{\sin\left[\lambda(\tau)\xi^{1/2}\int_\tau^\sigma \lambda^{-1}(u)du\right]}{\xi^{1/2}} g\Big(\sigma, \frac{\lambda^2(\tau)}{\lambda^2(\sigma)}\xi\Big)d\sigma.$$

To control the $\tilde{S}$ norm of the low-frequencies component of the k-th increment $(\triangle x^{(k)}, \mathcal{D}_\tau \triangle x^{(k)})$ one splits $I[\mathcal{R}\triangle x^{(k-1)} + \triangle f^{(k-1)}]$, the inhomogeneous parametrix with vanishing initial data at $\tau = \tau_0$, into $I_{>\tau}[\mathcal{R}\triangle x^{(k-1)} + \triangle f^{(k-1)}]$, an inhomogeneous parametrix with vanishing initial data at $\tau = +\infty$, plus $S[\tilde{\Delta}\tilde{x}_0^{(k)}, \tilde{\Delta}\tilde{x}_1^{(k)}]$, a homogeneous solutions with non-vanishing initial data at $\tau = \tau_0$. Therefore we obtain

$$\triangle x^{(k)} = I_{>\tau}[\mathcal{R}\triangle x^{(k-1)} + \triangle f^{(k-1)}] + S[\tilde{\Delta}\tilde{x}_0^{(k)} + \Delta\tilde{\tilde{x}}_0^{(k)}, \tilde{\Delta}\tilde{x}_1^{(k)} + \Delta\tilde{\tilde{x}}_1^{(k)}].$$

The corrections $\Delta\tilde{\tilde{x}}_{0,1}^{(k)}$ ensure that the small error introduced in the initial data will preserve the vanishing conditions (19), leading to an approximation $\varepsilon^{(j)}$ on the physical side with controlled growth. Therefore to guarantee that the vanishing conditions holds throughout each step one needs to adjust the initial data by adding a small correction.

The final portion of [11] consists in proving that such iteration scheme converges by picking τ_0 sufficiently large. This is achieved via a re-iteration argument of the inhomogeneous parametrix which allows to gain enough smallness and to obtain a convergent series. A similar procedure was employed in [15] and [17]. Once the convergence is established, we obtain a solution of system (15) that fulfills the initial data requirements where (x_0, x_1) have been replaced by $(x_0 + \triangle x_0, x_1 + \triangle x_1)$. The corrections $\triangle x_{0,1}$ are obtained by summing up all the k-th step corrections $\triangle \tilde{\tilde{x}}_{0,1}^{(k)}$. Moreover, they are small with respect to the $\tilde{S}$ norm when compared to the original initial data (x_0, x_1) and they depend in Lipschitz continuous fashion on $x_{0,1}$.

3.2 *Optimal Stability Result*

The elimination of the extra-vanishing conditions (19) imposed on the perturbation accomplished in [3] is attained in a four steps argument. Firstly, notice that one cannot time translate the solution u_ν without introducing an error of regularity $H^{1+\nu/2-}_{rad,loc}(\mathbb{R}^3)$ on each time-slice, that is too weak since the tollerate regularity of the perturbations is $H^{3/2+}_{rad,loc}(\mathbb{R}^3)$. Therefore a subtle modulation of the scaling law $\lambda(t) = t^{-1-\nu}$ is required. Precisely, one needs to work with a more flexible scaling law depending on two additional parameters γ_1 and γ_2. We stipulate the following ansatz:

$$\lambda^{(\gamma)}(t) = \left(1 + \gamma_1 \frac{t^N}{\langle t^N \rangle} + \gamma_2 \log t \frac{t^N}{\langle t^N \rangle}\right) t^{-1-\nu}, \tag{20}$$

here $N \gg 1$ is sufficiently large. Notice that $\lambda^{(\gamma)}$ asymptotically equals to λ as $t \to 0$, and such alteration implies a corresponding adjustment of renormalized coordinates (τ, R): let us introduce

$$\tau^{(\gamma)}(t) = \int_t^{t_0} \lambda^{(\gamma)}(s)ds + \nu^{-1} t_0^{-\nu}, \qquad R^{(\gamma)}(t, r) = \lambda^{(\gamma)}(t) r.$$

A similar iterative procedure that gave rise to the approximate solutions in [16] and [13] can be applied for the more general scaling law (20) to build approximate solutions of the form

$$u^{(\gamma)}_{app}(t, r) = W_{\lambda^{\gamma}(t)}(r) + \sum_{l=1}^{2k-1} v_l(t, r) + \sum_{a=1,2} v_{smooth,a}(t, r) + v(t, r)$$

which solves $\Box u_{app}^{(\gamma)} + (u_{app}^{(\gamma)})^5 = e_{app}^{(\gamma)}$ and where the error satisfies

$$e_{app}^{(\gamma)} = (|\gamma_1| + |\gamma_2|)\Big[O\Big(\log t \frac{\lambda^{1/2} R}{(t\lambda)^{k_0+4}}(1+(1-a)^{1/2+\nu/2})\Big) + O\Big(\log t \frac{\lambda^{1/2} R^{-1}}{(t\lambda)^{k_0+2}}(1+(1-a)^{1/2+\nu/2})\Big)\Big].$$

The main novelty is that we perturb around $W_{\lambda^{(\gamma)}}$ as opposed to W_λ, which when inserted into Eq. (4), generates additional error terms. We isolate the terms of the error which depend on $\gamma_{1,2}$ from the part which do not depend on $\gamma_{1,2}$. The former error terms are treated by adding a finite number of corrections v_l following the iterative scheme in [16] and [13]. On the other hand, the latter error terms are decimated by the two corrections $v_{smooth,a}$ which have better regularity property than the previous corrections. The final correction v is introduced to further improve the overall regularity to the error term.

Next, in the modulation step, one shows how to tune the parameters γ_1 and γ_2 such that a comparable procedure from [11] can be applied. Precisely, our point of departure is a singular type II solution constructed in the previous papers [16] and [13] which has the form $u_\nu = u_{2k-1} + \varepsilon$. Denote the associated initial data on the $t = t_0$ time slice by $(\varepsilon_1, \varepsilon_2)$ and consider (x_0, x_1) the corresponding initial data at $\tau = \tau_0$ on the distorted Fourier side (with respect to R) computed via the relations (17). The point is that the initial data (x_0, x_1) do not satisfy the vanishing conditions (19) with respect to scaling law λ anymore, thence we can not directly apply the argument of [11] as outlined in the previous section. To circumvent this impasse, we shall seek to complete the approximation $u_{app}^{(\gamma)}$ to an exact solution to the critical focusing wave equation (4) by introducing the function $\bar{\varepsilon}$:

$$u = u_{app}^{(\gamma)} + \bar{\varepsilon}. \tag{21}$$

Denote $(\bar{\varepsilon}_1, \bar{\varepsilon}_2) = \bar{\varepsilon}[t_0]$ the associated initial data of the new perturbation on the $t = t_0$ time slice and consider $(x_0^{(\gamma)}, x_1^{(\gamma)})$ the corresponding initial data at $\tau = \tau_0$ on the distorted Fourier side with respect to $R^{(\gamma)}$. We impose the following relations on the $t = t_0$ time slice

$$\bar{\varepsilon}_0 = \chi_{r \lesssim t_0}\big[W_\lambda(r) - W_{\lambda^{(\gamma)}}(r) - v_{smooth,1,2} - v\big] + \varepsilon_0,$$

as well as

$$\bar{\varepsilon}_1 = \chi_{r \lesssim t_0}\big[\partial_t\big[W_\lambda(r) - W_{\lambda^{(\gamma)}}(r)\big] - \partial_t v_{smooth,1,2} - \partial_t v\big] + \varepsilon_1.$$

Then one proves that there exists a unique choice of the parameters $\gamma_{1,2}$ such that the corresponding vanishing conditions (19) for $(x_0^{(\gamma)}, x_1^{(\gamma)})$ with respect to the scaling law (20) are satisfied.

Subsequently, we plug the ansatz (21) into (4) to find a corresponding equation for the perturbation $\tilde{\varepsilon}$. Proceeding as in the previous section we solve such equation by passing to the distorted Fourier side with respect to $R^{(\gamma)}$. Let us denote the distorted Fourier transform of $\tilde{\varepsilon}$ by $x^{(\gamma)}$, then we obtain the corresponding transport equation on the distorted Fourier side:

$$(\mathcal{D}_\tau^2 + \beta(\tau)\mathcal{D}_\tau + \xi)x^{(\gamma)}(\tau,\xi) = \mathcal{R}x^{(\gamma)}(\tau,\xi) + f[\tilde{\varepsilon}^{(\gamma)}](\tau,\xi) \tag{22}$$

where the linear source terms $\mathcal{R}$ are as in (11) and the nonlinear terms are defined by

$$\mathbf{f}[\tilde{\varepsilon}^{(\gamma)}](\tau,\xi) = \begin{pmatrix} f_d[\tilde{\varepsilon}^{(\gamma)}](\tau) \\ f[\tilde{\varepsilon}^{(\gamma)}](\tau,\xi) \end{pmatrix} = \begin{pmatrix} \lambda^{-2}(\tau)\langle \phi_d, R^{(\gamma)}e_{app}^{(\gamma)} + N(u_{app}^{(\gamma)}, \tilde{\varepsilon}^{(\gamma)})\rangle_{L^2_{dR}} \\ \lambda^{-2}(\tau)\mathcal{F}\Big(R^{(\gamma)}e_{app}^{(\gamma)} + N(u_{app}^{(\gamma)}, \tilde{\varepsilon}^{(\gamma)})\Big)(\tau,\xi) \end{pmatrix}.$$

The system (22) is coupled with initial data $(x_0^{(\gamma)}, x_1^{(\gamma)})$ which satisfy the vanishing relations (19) with respect to the scaling law (20). Thus we can apply a similar iterative scheme as in [11] to show that there exist corrections $\triangle x_{0,1}^{(\gamma)}$ such that a solution $x^{(\gamma)}$ to (22) with perturbed initial data $(x_0^{(\gamma)} + \triangle x_0^{(\gamma)}, x_1^{(\gamma)} + \triangle x_1^{(\gamma)})$ exists.

The last step consists to estimate the error induced by the small correction terms $\triangle x_{0,1}^{(\gamma)}$ which have been introduced in the iterative scheme in terms of the original variable R. Hence we analyse $\triangle\varepsilon_{0,1}^{(\gamma)}$ the inverse distorted Fourier transform of $\triangle x_{0,1}^{(\gamma)}$, with respect to the variable $R^{(\gamma)}$, and we prove that such errors are small when compared to the initial data perturbation. To show smallness one needs to compute the Fourier transform of $\triangle\varepsilon_{0,1}^{(\gamma)}$ with respect to the original variable R yielding to corrections denoted $\triangle x_{0,1}$, and prove that the latter corrections are small in the $\tilde{S}$ norm when compared to the original initial data $x_{0,1}$.

Finally, the investigation of the Lipschitz dependence of the corrections $\triangle x_{0,1}^{(\gamma)}$ with respect to the original data $x_{0,1}^{(\gamma)}$ is carried out in details in [3] by carefully analyzing the dependence of the error $e_{app}^{(\gamma)}$ from the parameters $\gamma_{1,2}$.

Acknowledgments The author would like to thank Joachim Krieger and Luigi Forcella for providing insightful comments and numerous corrections.

References

1. Aubin, T.: Équations différentielles non linéaires et problème de Yamabe concernant la courbure scalaire. J. Math. Pures Appl. **55**, 269–296 (1976)
2. Bizoń, P., Chmaj, T., Tabor, Z.: On blowup for semilinear wave equations with a focusing nonlinearity. Nonlinearity **17**(6), 2187 (2004)

3. Burzio, S., Krieger, J.: Type II blow up solutions with optimal stability properties for the critical focussing nonlinear wave equation on $\mathbb{R}^{3+1}$. Mem. Am. Math. Soc. (2017, accepted). arXiv preprint math/1709.06408
4. Donninger, R.: Strichartz estimates in similarity coordinates and stable blowup for the critical wave equation. Duke Math. J. **166**(9), 1627–1683 (2017)
5. Donninger, R., Schörkhuber, B.: Stable blow up dynamics for energy supercritical wave equations. Trans. Am. Math. Soc. **366**(4), 2167–2189 (2014)
6. Duyckaerts, T., Kenig, C., Merle, F.: Universality of blow-up profile for small radial type II blow-up solutions of the energy-critical wave equation. J. Eur. Math. Soc. (JEMS) **13**(3), 533–599 (2011)
7. Duyckaerts, T., Kenig, C., Merle, F.: Profiles of bounded radial solutions of the focusing, energy-critical wave equation. Geom. Funct. Anal. **22**(3), 639–698 (2012)
8. Duyckaerts, T., Kenig, C., Merle, F.: Universality of the blow-up profile for small type II blow-up solutions of the energy-critical wave equation: the nonradial case. J. Eur. Math. Soc. (JEMS) **14**(5), 1389–1454 (2012)
9. Duyckaerts, T., Kenig, C., Merle, F.: Classification of radial solutions of the focusing, energy-critical wave equation. Camb. J. Math. **1**, 75–144 (2013)
10. Hillairet, M., Raphaël, P.: Smooth type II blow-up solutions to the four-dimensional energy-critical wave equation. Anal. PDE **5**(4), 777–829 (2012)
11. Krieger, J.: On the stability of type II blow up for the critical NLW on $\mathbb{R}^{3+1}$. Mem. Am. Math. Soc. (2018). Preprint. arXiv:1705.03907
12. Krieger, J., Miao, S.: On stability of blow up solutions for the critical co-rotational wave maps problem (2018). Preprint. arXiv:1803.02706
13. Krieger, J., Schlag, W.: Full range of blow up exponents for the quintic wave equation in three dimensions. J. Math. Pures Appl. **101**(6), 873–900 (2014)
14. Krieger, J., Schlag, W., Tataru, D.: Renormalization and blow up for charge one equivariant critical wave maps. Invent. Math. **171**(3), 543–615 (2008)
15. Krieger, J., Schlag, W., Tataru, D.: Renormalization and blow up for the critical Yang–Mills problem. Adv. Math. **221**(5), 1445–1521 (2009)
16. Krieger, J., Schlag, W., Tataru, D.: Slow blow-up solutions for the $H^1(\mathbb{R}^3)$ critical focusing semilinear wave equation. Duke Math. J. **147**(1), 1–53 (2009)
17. Krieger, J., Donninger, R., Huang, M., Schlag, W.: Exotic blow up solutions for the u^5-focussing wave equation in $\mathbb{R}^3$. Mich. Math. J. **63**, 451–501 (2014)
18. Krieger, J., Nakanishi, K., Schlag, W.: Center-stable manifold of the ground state in the energy space for the critical wave equation. Math. Ann. **361**(1–2), 1–50 (2015)
19. Levine, H.A.: Instability and nonexistence of global solutions to nonlinear wave equations of the form $Pu_{tt} = -Au + \mathcal{F}(u)$. Trans. Am. Math. Soc. **192**, 1–21 (1974)
20. Lindblad, H., Sogge, C.D.: On existence and scattering with minimal regularity for semilinear wave equations. J. Funct. Anal. **130**(2), 357–426 (1995)
21. Martel, Y., Merle, F., Raphaël, P.: Blow up for the critical gKdV equation III: exotic regimes. Ann. Sc. Norm. Super. Pisa Cl. Sci. (5) **14**(2), 575–631 (2015)
22. Raphaël, P., Rodnianski, I.: Stable blow up dynamics for the critical co-rotational wave maps and equivariant Yang-Mills problems. Publ. Math. l'IHÉS **115**(1), 1–122 (2012)
23. Raphaël, P., Schweyer, R.: Quantized slow blow-up dynamics for the corotational energy-critical harmonic heat flow. Anal. PDE **7**(8), 1713–1805 (2014)
24. Sogge, C.D.: Lectures on Non-linear Wave Equations, vol. 2. International Press, Boston (1995)
25. Talenti, G.: Best constant in Sobolev inequality. Ann. Mat. Pura Appl. **110**(1), 353–372 (1976)
26. van den Berg, J.B., Hulshof, J., King, J.R.: Formal asymptotics of bubbling in the harmonic map heat flow. SIAM J. Appl. Math. **63**(5), 1682–1717 (2003)

Local Well-Posedness for the Nonlinear Schrödinger Equation in the Intersection of Modulation Spaces $M^s_{p,q}(\mathbb{R}^d) \cap M_{\infty,1}(\mathbb{R}^d)$

Leonid Chaichenets, Dirk Hundertmark, Peer Christian Kunstmann, and Nikolaos Pattakos

Abstract We introduce a Littlewood–Paley characterization of modulation spaces and use it to give an alternative proof of the algebra property, somehow implicitly contained in Sugimoto et al. (2011), of the intersection $M^s_{p,q}(\mathbb{R}^d) \cap M_{\infty,1}(\mathbb{R}^d)$ for $d \in \mathbb{N}$, $p, q \in [1, \infty]$ and $s \geq 0$. We employ this algebra property to show the local well-posedness of the Cauchy problem for the cubic nonlinear Schrödinger equation in the above intersection. This improves a theorem by Bényi and Okoudjou (2009), where only the case $q = 1$ is considered, and closes a gap in the literature. If $q > 1$ and $s > d\left(1 - \frac{1}{q}\right)$ or if $q = 1$ and $s \geq 0$ then $M^s_{p,q}(\mathbb{R}^d) \hookrightarrow M_{\infty,1}(\mathbb{R}^d)$ and the above intersection is superfluous. For this case we also reobtain a Hölder-type inequality for modulation spaces.

The original version of this chapter was revised. A correction to this chapter can be found at https://doi.org/10.1007/978-3-030-47174-3_20

L. Chaichenets (✉)
Department of Mathematics, Institute for Analysis, Karlsruhe Institute of Technology, Karlsruhe, Germany

Technical University of Dresden, Institute for Analysis, Dresden, Germany
e-mail: leonid.chaichenets@kit.edu; leonid.chaichenets@tu-dresden.de

D. Hundertmark · P. C. Kunstmann · N. Pattakos
Department of Mathematics, Institute for Analysis, Karlsruhe Institute of Technology, Karlsruhe, Germany
e-mail: dirk.hundertmark@kit.edu; peer.kunstmann@kit.edu; nikolaos.pattakos@kit.edu

W. Dörfler et al. (eds.), *Mathematics of Wave Phenomena*, Trends in Mathematics,
https://doi.org/10.1007/978-3-030-47174-3_6

1 Introduction

In this paper we contribute to the general theory of modulation spaces. Modulation spaces $M^s_{p,q}(\mathbb{R}^d)$ were introduced by Feichtinger in [10]. Here, we only briefly recall their definition and refer to Sect. 2 and the literature mentioned there for more information. Fix a so-called *window function* $g \in \mathcal{S}(\mathbb{R}^d) \setminus \{0\}$. The *short-time Fourier transform* $V_g f$ of a tempered distribution $f \in \mathcal{S}'(\mathbb{R}^d)$ with respect to the window g is defined by

$$(V_g f)(x,\xi) = \frac{1}{(2\pi)^{\frac{d}{2}}}\overline{\langle f, M_\xi S_x g\rangle} \qquad \forall x,\xi \in \mathbb{R}^d, \tag{1}$$

where $S_x g(y) = g(y-x)$ denotes the *right-shift* by $x \in \mathbb{R}^d$, $\left(M_\xi g\right)(y) = e^{ik\cdot y} g(y)$ the *modulation* by $\xi \in \mathbb{R}^d$ and $\langle f, g\rangle = \int_{\mathbb{R}^d} \overline{f}(x) g(x) \mathrm{d}x$ for $f \in L^1_{\mathrm{loc}}(\mathbb{R}^d)$, $g \in \mathcal{S}(\mathbb{R})^d$. We define

$$M^s_{p,q}(\mathbb{R}^d) = \left\{ f \in \mathcal{S}'(\mathbb{R}^d) \middle| \, \|f\|_{M^s_{p,q}(\mathbb{R}^d)} < \infty \right\}, \text{ where}$$

$$\|f\|_{M^s_{p,q}(\mathbb{R}^d)} = \left\| \xi \mapsto \langle \xi \rangle^s \left\| V_g f(\cdot, \xi) \right\|_p \right\|_q$$

for $s \in \mathbb{R}$, $p, q \in [1,\infty]$. As usual in the literature, we set $M_{p,q}(\mathbb{R}^d) := M^0_{p,q}(\mathbb{R}^d)$ and often shorten the notation for $M^s_{p,q}(\mathbb{R}^d)$ to $M^s_{p,q}$. It can be shown, that the $M^s_{p,q}(\mathbb{R}^d)$ are Banach spaces and that different choices of the window function g lead to equivalent norms.

To state our first result, let us recall the definition of the Littlewood–Paley decomposition. Consider a smooth radial function $\phi_0 \in C^\infty_c(\mathbb{R}^d)$ with $\phi_0(\xi) = 1$ for all $|\xi| \leq \frac{1}{2}$ and $\mathrm{supp}(\phi_0) \subseteq B_1(0)$. Set $\phi_1 = \phi_0\left(\frac{\cdot}{2}\right) - \phi_0$ and $\phi_l = \phi_1\left(\frac{\cdot}{2^{l-1}}\right)$ for all $l \in \mathbb{N}$. The multiplier operators defined by

$$\Delta_l f := \frac{1}{(2\pi)^{\frac{d}{2}}} \check{\phi}_l * f = \mathcal{F}^{(-1)} \phi_l \mathcal{F} f \qquad \forall l \in \mathbb{N}_0 \, \forall f \in \mathcal{S}'(\mathbb{R}^d)$$

are called *dyadic decomposition operators* and the sequence $(\Delta_l f)_{l\in\mathbb{N}_0}$ is called the *Littlewood–Paley decomposition* of $f \in \mathcal{S}'(\mathbb{R}^d)$. Above, $\mathcal{F}$ denotes the usual *Fourier transform* and $\mathcal{F}^{(-1)}$ its inverse.

Our first result is

Theorem 1 (Littlewood–Paley Characterization) *Let $d \in \mathbb{N}$, $p, q \in [1,\infty]$ and $s \in \mathbb{R}$. Then*

$$\|f\| := \left\| \left(2^{ls} \, \|\Delta_l f\|_{M_{p,q}(\mathbb{R}^d)} \right)_{l\in\mathbb{N}_0} \right\|_q \qquad \forall f \in \mathcal{S}'(\mathbb{R}^d)$$

defines an equivalent norm on $M^s_{p,q}(\mathbb{R}^d)$*. The constants of the norm equivalence depend only on d and s.*

The above characterization of modulation spaces is new and we shall use it to prove that the intersections $M^s_{p,q}(\mathbb{R}^d) \cap M_{\infty,1}(\mathbb{R}^d)$ are *Banach *-algebras.*[1] To state this second result, let us denote by $C_b(\mathbb{R}^d)$ the space of bounded complex-valued continuous functions on $\mathbb{R}^d$, where $d \in \mathbb{N}$. We then have

Theorem 2 (Algebra Property) *Let* $d \in \mathbb{N}$, $p, q \in [1, \infty]$ *and* $s \geq 0$. *Then* $M^s_{p,q}(\mathbb{R}^d) \cap M_{\infty,1}(\mathbb{R}^d)$ *is a Banach *-algebra with respect to pointwise multiplication and complex conjugation. These operations are well-defined due to the embedding* $M_{\infty,1}(\mathbb{R}^d) \hookrightarrow C_b(\mathbb{R}^d)$ *Furthermore, if* $q > 1$ *and* $s > d\left(1 - \frac{1}{q}\right)$ *or if* $q = 1$*, then* $M^s_{p,q}(\mathbb{R}^d) \hookrightarrow M_{\infty,1}(\mathbb{R}^d)$*, so in particular* $M^s_{p,q}(\mathbb{R}^d)$ *is a Banach *-algebra, in that case.*

The latter case of Theorem 2 had been observed already in 1983 by Feichtinger in his aforementioned pioneering work on modulation spaces (cf. [10, Proposition 6.9]), where it is presented as a consequence of convolution relations between Wiener amalgam spaces, as introduced in [9], combined with the fact that the Fourier image of a modulation space can be characterized as Wiener amalgam space with a local $\mathcal{F}L^p$ component. The case $q > 1$ and $s \in \left[0, d\left(1 - \frac{1}{q}\right)\right]$ seems to be new, at least as a statement. A different Proof of Theorem 2 can be given following the idea of proof of [18, Proposition 3.2], see [5, Proposition 4.2].

Next, we present a Hölder-type inequality for modulation spaces.

Theorem 3 (Hölder-Type Inequality) *Let* $d \in \mathbb{N}$ *and* $p, p_1, p_2, q \in [1, \infty]$ *be such that* $\frac{1}{p} = \frac{1}{p_1} + \frac{1}{p_2}$. *For* $q > 1$ *let* $s > d\left(1 - \frac{1}{q}\right)$ *and for* $q = 1$ *let* $s \geq 0$. *Then there is a* $C > 0$ *such that for any* $f \in M^s_{p_1,q}(\mathbb{R}^d)$ *and any* $g \in M^s_{p_2,q}(\mathbb{R}^d)$ *one has* $fg \in M^s_{p,q}(\mathbb{R}^d)$ *and*

$$\|fg\|_{M^s_{p,q}(\mathbb{R}^d)} \leq C \|f\|_{M^s_{p_1,q}(\mathbb{R}^d)} \|g\|_{M^s_{p_2,q}(\mathbb{R}^d)} . \tag{2}$$

The above pointwise multiplication fg is well-defined due to the embedding formulated in Theorem 2. The constant C does *not* depend on p, p_1 or p_2.

Theorem 3 easily generalizes to $m \in \mathbb{N}$ factors and $p, p_1, \ldots, p_m \in (0, \infty]$. Hence, it extends the multilinear estimate from [3, Equation 2.4] to the case $q_0 = \ldots = q_m > 1$. The case $q \neq \infty$ and $p \neq \infty$ of our Theorem 3 is a special case of [13, Theorem 1.4]. That theorem also establishes that the condition $q = 1$ and $s \geq 0$ or $q > 1$ and $s > \frac{d}{q'}$ is necessary for the space $M^s_{p,q}(\mathbb{R}^d)$ to be a Banach *-algebra.

[1] For us, a Banach *-algebra X is a Banach algebra over $\mathbb{C}$ on which a continuous *involution* $*$ is defined, i.e. $(x + y)^* = x^* + y^*$, $(\lambda x)^* = \overline{\lambda} x^*$, $(xy)^* = y^* x^*$ and $(x^*)^* = x$ for any $x, y \in X$ and $\lambda \in \mathbb{C}$. We neither require X to have a unit nor $C = 1$ in the estimates $\|x \cdot y\| \leq C \|x\| \|y\|$, $\|x^*\| \leq C \|x\|$.

Let us also mention, that other results on Hölder-type inequalities in modulation spaces, i.e.

$$\|fg\|_{M^s_{p,q}} \lesssim \|f\|_{M^{s_1}_{p_1,q_1}} \|g\|_{M^{s_2}_{p_2,q_2}}$$

(in contrast to (2) the q is allowed to vary), include [19, Theorem 2.4] and [7, Proposition 3.5] (but see also [21, Lemma 4.1]). In the case $q = q_1 = q_2$, the former result requires $q \leq 2$ and $p \neq \infty$, while the latter result only covers the case $q = 1$.

Here we present a direct Proof of Theorem 3, close to the approach found in [21, Corollary 4.2] and involving an application of Theorem 2. For a proof avoiding the Littewood-Paley characterization see the proof of [5, Theorem 4.3]. Yet another and more abstract proof could be given by invoking [9, Theorem 3] for a specific choice of Banach convolution triples.

Lastly, we employ Theorem 2 to study the Cauchy problem for the cubic nonlinear Schrödinger equation (*NLS*)

$$\begin{cases} \mathrm{i}\dfrac{\partial u}{\partial t}(x,t) + \Delta u(x,t) \pm |u|^2 u(x,t) = 0 & (x,t) \in \mathbb{R}^d \times \mathbb{R}, \\ u(x,0) = u_0(x) & x \in \mathbb{R}^d, \end{cases} \tag{3}$$

where the initial data u_0 is in an intersection of modulation spaces $M^s_{p,q}(\mathbb{R}^d) \cap M_{\infty,1}(\mathbb{R}^d)$. We are interested in *mild solutions* u of (3), i.e.

$$u \in C\left([0,T), M^s_{p,q}(\mathbb{R}^d) \cap M_{\infty,1}(\mathbb{R}^d)\right)$$

for some $T > 0$ which satisfy the corresponding integral equation

$$u(\cdot,t) = e^{\mathrm{i}t\Delta}u_0 \pm \mathrm{i}\int_0^t e^{\mathrm{i}(t-\tau)\Delta}\left(|u|^2 u(\cdot,\tau)\right)\mathrm{d}\tau \qquad \forall t \in [0,T). \tag{4}$$

Our last result is stated in

Theorem 4 (Local Well-Posedness) *Let* $d \in \mathbb{N}$, $p \in [1,\infty]$, $q \in [1,\infty)$ *and* $s \geq 0$. *Set* $X = M^s_{p,q}(\mathbb{R}^d) \cap M_{\infty,1}(\mathbb{R}^d)$ *and* $X(T) = C([0,T],X)$, $X_*(T) = C([0,T),X)$ *for any* $T > 0$. *Assume that* $u_0 \in X$. *Then, there exists a unique maximal mild solution* $u \in X_*(T_*)$ *of* (3) *and the* blow-up alternative

$$T_* < \infty \qquad \Rightarrow \qquad \limsup_{t \to T_*-} \|u(\cdot,t)\|_X = \infty$$

holds. Moreover, for any $T' \in (0,T_*)$ *there exists a neighborhood* V *of* u_0 *in* X, *such that the initial-data-to-solution-map* $V \to X(T')$, $v_0 \mapsto v$ *is Lipschitz continuous.*

As already stated in Theorem 2 one has that, if $q > 1$ and $s > d\left(1 - \frac{1}{q}\right)$ or if $q = 1$, then $M^s_{p,q}(\mathbb{R}^d) \hookrightarrow M_{\infty,1}(\mathbb{R}^d)$ and so $X = M^s_{p,q}(\mathbb{R}^d)$, in that case.

In the case $q = \infty$ excluded in Theorem 4, the situation is more subtle. Following our proof, one obtains local well-posedness in the larger space

$$L^\infty([0, T), M^s_{p,\infty}(\mathbb{R}^d) \cap M_{\infty,1}(\mathbb{R}^d)).$$

The missing continuity in time is due to the properties of the free Schrödinger evolution and we refer to the remarks after Theorem 6.

The precursors of Theorem 4 are [21, Theorem 1.1] by Wang, Zhao and Guo for the space $M^0_{2,1}(\mathbb{R}^d)$ and [3, Theorem 1.1] due to Bényi and Okoudjou for the space $M^s_{p,1}(\mathbb{R}^d)$ with $p \in [1, \infty]$ and $s \geq 0$. In fact, Theorem 4 generalizes [3, Theorem 1.1] to $q \geq 1$: Although our theorem is stated for the cubic nonlinearity, this is for simplicity of the presentation only. The proof allows for an easy generalization to *algebraic nonlinearities* considered in [3], which are of the form

$$f(u) = g(|u|^2)u = \sum_{k=0}^{\infty} c_k \, |u|^{2k} \, u, \tag{5}$$

where g is an entire function. Also, [3, Theorems 1.2 and 1.3], which concern the nonlinear wave and the nonlinear Klein-Gordon equation respectively, can be generalized in the same spirit. The reason for this is that the proof of these results is based on the well-known Banach's contraction principle, on the fact that the free propagator is a C_0-group, and on the algebra property of the spaces under consideration. Although the ingredients seem to be known in the community, the results to be found in the literature (e.g. [22, Theorem 6.2]) are not as general as Theorem 4. In this sense, it closes a gap in the literature.

Let us remark that local well-posedness results in the case of modulation spaces that are not Banach *-algebras are [12, Theorem 1.4] for $u_0 \in M_{2,q}(\mathbb{R})$ with $q \in [2, \infty)$ and [6, Theorem 6] with $u_0 \in M^s_{p,q}(\mathbb{R})$ with either $p \in [2, 3]$, $q \in \left[1, \frac{3}{2}\right]$ and $s \geq 0$ or $p \in [2, 3]$, $q \in \left(\frac{3}{2}, \frac{18}{11}\right]$ and $s > \frac{2}{3} - \frac{1}{q}$ or $q \in \left(\frac{18}{11}, 2\right]$, $p \in \left[2, \frac{10q}{7q-6}\right)$ and $s > \frac{2}{3} - \frac{1}{q}$ (see also [15, Theorem 4]).

The remainder of our paper is structured as follows. We start with Sect. 2 providing an overview over modulation spaces and the free Schrödinger propagator on them. In Sect. 3 we apply methods from the Littlewood–Paley theory to prove Theorem 1. In the subsequent Sect. 4 we prove the algebra property from Theorem 2, notice the resulting similar property for weighted sequence spaces in Lemma 6, and deduce the Hölder-type inequality stated in Theorem 3. Finally, we prove Theorem 4 on the local well-posedness in Sect. 5.

Notation

We denote generic constants by C. To emphasize on which quantities a constant depends we write e.g. $C = C(d)$ or $C = C(d, s)$. Sometimes we omit a positive constant from an inequality by writing "$\lesssim$", e.g. $A \lesssim_d B$ instead of $A \le C(d)B$. By $A \approx B$ we mean $A \lesssim B$ and $B \lesssim A$. Special constants are $d \in \mathbb{N}$ for the *dimension*, $p, q \in [1, \infty]$ for the *Lebesgue* exponents and $s \in \mathbb{R}$ for the *regularity* exponent. By p' we mean the *dual* exponent of p, that is the number satisfying $\frac{1}{p} + \frac{1}{p'} = 1$.

We denote by $\mathcal{S}(\mathbb{R}^d)$ the set of *Schwartz functions* and by $\mathcal{S}'(\mathbb{R}^d)$ the space of *tempered distributions*. Furthermore, we denote the *Bessel potential spaces* or simply L^2-based *Sobolev spaces* by $H^s = H^s(\mathbb{R}^d)$. For the space of smooth functions with compact support we write C_c^∞. The letters f, g, h denote either generic functions $\mathbb{R}^d \to \mathbb{C}$ or generic tempered distributions and $(a_k)_{k\in\mathbb{Z}^d} = (a_k)_k = (a_k)$, $(b_k)_{k\in\mathbb{Z}^d} = (b_k)_k = (b_k)$ denote generic complex-valued sequences. By $\langle\cdot\rangle = \sqrt{1 + |\cdot|^2}$ we mean the *Japanese bracket*.

For a Banach space X we write X^* for its dual and $\|\cdot\|_X$ for the norm it is canonically equipped with. By $\mathcal{L}(X, Y)$ we denote the space of all bounded linear maps from X to Y, where Y is another Banach space, and set $\mathcal{L}(X) = \mathcal{L}(X, X)$. By $[X, Y]_\theta$ we mean complex interpolation between X and Y, if (X, Y) is an interpolation couple. For brevity we write $\|\cdot\|_p$ for the p-norm on the *Lebesgue space* $L^p = L^p(\mathbb{R}^d)$, the *sequence space* $l^p = l^p(\mathbb{Z}^d)$ or $l^p = l^p(\mathbb{N}_0)$ and $\|(a_k)\|_{q,s} := \|(\langle k\rangle^s a_k)\|_q$ for the norm on $\langle\cdot\rangle^s$-weighted sequence spaces $l_s^q = l_s^q(\mathbb{Z}^d)$. If the norm is apparent from the context, we write $B_r(x)$ for a ball of radius r around $x \in X$.

We use the symmetric choice of constants for the Fourier transform and also write

$$\hat{f}(\xi) := (\mathcal{F}f)(\xi) = \frac{1}{(2\pi)^{\frac{d}{2}}} \int_{\mathbb{R}^d} e^{-\mathrm{i}\xi\cdot x} f(x)\mathrm{d}x,$$

$$\check{g}(x) := \left(\mathcal{F}^{(-1)}g\right)(x) = \frac{1}{(2\pi)^{\frac{d}{2}}} \int_{\mathbb{R}^d} e^{\mathrm{i}\xi\cdot x} g(\xi)\mathrm{d}\xi.$$

2 Preliminaries

As already mentioned in the introduction, modulation spaces were introduced by Feichtinger in [10] in the setting of locally compact Abelian groups. A thorough introduction is given in the textbook [11] by Gröchenig. A presentation incorporating the characterization of modulation spaces via *isometric decomposition operators*, which we are going to use below, is contained in the paper [20, Section 2, 3] by Wang and Hudzik. A survey on modulation spaces and nonlinear evolution equations is given in [17].

A convenient equivalent norm on modulation spaces which we are going to use is constructed as follows (cf. [20, Propostition 2.1]): Set $Q_0 := \left[-\frac{1}{2}, \frac{1}{2}\right)^d$ and $Q_k := Q_0 + k$ for all $k \in \mathbb{Z}^d$. Consider a smooth partition of unity $(\sigma_k)_{k\in\mathbb{Z}^d} \in \left(C_c^\infty(\mathbb{R}^d)\right)^{\mathbb{Z}^d}$ satisfying

- $\exists c > 0 : \forall k \in \mathbb{Z}^d : \forall \eta \in Q_k : |\sigma_k(\eta)| \geq c$,
- $\forall k \in \mathbb{Z}^d : \operatorname{supp}(\sigma_k) \subseteq B_{\sqrt{d}}(k)$,
- $\sum_{k\in\mathbb{Z}^d} \sigma_k = 1$,
- $\forall m \in \mathbb{N}_0 : \exists C_m > 0 : \forall k \in \mathbb{Z}^d : \forall \alpha \in \mathbb{N}_0^d : |\alpha| \leq m \Rightarrow \|D^\alpha \sigma_k\|_\infty \leq C_m$

and define the *isometric decomposition operators* $\Box_k := \mathcal{F}^{(-1)}\sigma_k\mathcal{F}$. We need the following often used (cf. [20, Proposition 1.9])

Lemma 1 (Bernstein Multiplier Estimate) *Let $d \in \mathbb{N}$, $\sigma \in \mathcal{S}(\mathbb{R}^d)$ and $r, p_1, p_2 \in [1, \infty]$ such that $1 + \frac{1}{p_2} = \frac{1}{r} + \frac{1}{p_1}$. Consider the* multiplier operator *$T_\sigma : \mathcal{S}'(\mathbb{R}^d) \to \mathcal{S}'(\mathbb{R}^d)$ with* symbol *σ defined by*

$$T_\sigma f = \mathcal{F}^{(-1)}\sigma\mathcal{F}f = \frac{1}{(2\pi)^{\frac{d}{2}}}\check{\sigma} * f \qquad \forall f \in \mathcal{S}'(\mathbb{R}^d).$$

Then, for any $f \in \mathcal{S}'(\mathbb{R}^d)$, every derivative of $T_\sigma f \in C^\infty(\mathbb{R}^d)$ (including $T_\sigma f$) has at most polynomial growth. Furthermore $\|T_\sigma f\|_{p_2} \leq \frac{\|\hat{\sigma}\|_r}{(2\pi)^{\frac{d}{2}}}\|f\|_{p_1}$ for any $f \in L^{p_1}(\mathbb{R}^d)$.

Putting $r = 1$ and $p_1 = p_2 = p$ in Lemma 1, shows that $\Box_k f \in C^\infty(\mathbb{R}^d)$ for $f \in \mathcal{S}'(\mathbb{R}^d)$ and $\|\Box_k\|_{\mathcal{L}(L^p(\mathbb{R}^d))}$ is bounded independently of k and p. The aforementioned equivalent norm for the modulation space $M^s_{p,q}(\mathbb{R}^d)$ is given by (see [20, Proposition 2.1])

$$\|f\|_{M^s_{p,q}} \approx \left\| \left(\langle k\rangle^s \|\Box_k f\|_p\right)_{k\in\mathbb{Z}^d} \right\|_q. \tag{6}$$

Choosing a different partition of unity (σ_k) yields yet another equivalent norm.

Lemma 2 (Continuous Embeddings) *Let $s_1 \geq s_2$, $1 \leq p_1 \leq p_2 \leq \infty$, $1 \leq q_1 \leq q_2 \leq \infty$, $q > 1$ and $s > \frac{d}{q'}$. Then*

1. *$M^{s_1}_{p_1,q_1}(\mathbb{R}^d) \subseteq M^{s_2}_{p_2,q_2}(\mathbb{R}^d)$ and the embedding is continuous,*
2. *$M^s_{p_1,q}(\mathbb{R}^d) \subseteq M_{p_1,1}(\mathbb{R}^d)$ and the embedding is continuous,*
3. *$M_{p_1,1}(\mathbb{R}^d) \hookrightarrow C_b(\mathbb{R}^d)$.*

Lemma 2 is well-known (cf. [20, Proposition 2.5, 2.7]), but for convenience we sketch a

Proof

1. One can change indices one by one. The inclusion for "s" is by monotonicity and the inclusion for "q" is by the embeddings of the l^q spaces. For the "p"-embedding consider $\tau \in C_c^\infty(\mathbb{R}^d)$ such that $\tau|_{B_{\sqrt{d}}} \equiv 1$ and $\operatorname{supp}(\tau) \subseteq B_d$. For every $k \in \mathbb{Z}^d$, consider the shifted symbol $\tau_k = S_k\tau$, define the corresponding multiplier operator $\tilde{\Box}_k = \mathcal{F}^{(-1)}\tau_k\mathcal{F}$ and observe, that $\hat{\tau}_k = M_k\hat{\tau}$. Hence, by Lemma 1, the family $\left(\tilde{\Box}_k\right)_{k\in\mathbb{Z}^d}$ is bounded in $\mathcal{L}(L^{p_1}(\mathbb{R}^d), L^{p_2}(\mathbb{R}^d))$. So, $\|\Box_k f\|_{p_2} = \left\|\tilde{\Box}_k\Box_k f\right\|_{p_2} \lesssim_d \|\Box_k f\|_{p_1}$ for any $k \in \mathbb{Z}^d$. Recalling (6) completes the argument.
2. By Hölder's inequality we immediately have

$$\|f\|_{p_1,1} \approx \sum_{k\in\mathbb{Z}^d} \|\Box_k f\|_{p_1} \le \left(\sum_{k\in\mathbb{Z}^d} \langle k\rangle^{-sq'}\right)^{\frac{1}{q'}} \left(\sum_{k\in\mathbb{Z}^d} \langle k\rangle^{sq} \|\Box_k f\|_p^q\right)^{\frac{1}{q}}$$
$$\approx \left(\sum_{k\in\mathbb{Z}^d} \langle k\rangle^{-sq'}\right)^{\frac{1}{q'}} \|f\|_{M^s_{p_1,q}}$$

   and the first factor is finite for $s > \frac{d}{q'}$ by comparison with the integral $\int_{\mathbb{R}^d} \langle x\rangle^{-sq'} \mathrm{d}x$.
3. By part (1) it is enough to show that $M_{\infty,1} \hookrightarrow C_b$. For any $f \in M_{\infty,1}$ we have $\underbrace{\sum_{|k|\le N} \Box_k f}_{\in C^\infty} \to f$ in $\mathcal{S}'$ as $N \to \infty$. But simultaneously, the series $\sum_{k\in\mathbb{Z}^d} \Box_k f$ is absolutely convergent in L^∞ to, say, $g \in C_b$. As $M_{\infty,1} \hookrightarrow \mathcal{S}'$ (see [10, Thm. 6.1 (B)]), we have $f = g$.

□

For the Proof of Theorem 2 we will need the following (cf. [3, eqn. (2.4)]).

Lemma 3 (Bilinear Estimate) *Let $d \in \mathbb{N}$ and $1 \le p \le \infty$. Assume $f \in M_{p,q}(\mathbb{R}^d)$ and $g \in M_{\infty,1}(\mathbb{R}^d)$. Then*

$$\|fg\|_{M_{p,q}(\mathbb{R}^d)} \lesssim \|f\|_{M_{p,q}(\mathbb{R}^d)} \|g\|_{M_{\infty,1}(\mathbb{R}^d)},$$

where the implicit constant does not *depend on p or q.*

For convenience, and because we will generalize Lemma 3 to Theorem 3, we present a proof close to the one of [21, Corollary 4.2].

Proof We use (6) to estimate the modulation space norm of the left-hand side. Fix a $k \in \mathbb{Z}^d$. By the definition of the operator $\Box_k$ we have

$$\Box_k(fg) = \frac{1}{(2\pi)^{\frac{d}{2}}}\mathcal{F}^{(-1)}\left(\sigma_k(\hat{f} * \hat{g})\right) = \frac{1}{(2\pi)^{\frac{d}{2}}}\sum_{l,m\in\mathbb{Z}^d}\mathcal{F}^{(-1)}\left(\sigma_k((\sigma_l\hat{f}) * (\sigma_m\hat{g}))\right).$$

As the supports of the partition of unity are compact, many summands vanish. Indeed, for any $k, l, m \in \mathbb{Z}^d$

$$\begin{aligned}\operatorname{supp}\left(\sigma_k\left((\sigma_l\hat{f}) * (\sigma_m\hat{g})\right)\right) &\subseteq \operatorname{supp}(\sigma_k) \cap (\operatorname{supp}(\sigma_l) + \operatorname{supp}(\sigma_m))\\ &\subseteq B_{\sqrt{d}}(k) \cap B_{2\sqrt{d}}(l+m)\end{aligned}$$

and so $\sigma_k\left((\sigma_l\hat{f}) * (\sigma_m\hat{g})\right) \equiv 0$ if $|(k-l)-m| > 3\sqrt{d}$. Hence, the double series over $l, m \in \mathbb{Z}^d$ boils down to a finite sum of discrete convolutions

$$\begin{aligned}\Box_k(fg) &= \frac{1}{(2\pi)^{\frac{d}{2}}}\mathcal{F}^{(-1)}\left(\sigma_k\sum_{m\in M}\sum_{l\in\mathbb{Z}^d}(\sigma_l\hat{f}) * (\sigma_{k-l+m}\hat{g})\right)\\ &= \Box_k\sum_{m\in M}\sum_{l\in\mathbb{Z}^d}(\Box_l f)\cdot(\Box_{k+m-l}g),\end{aligned}$$

where $M = \left\{m \in \mathbb{Z}^d \,\middle|\, |m| \leq 3\sqrt{d}\right\}$ and $\#M \leq \left(6\sqrt{d}+1\right)^d < \infty$. That was the job of $\Box_k$ and we now get rid of it,

$$\|\Box_k(fg)\|_p \lesssim \sum_{m\in M}\sum_{l\in\mathbb{Z}^d}\|(\Box_l f)\cdot(\Box_{k+m-l}g)\|_p,$$

using the Bernstein multiplier estimate from Lemma 1.

Invoking Hölder's inequality we further estimate

$$\|\Box_k(fg)\|_p \lesssim \sum_{m\in M}\left(\left(\|\Box_l(f)\|_p\right)_l * \left(\|\Box_{n+m}(g)\|_\infty\right)_n\right)(k) \tag{7}$$

pointwise in $k \in \mathbb{Z}^d$, where $*$ denotes the convolution of sequences, and hence obtain

$$\|fg\|_{M_{p,q}} \lesssim \left\|\left(\|\Box_l f\|_p\right)_l\right\|_q \left\|\left(\|\Box_n g\|_\infty\right)_n\right\|_1$$

by Young's inequality. $\square$

Lemma 4 (Dual Space) *For $s \in \mathbb{R}$, $p, q \in [1, \infty)$ we have*

$$\left(M^s_{p,q}(\mathbb{R}^d)\right)^* = M^{-s}_{p',q'}(\mathbb{R}^d),$$

with the duality pairing given by

$$\langle v, u\rangle_{M^{-s}_{p',q'} \times M^s_{p,q}} = \sum_{k\in\mathbb{Z}^d} \sum_{l\in M} \int_{\mathbb{R}^d} \overline{\square_{k+l} v} \square_k u \mathrm{d}x$$

for any $v \in M^{-s}_{p',q'}(\mathbb{R}^d)$ and any $u \in M^s_{p,q}(\mathbb{R}^d)$, and where M is as in the proof of Lemma 3 (cf. [20, Theorem 3.1]).

Theorem 5 (Complex Interpolation) *For $p_1, q_1 \in [1, \infty)$, $p_2, q_2 \in [1, \infty]$, $s_1, s_2 \in \mathbb{R}$ and $\theta \in (0, 1)$ one has*

$$\left[M^{s_1}_{p_1,q_1}(\mathbb{R}^d), M^{s_2}_{p_2,q_2}(\mathbb{R}^d)\right]_\theta = M^s_{p,q}(\mathbb{R}^d),$$

with

$$\frac{1}{p} = \frac{1-\theta}{p_1} + \frac{\theta}{p_2}, \quad \frac{1}{q} = \frac{1-\theta}{q_1} + \frac{\theta}{q_2}, \quad s = (1-\theta)s_1 + \theta s_2$$

(see [10, Theorem 6.1 (D)]).

We are now ready to state and prove the following.

Theorem 6 (Schrödinger Propagator Bound) *There is a constant $C > 0$ such that for any $d \in \mathbb{N}$, $p, q \in [1, \infty]$ and $s \in \mathbb{R}$ the inequality*

$$\left\| e^{\mathrm{i}t\Delta} \right\|_{\mathcal{L}(M^s_{p,q}(\mathbb{R}^d))} \le C^d (1 + |t|)^{d\left|\frac{1}{2} - \frac{1}{p}\right|} \tag{8}$$

holds for all $t \in \mathbb{R}$. Furthermore, the exponent of the time dependence is sharp.

The boundedness has been obtained e.g. in [2, Theorem 1] whereas the sharpness was proven in [7, Proposition 4.1]. If $q < \infty$, then $(e^{\mathrm{i}t\Delta})_{t\in\mathbb{R}}$ is a C_0-group on $M^s_{p,q}$, i.e.,

$$\lim_{t\to 0} \left\| e^{\mathrm{i}t\Delta} f - f \right\|_{M^s_{p,q}} = 0 \qquad \forall f \in M^s_{p,q}$$

(see e.g. [5, Proposition 3.5]). This is not true for $q = \infty$ and we refer to [14] for this more subtle case.

Proof of Theorem 6 By definition, we have

$$(V_g e^{\mathrm{i}t\Delta} f)(x, \xi) = e^{-\mathrm{i}t|\xi|^2} (V_{e^{\mathrm{i}t\Delta} g} f)(x + 2t\xi, \xi)$$

for any $f \in \mathcal{S}'(\mathbb{R}^d)$, any $(x, \xi) \in \mathbb{R}^d \times \mathbb{R}^d$, and any $t \in \mathbb{R}$, i.e. the Schrödinger time evolution of the initial data can be interpreted as the time evolution of the window function. The price for changing from window g_0 to window g_1 is $\|V_{g_0} g_1\|_{L^1(\mathbb{R}^d \times \mathbb{R}^d)}$ by Gröchenig [11, Proposition 11.3.2 (c)]. For $g(x) = e^{-|x|^2}$ one explicitly calculates

$$\left\| V_{e^{-it\Delta}g} g \right\|_{L^1(\mathbb{R}^d \times \mathbb{R}^d)} = C^d \, (1 + |t|)^{\frac{d}{2}} ,$$

which proves the claimed bound for $p \in \{1, \infty\}$. Conservation for $p = 2$ is easily seen from (6). Complex interpolation between the cases $p = 2$ and $p = \infty$ yields (8) for $p \in [2, \infty]$. The remaining case $p \in (1, 2)$ is covered by duality.

Optimality in the case $p \in [1, 2]$ is proven by choosing the window g and the argument f to be a Gaussian and explicitly calculating $\|e^{it\Delta} f\|_{M^s_{p,q}} \approx (1 + |t|)^{d\left(\frac{1}{p} - \frac{1}{2}\right)}$. This implies the optimality for $p \in (2, \infty]$ by duality. □

3 Littlewood–Paley Theory

In this section we extend some ideas of the Littlewood–Paley decomposition from Sobolev spaces $H^s(\mathbb{R}^d)$ to modulation spaces $M^s_{p,q}(\mathbb{R}^d)$. The inspiration for this was [1, Chapter II].

Observe, that for any $\xi \in \mathbb{R}^d$ one has

$$\sum_{l=0}^{\infty} \phi_l(\xi) = \phi_0(\xi) + \lim_{N \to \infty} \sum_{l=1}^{N} \left[\phi_1\left(\frac{\xi}{2^l}\right) - \phi_1\left(\frac{\xi}{2^{l-1}}\right) \right] = \lim_{N \to \infty} \phi_0\left(\frac{\xi}{2^N}\right) = 1,$$

i.e. $\{\phi_0, \phi_1, \phi_2, \dots\}$ is a smooth partition of unity. Moreover, $\operatorname{supp}(\phi_l) \subseteq A_l$ for any $l \in \mathbb{N}_0$, where

$$A_0 := \left\{ \xi \in \mathbb{R}^d \,|\, |\xi| \le 1 \right\} \quad \text{and} \quad A_l := \left\{ \xi \in \mathbb{R}^d \,|\, 2^{l-2} \le |\xi| \le 2^l \right\} \qquad \forall l \in \mathbb{N}.$$

The symbols of the dyadic decomposition operators satisfy

$$\left\| \hat{\phi}_l \right\|_1 = \left\| \mathcal{F}\left[\phi_1\left(\frac{\cdot}{2^{l-1}}\right) \right] \right\|_1 = \left\| 2^{l-1} \hat{\phi}_1(2^{l-1} \cdot) \right\|_1 = \left\| \hat{\phi}_1 \right\|_1 \le 2 \left\| \hat{\phi}_0 \right\|_1$$

for all $l \in \mathbb{N}$. Applying Lemma 1 shows that for any $l \in \mathbb{N}_0$ and any $f \in \mathcal{S}'(\mathbb{R}^d)$ one has that $\Delta_l f \in C^\infty$ and any of its derivatives has at most polynomial growth. Furthermore, $\|\Delta_l\|_{\mathcal{L}(L^p(\mathbb{R}^d))}$ is bounded independently of $l \in \mathbb{N}_0$ and $p \in [1, \infty]$.

Proof of Theorem 1 We start by gathering some useful facts. Fix $l \in \mathbb{N}_0$ and $k \in \mathbb{Z}^d$. Recall, that $\operatorname{supp}(\phi_l) \subseteq A_l$ and $\operatorname{supp}(\sigma_k) \subseteq B_{\sqrt{d}}(k)$. Hence,

$$\Box_k \Delta_l \not\equiv 0 \Rightarrow k \in A_l' := \left\{ k' \in \mathbb{Z}^d \,\middle|\, 2^{l-2} - \sqrt{d} \leq |k'| \leq 2^l + \sqrt{d} \right\}. \tag{9}$$

On A_l' the Japanese bracket can be controlled. In fact, for all $t \in \mathbb{R}$ we have

$$\langle k \rangle^t \approx 2^{lt}, \tag{10}$$

where the implicit constant does not depend on l.

Finally, observe that $k \in A_l'$ is satisfied for only finitely many $l \in \mathbb{N}_0$, whose number is independent of $k \in \mathbb{Z}^d$, i.e.

$$\sum_{l=0}^{\infty} \mathbb{1}_{A_l'}(k) \lesssim 1, \tag{11}$$

where the implicit constant depends on d only.

- $\gtrsim$: Consider $q < \infty$ first. By (6) and (9), Bernstein multiplier estimate, (10) and (11) we have

$$\begin{aligned}
&\left\| \left(2^{ls} \| \Delta_l f \|_{M_{p,q}} \right)_l \right\|_q \\
&\approx \left(\sum_{l=0}^{\infty} 2^{lsq} \sum_{k \in \mathbb{Z}^d} \| \Box_k \Delta_l f \|_p^q \right)^{\frac{1}{q}} \lesssim \left(\sum_{l=0}^{\infty} \sum_{k \in A_l'} 2^{lsq} \| \Box_k f \|_p^q \right)^{\frac{1}{q}} \\
&\approx \left(\sum_{l=0}^{\infty} \sum_{k \in \mathbb{Z}^d} \mathbb{1}_{A_l'}(k) \langle k \rangle^{qs} \| \Box_k f \|_p^q \right)^{\frac{1}{q}} \lesssim \| f \|_{M_{p,q}^s}.
\end{aligned}$$

  Similarly, for $q = \infty$, we have

$$\begin{aligned}
\left\| \left(2^{ls} \| \Delta_l f \|_{M_{p,\infty}} \right)_l \right\|_\infty &= \sup_{l \in \mathbb{N}_0} 2^{ls} \sup_{k \in \mathbb{Z}^d} \| \Box_k \Delta_l f \|_p \\
&\lesssim \sup_{l \in \mathbb{N}_0} \sup_{k \in A_l'} \langle k \rangle^s \| \Box_k f \|_p \approx \| f \|_{M_{p,\infty}^s}.
\end{aligned}$$

- $\lesssim$: Again, consider $q < \infty$ first. By (6), $f = \sum_{l=0}^{\infty} \Delta_l f$ in $\mathcal{S}'$ and (9) we have

$$\|f\|_{M^s_{p,q}} \lesssim \left(\sum_{k\in\mathbb{Z}^d} \langle k\rangle^{qs} \left(\sum_{l=0}^{\infty} \|\Box_k \Delta_l f\|_p \right)^q \right)^{\frac{1}{q}}$$

$$\lesssim \left(\sum_{k\in\mathbb{Z}^d} \langle k\rangle^{qs} \left(\sum_{l=0}^{\infty} \mathbb{1}_{A'_l}(k)\, \|\Box_k \Delta_l f\|_p \right)^q \right)^{\frac{1}{q}}.$$

For each $k \in \mathbb{Z}^d$ the sum over l contains only finitely many non-vanishing summands and their number is independent of k by (11). Hölder's inequality estimates the last term against

$$\left(\sum_{k\in\mathbb{Z}^d} \langle k\rangle^{qs} \sum_{l=0}^{\infty} \mathbb{1}_{A'_l}(k)\, \|\Box_k \Delta_l f\|_p^q \right)^{\frac{1}{q}} \approx \left(\sum_{l=0}^{\infty} 2^{lsq} \sum_{k\in\mathbb{Z}^d} \mathbb{1}_{A'_l}(k)\, \|\Delta_l \Box_k f\|_p^q \right)^{\frac{1}{q}}$$

$$\leq \left\| \left(2^{ls} \|\Delta_l f\|_{M_{p,q}} \right)_l \right\|_q,$$

where we additionally used (10). The proof for $q = \infty$ is along the same lines.

□

The individual parts of the Littlewood–Paley decomposition had their Fourier transform supported in almost disjoint dyadic annuli. Theorem 1 characterized elements of modulation spaces by the decay of these parts. The following lemma provides a sufficient condition for the case of non-disjoint balls.

Lemma 5 (Sufficient Condition) *Let $1 \leq q \leq \infty$ and $s > 0$. For $m \in \mathbb{N}_0$ let $f_m \in \mathcal{S}'(\mathbb{R}^d)$ be such that*

$$\operatorname{supp}(\hat{f}_m) \subseteq B_m := \left\{ \xi \in \mathbb{R}^d \,\middle|\, |\xi| \leq 2^m \right\} \qquad \forall m \in \mathbb{N}_0.$$

Set $f := \sum_{m=0}^{\infty} f_m$ in $\mathcal{S}'(\mathbb{R}^d)$. Then

$$\|f\|_{M^s_{p,q}(\mathbb{R}^d)} \lesssim \left\| \left(2^{ms} \|f_m\|_{M_{p,q}(\mathbb{R}^d)} \right)_{m\in\mathbb{N}_0} \right\|_q,$$

where the implicit constant depends on d and s only.

Proof of Theorem 1 Observe, that $A_l \cap B_m = \emptyset$ if $l > m + 2$. Hence, we have

$$\|f\|_{M^s_{p,q}} \approx \left\| \left(2^{ls} \|\Delta_l f\|_{M_{p,q}} \right)_l \right\|_q \lesssim \left\| \left(2^{ls} \sum_{m=l}^{\infty} \|\Delta_l f_m\|_{M_{p,q}} \right)_l \right\|_q$$

$$\lesssim \left\| \left(2^{ls} \sum_{m=l}^{\infty} \|f_m\|_{M_{p,q}} \right)_l \right\|_q ,$$

where we additionally used Theorem 1 and Bernstein multiplier estimate. From now on, we assume $q \in (1, \infty)$, as the proof for the other cases is easier and follows the same lines. Hölder's inequality and geometric sum formula estimates the last term against

$$\left(\sum_{l=0}^{\infty} \left(\sum_{m=l}^{\infty} 2^{ls} \|f_m\|_{M_{p,q}} \right)^q \right)^{\frac{1}{q}}$$

$$= \left(\sum_{l=0}^{\infty} \left(\sum_{m=l}^{\infty} 2^{\frac{(l-m)s}{q'}} \times 2^{\frac{(l-m)s}{q}} 2^{ms} \|f_m\|_{M_{p,q}} \right)^q \right)^{\frac{1}{q}}$$

$$\leq \left(\sum_{l=0}^{\infty} \left(\sum_{m=l}^{\infty} 2^{(l-m)s} \right)^{\frac{q}{q'}} \left(\sum_{m=l}^{\infty} 2^{(l-m)s} 2^{msq} \|f_m\|^q_{M_{p,q}} \right) \right)^{\frac{1}{q}}$$

$$\approx \left(\sum_{m=0}^{\infty} \sum_{l=0}^{m} 2^{(l-m)s} 2^{msq} \|f_m\|^q_{M_{p,q}} \right)^{\frac{1}{q}}$$

$$\approx \left\| \left(2^{ms} \|f_m\|_{M_{p,q}} \right)_m \right\|_q ,$$

finishing the proof. □

4 Algebra Property and Hölder-Type Inequality

Main goal of this section is to prove Theorem 2, which was inspired by the fact that $H^s(\mathbb{R}^d) \cap L^\infty(\mathbb{R}^d)$ is a Banach *-algebra with respect to pointwise multiplication for $s \geq 0$.

Proof of Theorem 2 Parts 2 and 3 of Lemma 2 prove the claimed embedding. Continuity of complex conjugation is obvious from (6). Continuity of multiplication follows by the paraproduct argument

$$fg = \left(\sum_{l=0}^{\infty} \Delta_l f\right)\left(\sum_{m=0}^{\infty} \Delta_m g\right) = \sum_{l=0}^{\infty} \underbrace{\left(\Delta_l f \sum_{m=0}^{l} \Delta_m g\right)}_{=:u_l} + \sum_{m=1}^{\infty} \underbrace{\left(\Delta_m g \sum_{l=0}^{m-1} \Delta_l f\right)}_{=:v_m}.$$

Observe, that for any $l, m \in \mathbb{N}_0$ we have $\mathrm{supp}(\hat{u}_l) \subseteq B_{l+1}$ and $\mathrm{supp}(\hat{v}_m) \subseteq B_m$ by the properties of convolution. Hence, Lemma 5 could be applied. Bilinear estimate from Lemma 3 and Theorem 1 show

$$\left\| \left(2^{ls} \|u_l\|_{M_{p,q}}\right)_l \right\|_q \le \left\| \left(2^{ls} \|\Delta_l f\|_{M_{p,q}}\right)_l \right\|_q \sum_{m=0}^{\infty} \|\Delta_m g\|_{M_{\infty,1}} \approx \|f\|_{M^s_{p,q}} \|g\|_{M_{\infty,1}}.$$

The same argument yields $\left\|\sum_{m=1}^{\infty} v_m\right\|_{M^s_{p,q}} \lesssim \|f\|_{M_{\infty,1}} \|g\|_{M^s_{p,q}}$ and finishes the proof. □

The analogon of Theorem 2 for sequence spaces is stated in

Lemma 6 (Algebra Property) *Let $1 \le q \le \infty$ and $s \ge 0$. Then $l^q_s(\mathbb{Z}^d) \cap l^1(\mathbb{Z}^d)$ is a Banach algebra with respect to convolution*

$$(a_l) * (b_m) = \left(\sum_{m \in \mathbb{Z}^d} a_{k-m} b_m\right)_{k \in \mathbb{Z}^d}, \tag{12}$$

which is well-defined, as the series above always converge absolutely.

Furthermore, if $q > 1$ and $s > d\left(1 - \frac{1}{q}\right)$ or $q = 1$, then $l^q_s(\mathbb{Z}^d) \hookrightarrow l^1(\mathbb{Z}^d)$, so in particular $l^q_s(\mathbb{Z}^d)$ is a Banach algebra, in that case.

This result is not new, see e.g. [8, Satz 3.7]. The proof given there works in much more general situations and relies on the fact that for $s \ge 0$ the weight function $x \mapsto \langle x \rangle^s$ is *weakly sub-additive* (a notion going back to [4], at least), i.e.,

$$\langle x + y \rangle^s \lesssim \langle x \rangle^s + \langle y \rangle^s \qquad \forall x, y \in \mathbb{R}.$$

For the sake of a self-contained presentation, let us remark that another proof can be given using the same techniques as for the proof of Theorem 2, i.e. by proving analoga of Theorem 1 and Lemma 5 for the weighted sequence spaces. Yet another approach is to notice that by definition

$$\left\| \sum_{k\in\mathbb{Z}^d} a_k e^{ikx} \right\|_{M^s_{\infty,q}} \approx \|(a_k)\|_{l^q_s}$$

and hence, by Theorem 2, one has

$$\begin{aligned}
&\|(a_k) * (b_k)\|_{l^q_s} \\
&\approx \left\| \left(\sum_{k\in\mathbb{Z}^d} a_k e^{ikx} \right) \cdot \left(\sum_{k\in\mathbb{Z}^d} b_k e^{ikx} \right) \right\|_{M^s_{\infty,q}} \\
&\lesssim \left\| \sum_{k\in\mathbb{Z}^d} a_k e^{ikx} \right\|_{M^s_{\infty,q}} \left\| \sum_{k\in\mathbb{Z}^d} b_k e^{ikx} \right\|_{M_{\infty,1}} + \left\| \sum_{k\in\mathbb{Z}^d} a_k e^{ikx} \right\|_{M_{\infty,1}} \left\| \sum_{k\in\mathbb{Z}^d} b_k e^{ikx} \right\|_{M^s_{\infty,q}} \\
&\approx \|(a_k)\|_{l^q_s} \|(b_k)\|_{l^1} + \|(a_k)\|_{l^1} \|(b_k)\|_{l^q_s} .
\end{aligned}$$

We are now ready to give a

Proof of Theorem 3 We arrive, as for Eq. (7) in the proof of Lemma 3, at

$$\|\square_k(fg)\|_p \lesssim \sum_{m\in M} \left(\left(\|\square_l(f)\|_{p_1}\right)_l * \left(\|\square_{n+m}(g)\|_{p_2}\right)_n \right)(k)$$

pointwise in $k \in \mathbb{Z}^d$. By the algebra property from Lemma 6, it follows that

$$\|fg\|_{M^s_{p,q}} \lesssim \left\| \left(\|\square_l f\|_{p_1}\right)_l \right\|_{q,s} \left(\sum_{m\in M} \left\| \left(\|\square_{n+m} g\|_{p_2}\right)_n \right\|_{q,s} \right)$$

and the first factor is already $\|f\|_{M^s_{p,q}}$. Finally, we remove the sum over m in the second factor

$$\sum_{m\in M} \left\| \left(\|\square_{n+m} g\|_{p_2}\right)_n \right\|_{q,s} \lesssim \|g\|_{M^s_{p_2,q}}$$

applying Peetre's inequality $\langle k+l\rangle^s \leq 2^{|s|}\langle k\rangle^s \langle l\rangle^{|s|}$ (see e.g. [16, Proposition 3.3.31]).

Let us finish the proof remarking that the only estimate involving "p"s we used was Hölder's inequality and thus the implicit constant indeed does not depend on p, p_1 or p_2. □

5 Proof of the Local Well-Posedness, Theorem 4

Theorem 2 immediately implies that $X(T)$ is a Banach *-algebra, i.e.

$$\|uv\|_{X(T)} = \sup_{0\le t\le T} \|uv(\cdot,t)\|_X \lesssim \left(\sup_{0\le s\le T} \|u(\cdot,s)\|_X\right)\left(\sup_{0\le t\le T} \|v(\cdot,t)\|_X\right)$$
$$= \|u\|_{X(T)}\, \|v\|_{X(T)}\,.$$

For $R>0$ we denote by $M(R,T) = \left\{u \in X(T) \middle| \, \|u\|_{X(T)} \le R\right\}$ the closed ball of radius R in $X(T)$ centered at the origin. We show that for some $R, T > 0$ the right-hand side of (4),

$$(\mathcal{T}u)(\cdot,t) := e^{\mathrm{i}t\Delta}u_0 \pm \mathrm{i}\int_0^t e^{\mathrm{i}(t-\tau)\Delta}\left(|u|^2 u(\cdot,\tau)\right)\mathrm{d}\tau \qquad (\forall t \in [0,T]), \tag{13}$$

defines a contractive self-mapping $\mathcal{T} = \mathcal{T}(u_0) : M_{R,T} \to M_{R,T}$.

To that end, let us observe that Theorem 6 implies the *homogeneous estimate*

$$\left\|t \mapsto e^{\mathrm{i}t\Delta}v\right\|_X \le C_0(1+T)^{\frac{d}{2}}\,\|v\|_X \qquad (\forall v \in X),$$

which, together with the algebra property of $X(T)$, proves the *inhomogeneous estimate*

$$\left\|\int_0^t e^{\mathrm{i}(t-\tau)\Delta}\left(|u|^2 u(\cdot,\tau)\right)\mathrm{d}\tau\right\|_X$$
$$\le C_0(1+T)^{\frac{d}{2}}\int_0^t \left\||u|^2 u(\cdot,\tau)\right\|_X \mathrm{d}\tau \le C_0C_1T(1+T)^{\frac{d}{2}}\,\|u\|_X^3\,,$$

holding for $0 \le t \le T$ and $u \in X(T)$.

Applying the triangle inequality in (13) yields

$$\|\mathcal{T}u\|_X \le C_0(1+T)^{\frac{d}{2}}(\|u_0\|_X + C_1TR^3)$$

for any $u \in M(R,T)$. Thus, $\mathcal{T}$ maps $M(R,T)$ into itself for $R = 2C_0C_1\,\|u_0\|_X$ and T small enough. Furthermore,

$$|u|^2 u - |v|^2 v = (u-v)\,|u|^2 + (\overline{u}u - \overline{v}v)v = (u-v)(|u|^2 + \overline{u}v) + (\overline{u}-\overline{v})v^2$$

and hence

$$\|\mathcal{T}u - \mathcal{T}v\|_{X(T)} \lesssim T(1+T)^{\frac{d}{2}}R^2\,\|u-v\|_{X(T)}$$

for $u, v \in M(R, T)$, where we additionally used the algebra property of $X(T)$ and the homogeneous estimate. Taking T sufficiently small makes $\mathcal{T}$ a contraction.

Banach's fixed-point theorem implies the existence and uniqueness of a mild solution up to the *guaranteed time of existence* $T_0 = T_0(\|u_0\|_X) \approx \|u_0\|_X^{-2} > 0$. Uniqueness of the maximal solution and the blow-up alternative now follow easily by the usual contradiction argument.

For the proof of the Lipschitz continuity, let us notice that for any $r > \|u_0\|_X$, $v_0 \in B_r(0)$ and $0 < T \leq T_0(r)$ we have

$$\begin{aligned}\|u - v\|_{X(T)} &= \|\mathcal{T}(u_0)u - \mathcal{T}(v_0)v\|_{X(T)} \\ &\lesssim (1+T)^{\frac{d}{2}} \|u_0 - v_0\|_X + T(1+T)^{\frac{d}{2}} R^2 \|u - v\|_{X(T)},\end{aligned}$$

where v is the mild solution corresponding to the initial data v_0 and $R = 2Cr$, similar to the above. Collecting terms containing $\|u - v\|_{X(T)}$ shows Lipschitz continuity with constant $L = L(r)$ for sufficiently small T, say $T_l = T_l(r)$. For arbitrary $0 < T' < T_*$ put $r = 2\|u\|_{X(T')}$ and divide $[0, T']$ into n subintervals of length $\leq T_l$. The claim follows for $V = B_\delta(u_0)$ where $\delta = \frac{\|u_0\|_X}{L^n}$ by iteration. This concludes the proof. □

Acknowledgments Funded by the Deutsche Forschungsgemeinschaft (DFG, German Research Foundation) project-id 258734477 SFB 1173.

The authors would like to thank Professor H.G. Feichtinger for comments and remarks on an earlier version of this paper.

References

1. Alinhac, S., Gérard, P.: Pseudo-differential operators and the Nash-Moser theorem. In: Graduate Studies in Mathematics, vol. 82. American Mathematical Society, Providence (2007). https://doi.org/10.1090/gsm/082
2. Bényi, A., Gröchenig, K., Okoudjou, K.A., Rogers, L.G.: Unimodular Fourier multipliers for modulation spaces. J. Funct. Anal. **246**(2), 366–384 (2007). https://doi.org/10.1016/j.jfa.2006.12.019
3. Bényi, A., Okoudjou, K.A.: Local well-posedness of nonlinear dispersive equations on modulation spaces. Bull. Lond. Math. Soc. **41**(3), 549–558 (2009). https://doi.org/10.1112/blms/bdp027
4. Brandenburg, L.H.: On identifying the maximal ideals in Banach algebras. J. Math. Anal. Appl. **50**, 489–510 (1975). https://doi.org/10.1016/0022-247X(75)90006-2
5. Chaichenets, L.: Modulation spaces and nonlinear Schrödinger equations. Ph.D. Thesis, Karlsruhe Institute of Technology (KIT) (2018). https://doi.org/10.5445/IR/1000088173
6. Chaichenets, L., Hundertmark, D., Kunstmann, P., Pattakos, N.: Nonlinear Schrödinger equation, differentiation by parts and modulation spaces. J. Evol. Eq. **19**, 803–843 (2019). https://doi.org/10.1007/s00028-019-00501-z
7. Cordero, E., Nicola, F.: Sharpness of some properties of Wiener amalgam and modulation spaces. Bull. Aust. Math. Soc. **80**(1), 105–116 (2009). https://doi.org/10.1017/S0004972709000070

8. Feichtinger, H.G.: Gewichtsfunktionen auf lokalkompakten Gruppen. Österreich. Akad. Wiss. Math.-Natur. Kl. Sitzungsber. II **188**(8–10), 451–471 (1979). https://www.univie.ac.at/nuhag-php/bibtex/open_files/fe79-2_gewfkta.pdf
9. Feichtinger, H.G.: Banach convolution algebras of Wiener type. In: Proceeding of the Conference on Functions, Series, Operators, Budapest 1980. Colloquia Mathematica Societatis János Bolyai, vol. 35, pp. 509 – 524. North-Holland, Amsterdam (1983). https://www.univie.ac.at/nuhag-php/bibtex/open_files/fe83_wientyp1.pdf
10. Feichtinger, H.G.: Modulation spaces on locally compact abelian groups. University Vienna (1983). https://www.univie.ac.at/nuhag-php/bibtex/open_files/120_ModICWA.pdf
11. Gröchenig, K.: Foundations of time-frequency analysis. In Applied and Numerical Harmonic Analysis. Birkhäuser, Boston (2001). https://doi.org/10.1007/978-1-4612-0003-1
12. Guo, S.: On the 1D cubic nonlinear Schrödinger equation in an almost critical space. J. Fourier Anal. Appl. **23**(1), 91–124 (2016). https://doi.org/10.1007/s00041-016-9464-z
13. Guo, W., Fan, D., Wu, H., Zhao, G.: Sharp weighted convolution inequalities and some applications. Studia Math. **241**(3), 201–239 (2018). https://doi.org/10.4064/sm8583-5-2017
14. Kunstmann, P.C.: Modulation type spaces for generators of polynomially bounded groups and Schrödinger equations. Semigroup Forum **98**, 645–668 (2019). https://doi.org/10.1007/s00233-019-10016-1
15. Pattakos, N.: NLS in the modulation space $M_{2,q}(\mathbb{R})$. J. Fourier Anal. Appl. **25**, 1447–1486 (2018). https://doi.org/10.1007/s00041-018-09655-9
16. Ruzhansky, M.V., Turunen, V.: Pseudo-differential operators and symmetries. In: Pseudo-Differential Operators: Theory and Applications, No. 2. Birkhäuser, Basel (2010). https://doi.org/10.1007/978-3-7643-8514-9
17. Ruzhansky, M., Sugimoto, M., Wang, B.: Modulation spaces and nonlinear evolution equations. In: Evolution Equations of Hyperbolic and Schrödinger Type. Progress in Mathematics, vol. 301, pp. 267–283. Springer, Basel (2012). https://doi.org/10.1007/978-3-0348-0454-7_14
18. Sugimoto, M., Tomita, N., Wang, B.: Remarks on nonlinear operations on modulation spaces. Integr. Trans. Spec. F. **22**(4–5), 351–358 (2011). https://doi.org/10.1080/10652469.2010.541054
19. Toft, J., Johansson, K., Pilipović, S., Teofanov, N.: Sharp convolution and multiplication estimates in weighted spaces. Anal. Appl. **13**(5), 457–480 (2015). https://doi.org/10.1142/S0219530514500523
20. Wang, B., Hudzik, H.: The global Cauchy problem for the NLS and NLKG with small rough data. J. Differ. Eq. **232**(1), 36–73 (2007). https://doi.org/10.1016/j.jde.2006.09.004
21. Wang, B., Zhao, L., Guo, B.: Isometric decomposition operators, function spaces $E^{\lambda}_{p,q}$ and applications to nonlinear evolution equations. J. Funct. Anal. **233**(1), 1–39 (2006). https://doi.org/10.1016/j.jfa.2005.06.018
22. Wang, B., Huo, Z., Hao, C., Guo, Z.: Harmonic Analysis Method for Nonlinear Evolution Equations, I. World Scientific, Singapore (2011). https://doi.org/10.1142/8209

FEM-BEM Coupling of Wave-Type Equations: From the Acoustic to the Elastic Wave Equation

Sarah Eberle

Abstract We consider the FEM-BEM coupling for two special wave-type equations: the acoustic and the elastic wave equation. In more detail, we take a look at the coupling of the interior and exterior problems of these wave-type equations and review a stable numerical method including the corresponding error estimates as well as the convergence. The intent of this paper is to thereby highlight the similarities, while at the same time presenting the challenges inherent in switching from the scalar acoustic to the vectorial elastic equation.

1 Introduction and Motivation

In this paper we review a stable numerical method which couples the interior and exterior problem of wave-type equations, specifically the acoustic wave equation given in [3] and the elastic wave equation analyzed in [4]. We summarize the results from the cited literature, where these were originally presented, and also compare their results with each other. Thus, this article gives an overview over the necessary steps to prove a convergence result. For the in-depth discussion however, the reader is referred to [3] and [4].

The framework is organized as follows: We start with the problem statement and introduce the partial differential equations for the acoustic and elastic wave equation as well as their Laplace transformation. In order to construct the so-called Calderón operator, we examine the boundary conditions and the corresponding boundary integral operators. One important result in this context is the coercivity of the Calderón operator. Here, we compare the operators for the elastic and acoustic wave equation and highlight the additional necessary steps to switch from the acoustic to the elastic case. Further on, the space and time discretization is considered, where

---

S. Eberle (✉)
Institute of Mathematics, Goethe-University Frankfurt, Frankfurt, Germany
e-mail: eberle@math.uni-frankfurt.de

W. Dörfler et al. (eds.), *Mathematics of Wave Phenomena*, Trends in Mathematics,
https://doi.org/10.1007/978-3-030-47174-3_7

we introduce the first order systems in time which build the basis for the FEM-BEM formulation. Based on this, we finally present the error estimates and the convergence result.

2 Problem Statement and Background

We start with the definition of the acoustic and elastic wave equation.

Definition 1 (Acoustic Wave Equation) The acoustic wave equation is given by

$$\partial_t^2 u_{\mathrm{a}}(x,t) = \Delta u_{\mathrm{a}}(x,t) + \partial_t f_{\mathrm{a}}(x,t) \text{ in } \mathbb{R}^3 \times (0,T], \tag{1}$$

$$u_{\mathrm{a}}(x,0) = u_{\mathrm{a}}^0 \text{ in } \mathbb{R}^3, \tag{2}$$

$$\partial_t u_{\mathrm{a}}(x,0) = v_{\mathrm{a}}^0 \text{ in } \mathbb{R}^3, \tag{3}$$

with u_{a} the scalar pressure and $\partial_t f_{\mathrm{a}}$ the inhomogeneity w.r.t. u_{a}.

Definition 2 (Elastic Wave Equation) The elastic wave equation is given by

$$\rho\partial_t^2 u_{\mathrm{e}}(x,t) = \mu\Delta u_{\mathrm{e}}(x,t) + (\lambda+\mu)\nabla(\nabla\cdot u_{\mathrm{e}}(x,t)) + \rho\partial_t f_{\mathrm{e}}(x,t) \text{ in } \mathbb{R}^3 \times (0,T], \tag{4}$$

$$u_{\mathrm{e}}(x,0) = u_{\mathrm{e}}^0 \text{ in } \mathbb{R}^3, \tag{5}$$

$$\partial_t u_{\mathrm{e}}(x,0) = v_{\mathrm{e}}^0 \text{ in } \mathbb{R}^3, \tag{6}$$

with u_{e} as the displacement vector, $\rho > 0$ the density, $\mu, \lambda > 0$ the Lamé parameters, and $\partial_t f_{\mathrm{e}}$ the inhomogeneity w.r.t. u_{e}.

The similarities between those equations can be made more clear by introducing the stress tensor

$$\begin{aligned}\sigma(u_{\mathrm{e}}(x,t)) &= \mu\left(\nabla u_{\mathrm{e}}(x,t) + (\nabla u_{\mathrm{e}}(x,t))^T\right) + \lambda\,(\nabla\cdot u_{\mathrm{e}}(x,t))\,I \\ &= 2\mu\varepsilon(u_{\mathrm{e}}(x,t)) + \lambda\,(\nabla\cdot u_{\mathrm{e}}(x,t))\,I\end{aligned}$$

with I as the 3×3-identity matrix and $\varepsilon(u)$ as the elastic strain tensor given by

$$\varepsilon(u_{\mathrm{e}}(x,t)) = \frac{1}{2}\left(\nabla u_{\mathrm{e}}(x,t) + (\nabla u_{\mathrm{e}}(x,t))^T\right).$$

This allows us to rewrite Eqs. (1) and (4) as

$$\partial_t^2 u_{\mathrm{a}}(x,t) = \nabla\cdot(\nabla u_{\mathrm{a}}(x,t)) + \partial_t f_{\mathrm{a}}(x,t) \text{ in } \mathbb{R}^3 \times (0,T] \tag{7}$$

and

$$\partial_t^2 u_e(x,t) = \nabla \cdot (\sigma\,(u_e(x,t))) + \rho \partial_t f_e(x,t) \text{ in } \mathbb{R}^3 \times (0,T], \tag{8}$$

respectively. Apart from the scalar and vectorial structure of the equations, we use the elastic strain tensor instead of the gradient in elasticity and the equation is extended by the term $\lambda\,(\nabla \cdot u_e(x,t))\,I$ as well.

In this paper, we are especially interested in the following transmission problems.

Definition 3 Let $\Omega \subset \mathbb{R}^3$ be a bounded Lipschitz domain with boundary $\partial\Omega$ and $\Omega^+ = \mathbb{R}^3 \setminus \Omega$. With the upper $-$ and $+$ sign indicating the interior and exterior, respectively, the corresponding transmission problem is represented by the interior problem given by

$$\begin{aligned}
\partial_t^2 u_a^-(x,t) &= \nabla \cdot \left(\nabla u_a^-(x,t)\right) + \partial_t f_a(x,t) \text{ in } \Omega \times (0,T],\\
u_a^-(x,0) &= u_a^0 \text{ in } \Omega,\\
\partial_t u_a^-(x,0) &= v_a^0 \text{ in } \Omega
\end{aligned}$$

for the acoustic case and

$$\begin{aligned}
\partial_t^2 u_e^-(x,t) &= \nabla \cdot \left(\sigma\left(u_e^-(x,t)\right)\right) + \rho \partial_t f_e(x,t) \text{ in } \Omega \times (0,T],\\
u_e^-(x,0) &= u_e^0 \text{ in } \Omega,\\
\partial_t u_e^-(x,0) &= v_e^0 \text{ in } \Omega
\end{aligned}$$

for the elastic case, coupled with the corresponding exterior problem as

$$\begin{aligned}
\partial_t^2 u_a^+(x,t) &= \nabla \cdot \left(\nabla u_a^+(x,t)\right) \text{ in } \Omega^+ \times (0,T],\\
u_a^+(x,0) &= 0 \text{ in } \Omega^+,\\
\partial_t u_a^+(x,0) &= 0 \text{ in } \Omega^+
\end{aligned}$$

for the acoustic case and

$$\begin{aligned}
\partial_t^2 u_e^+(x,t) &= \nabla \cdot \left(\sigma\left(u_e^+(x,t)\right)\right) \text{ in } \Omega^+ \times (0,T],\\
u_e^+(x,0) &= 0 \text{ in } \Omega^+,\\
\partial_t u_e^+(x,0) &= 0 \text{ in } \Omega^+
\end{aligned}$$

for the elastic wave equation.

For the analysis of the transmission problem, we also introduce the Laplace transformations exemplary for the interior case given by

$$s^2 \mathring{u}_a^- = \Delta \mathring{u}_a^- + s \mathring{f}_a + s \mathring{v}_a^0 + \mathring{u}_a^0 \tag{9}$$

for the acoustic wave equation and

$$\rho s^2 \mathring{u}_e^- = \mu \Delta \mathring{u}_e^- + (\lambda + \mu)\nabla(\nabla \cdot \mathring{u}_e^-) + \rho s \mathring{f}_e + s\mathring{v}_e^0 + \mathring{u}_e^0 \tag{10}$$

for the elastic wave equation. Here, the upper $\circ$ represents the Laplace transform of a given function.

Since the general approach to get the error estimates and coercivity is the same for both types of equations, we summarize both the scalar pressure distribution u_a as well as the vector valued displacement u_e by the place holder U.

The transmission conditions for the setup in Definition 3 including the Dirichlet and Neumann boundary conditions can be summarized as

$$\gamma_D^- \mathring{U}^- = \gamma_D^+ \mathring{U}^+,$$
$$\gamma_N^- \mathring{U}^- = \gamma_N^+ \mathring{U}^+.$$

The Neumann boundary conditions can be directly read from Eqs. (7) and (8) as the terms which are subjected to the divergence multiplied by the corresponding normal vector, so that the acoustic Neumann boundary condition is given by

$$\gamma_N \mathring{u}_a = \nabla \mathring{u}_a \cdot n = \partial_n \mathring{u}_a$$

and the elastic Neumann boundary condition by

$$\gamma_N \mathring{u}_e = \sigma(\mathring{u}_e) n.$$

In the following, we will consider the problems in the Laplace domain and will indicate this dependence by the parameter s, thereby neglecting to write $\mathring{U}$ instead of U.

3 Calderón Operator

As a first step, the fundamental solutions corresponding to both problems are required in order to obtain the potentials and boundary integral operators which can be found in [3] and [4]. We follow the standard approach, thus, we define the jumps over the trace Γ by

$$[U] = \gamma_D^-(U) - \gamma_D^+(U),$$
$$[\gamma_N(U)] = \gamma_N^-(U) - \gamma_N^+(U),$$

and consider the representation formula

$$U = sS(s)\phi + D(s)\psi,$$

where $S(s)$ is the single layer potential and $D(s)$ the double layer potential.

Further on, we introduce the boundary densities $(\phi, \psi)^T$ in a similar way as the place holder U and indicate the connection to the corresponding wave equation by the indices $\cdot_{\mathrm{a}}$ and $\cdot_{\mathrm{e}}$. In general the boundary densities are given by

$$\psi = -[U],$$

$$\phi = \frac{1}{s}[\gamma_N(U)],$$

where ψ and ϕ are scalar in case of $U = u_{\mathrm{a}}$ and vector valued in case of $U = u_{\mathrm{e}}$.

Finally, we take a look at the generalized boundary integral operators

$$J(s)\phi = \gamma_D S(s)\phi, \tag{11}$$

$$W(s)\psi = -\gamma_N D(s)\psi, \tag{12}$$

$$K(s)\psi = \frac{1}{2}\left(\gamma_D^+ D(s)\psi + \gamma_D^- D(s)\psi\right) = \{\{\gamma_D D(s)\psi\}\}, \tag{13}$$

$$K^T(s)\phi = \frac{1}{2}\left(\gamma_N^+ S(s)\phi + \gamma_N^- S(s)\phi\right) = \{\{\gamma_N S(s)\phi\}\}, \tag{14}$$

where we also introduced the notation $\{\{\cdot\}\}$ for averages $\{\{u\}\} = \frac{1}{2}(u^+ + u^-)$ of inner and outer limits at the boundary. This leads us to the definition of the Calderón operator

$$B(s) = \begin{pmatrix} sJ(s) & K(s) \\ -K^T(s) & \frac{1}{s}W(s) \end{pmatrix}. \tag{15}$$

Again, the details of the construction can be found in [3] and [4].

Remark 1 As we can see, at this point, the structure of the Calderón operator for our two problems (9) and (10) are the same but the boundary integral operators (11)–(14) themselves are not, due to the fact that the single layer potential $S(s)$ and double layer potential $D(s)$ depend on the fundamental solution, which substantially differs for the given problems. In addition, if we take a closer look at the coercivity results, we have to take into account that we deal with different boundary densities due to the different Neumann boundary conditions for the acoustic and elastic case and the scalar and vector valued nature of u_{a} and u_{e}, respectively.

All in all, this allows us to recapitulate the coercivity of $B(s)$ as the first main result from [3] and [4]:

Lemma 1 (Lemma 3.1 from [3]) *There exists a constant* $\tilde{\beta}_{\mathrm{a}} > 0$ *such that the Calderón operator* $B_{\mathrm{a}}(s)$ *of the* ***acoustic wave equation*** *satisfies*

$$\begin{aligned} &\mathrm{Re}\left\langle \begin{pmatrix} \phi_{\mathrm{a}} \\ \psi_{\mathrm{a}} \end{pmatrix}, B_{\mathrm{a}}(s) \begin{pmatrix} \phi_{\mathrm{a}} \\ \psi_{\mathrm{a}} \end{pmatrix} \right\rangle_{\Gamma} \\ &\geq \tilde{\beta}_{\mathrm{a}} \frac{\mathrm{Re}(s)}{|s|^2} \min\left(1, |s|^2\right) \left(\|\phi_{\mathrm{a}}\|^2_{H^{-\frac{1}{2}}(\Gamma)} + \|\psi_{\mathrm{a}}\|^2_{H^{\frac{1}{2}}(\Gamma)} \right) \end{aligned} \tag{16}$$

for $Re(s) > 0$ *and for all* $\phi_{\mathrm{a}} \in H^{-\frac{1}{2}}(\Gamma)$ *and* $\psi_{\mathrm{a}} \in H^{\frac{1}{2}}(\Gamma)$.

Sketch of the Proof We summarize the most important steps of the proof and refer to [3] for the details:

- Start with $\mathrm{Re}\left\langle \begin{pmatrix} \phi_{\mathrm{a}} \\ \psi_{\mathrm{a}} \end{pmatrix}, B_{\mathrm{a}}(s) \begin{pmatrix} \phi_{\mathrm{a}} \\ \psi_{\mathrm{a}} \end{pmatrix} \right\rangle_{\Gamma}$ and apply the corresponding Green's formula.
- Find estimates for $\|\phi_{\mathrm{a}}\|_{H^{-\frac{1}{2}}(\Gamma)}$ and $\|\psi_{\mathrm{a}}\|_{H^{\frac{1}{2}}(\Gamma)}$ by use of the trace inequalities.
- Finally, use these estimates to show the positivity of the Calderón operator.

□

The corresponding lemma for the elastic wave equation reads as follows.

Lemma 2 (Lemma 1 from [4]) *There exists a constant* $\tilde{\beta}_{\mathrm{e}} > 0$ *such that the Calderón operator* $B_{\mathrm{e}}(s)$ *of the* ***elastic wave equation*** *satisfies*

$$\begin{aligned} &\mathrm{Re}\left\langle \begin{pmatrix} \phi_{\mathrm{e}} \\ \psi_{\mathrm{e}} \end{pmatrix}, B_{\mathrm{e}}(s) \begin{pmatrix} \phi_{\mathrm{e}} \\ \psi_{\mathrm{e}} \end{pmatrix} \right\rangle_{\Gamma} \\ &\geq \tilde{\beta}_{\mathrm{e}} \frac{\mathrm{Re}(s)}{|s|^2} \min\left(1, |s|^2\right) \left(\|\phi_{\mathrm{e}}\|^2_{H^{-\frac{1}{2}}(\Gamma)^3} + \|\psi_{\mathrm{e}}\|^2_{H^{\frac{1}{2}}(\Gamma)^3} \right) \end{aligned} \tag{17}$$

for $Re(s) > 0$ *and for all* $\phi_{\mathrm{e}} \in H^{-\frac{1}{2}}(\Gamma)^3$ *and* $\psi_{\mathrm{e}} \in H^{\frac{1}{2}}(\Gamma)^3$.

Sketch of the Proof Again the whole proof can be found in [4] and we highlight the essential steps:

- First, take a look at $\mathrm{Re}\left\langle \begin{pmatrix} \phi_{\mathrm{e}} \\ \psi_{\mathrm{e}} \end{pmatrix}, B_{\mathrm{e}}(s) \begin{pmatrix} \phi_{\mathrm{e}} \\ \psi_{\mathrm{e}} \end{pmatrix} \right\rangle_{\Gamma}$ and apply the corresponding Green's formula for the elastic wave equation which is also known as Betti's formula.
- Determine estimates for $\|\phi_{\mathrm{e}}\|_{H^{-\frac{1}{2}}(\Gamma)^3}$ and $\|\psi_{\mathrm{e}}\|_{H^{\frac{1}{2}}(\Gamma)^3}$ by the application of estimates due to the energy norm for the elastic wave equation.
- For the last step, apply these estimates to prove the positivity of the Calderón operator.

□

Remark 2 As can be seen in the sketches of the proofs of the coercivity estimates, both follow a very similar procedure. However, the challenge in the elastic case is its vector valued setup which complicates the proof. Further on, an additional term has to be considered in such a way that the step from the estimate regarding the Calderón operator to the estimates of the boundary densities follow through.

These coercivity results can be transformed via the inverse Laplace transformation such that we end up with the estimates for the Calderón operator in time, i.e., $B(\partial_t)$ (see [3] and [4]).

4 Space Discretization

We start with the space discretization of our problems and transform them into first order systems w.r.t. the time t. Thus, we take a look at the first order system for the **acoustic wave equation**

$$\begin{aligned} \partial_t u_{\mathrm{a}} &= \nabla \cdot v + f_{\mathrm{a}}, \\ \partial_t v &= \nabla u_{\mathrm{a}}, \\ B_{\mathrm{a}}(\partial_t) \begin{pmatrix} \phi_{\mathrm{a}} \\ \psi_{\mathrm{a}} \end{pmatrix} &= \frac{1}{2} \begin{pmatrix} \gamma u_{\mathrm{a}} \\ -\gamma v \cdot n \end{pmatrix}. \end{aligned}$$

Compared to the acoustic wave equation, the elastic wave equation differs by the use of the symmetric gradient $\nabla u + (\nabla u)^T$ instead of ∇u. The additional term due to the λ part of the initial equation results in an additional equation in the first order formulation, so that the first order system for the **elastic wave equation** (see [4]) is given by

$$\begin{aligned} \rho \partial_t u_{\mathrm{e}} &= \mu \nabla \cdot V + \lambda \nabla \omega + \rho f_{\mathrm{e}}, \\ \partial_t V &= 2\varepsilon(u_{\mathrm{e}}) = (\nabla u_{\mathrm{e}} + (\nabla u_{\mathrm{e}})^T), \\ \partial_t \omega &= \nabla \cdot u_{\mathrm{e}}, \\ B_{\mathrm{e}}(\partial_t) \begin{pmatrix} \phi_{\mathrm{e}} \\ \psi_{\mathrm{e}} \end{pmatrix} &= \frac{1}{2} \begin{pmatrix} \gamma u_{\mathrm{e}} \\ -\gamma(\mu V + \lambda \omega I)n \end{pmatrix}. \end{aligned}$$

All in all, these first order systems build the starting point for the weak formulation. With $(\cdot, \cdot)$ and $\langle \cdot, \cdot \rangle$ as the classical L^2-scalar products in Ω and $\partial\Omega$ respectively, the following special choice of testing and application of partial integration (see, e.g., [1]) results in

$$(\dot{u}_{\mathrm{a}}\, u_{\mathrm{a}}) = -\frac{1}{2}(v, \nabla u_{\mathrm{a}}) + \frac{1}{2}(\nabla \cdot v, u_{\mathrm{a}}) + \frac{1}{2}\langle \gamma v \cdot n, \gamma u_{\mathrm{a}} \rangle_{\Gamma} + (f_{\mathrm{a}}, u_{\mathrm{a}}), \tag{18}$$

$$(\dot{v}, v) = \frac{1}{2}(\nabla u_{\text{a}}, v) - \frac{1}{2}(u_{\text{a}}, \nabla \cdot v) + \frac{1}{2}\langle \gamma u_{\text{a}}, \gamma v \cdot n\rangle_\Gamma, \tag{19}$$

$$\left\langle \begin{pmatrix} \phi_{\text{a}} \\ \psi_{\text{a}} \end{pmatrix}, B_{\text{a}}(\partial_t) \begin{pmatrix} \phi_{\text{a}} \\ \psi_{\text{a}} \end{pmatrix} \right\rangle_\Gamma = \frac{1}{2}\langle \phi_{\text{a}}, \gamma u_{\text{a}}\rangle_\Gamma - \frac{1}{2}\langle \psi_{\text{a}}, \gamma v \cdot n\rangle_\Gamma. \tag{20}$$

By summing up Eqs. (18)–(20), we obtain for the **acoustic wave equation**

$$\frac{d}{dt}\left(\frac{1}{2}\|u_{\text{a}}\|^2 + \frac{1}{2}\|v\|^2\right) + \left\langle \begin{pmatrix} \phi_{\text{a}} \\ \psi_{\text{a}} \end{pmatrix}, B_{\text{a}}(\partial_t) \begin{pmatrix} \phi_{\text{a}} \\ \psi_{\text{a}} \end{pmatrix} \right\rangle_\Gamma = (f_{\text{a}}, u_{\text{a}}).$$

Due to the coercivity of the Calderón operator $B_a(\partial_t)$ in the time-domain, we end up with the energy $E_{\text{a}}(t)$ given by

$$E_a(t) = \frac{1}{2}\left(\|u_{\text{a}}\|^2 + \|v\|^2\right). \tag{21}$$

We proceed in the same way with the weak formulation of the elastic wave equation so that we get

$$\begin{aligned} \rho(\dot{u}_{\text{e}}, u_{\text{e}}) = &- \mu\frac{1}{2}(V, \nabla u_{\text{e}}) + \mu\frac{1}{2}(\nabla \cdot V, u_{\text{e}}) - \lambda\frac{1}{2}(\omega, \nabla \cdot u_{\text{e}}) \\ &+ \lambda\frac{1}{2}(\nabla\omega, u_{\text{e}}) + \frac{1}{2}\langle \gamma(\mu V + \lambda\omega I)n, \gamma u_{\text{e}}\rangle_\Gamma + \rho(f_{\text{e}}, u_{\text{e}}), \end{aligned} \tag{22}$$

$$\mu(\dot{V}, \frac{1}{2}V) = \mu\frac{1}{2}(\nabla u_{\text{e}}, V) - \mu\frac{1}{2}(u_{\text{e}}, \nabla \cdot V) + \mu\frac{1}{2}\langle \gamma u_{\text{e}}, \gamma V n\rangle_\Gamma, \tag{23}$$

$$\lambda(\dot{\omega}, \omega) = -\lambda\frac{1}{2}(u_{\text{e}}, \nabla\omega) + \lambda\frac{1}{2}(\nabla \cdot u_{\text{e}}, \omega) + \lambda\frac{1}{2}\langle \gamma u_{\text{e}}, (\gamma\omega I)n\rangle_\Gamma, \tag{24}$$

$$\left\langle \begin{pmatrix} \phi_{\text{e}} \\ \psi_{\text{e}} \end{pmatrix}, B_{\text{e}}(\partial_t) \begin{pmatrix} \phi_{\text{e}} \\ \psi_{\text{e}} \end{pmatrix} \right\rangle_\Gamma = \frac{1}{2}\langle \phi, \gamma u_{\text{e}}\rangle_\Gamma - \frac{1}{2}\langle \psi_{\text{e}}, \gamma\mu V n\rangle_\Gamma - \frac{1}{2}\langle \psi_{\text{e}}, (\gamma\lambda\omega I)n\rangle_\Gamma, \tag{25}$$

and end up with the energy for the **elastic wave equation**

$$E_e(t) = \frac{1}{2}\left(\rho\|u_{\text{e}}\|^2 + \frac{1}{2}\mu\|V\|^2 + \lambda\|\omega\|^2\right). \tag{26}$$

For more details, the reader is again referred to [3] and [4].

Remark 3 It should be noted that the testing in the elastic case also involves the material parameters ρ, λ, μ. This is not necessary in the acoustic case since this equation has only one parameter, the wave speed c, which can be scaled to $c = 1$. Hence, the elastic energy depends on the elastic parameters whereas the acoustic energy contains none due to scaling.

In addition, these energies satisfy

$$E_a(T) + \tilde{\beta}_a c_{a,T} \int_0^T \left(\|\partial_t^{-1}\phi_a(\cdot,t)\|^2_{H^{-\frac{1}{2}}(\Gamma)} + \|\partial_t^{-1}\psi_a(\cdot,t)\|^2_{H^{\frac{1}{2}}(\Gamma)} \right) dt$$
$$\leq \exp(2)\, E_a(0),$$
$$E_e(T) + \tilde{\beta}_e c_{e,T} \int_0^T \left(\|\partial_t^{-1}\phi_e(\cdot,t)\|^2_{H^{-\frac{1}{2}}(\Gamma)^3} + \|\partial_t^{-1}\psi_e(\cdot,t)\|^2_{H^{\frac{1}{2}}(\Gamma)^3} \right) dt$$
$$\leq \exp(2)\, E_e(0),$$

which means that the systems have the highest energy $E(t)$ in $t = 0$ (see [3] and [4]).

Based on the weak formulation (18)–(20) as well as (23)–(25), we introduce the FEM-BEM discretization. The corresponding spaces and functions are introduced in accordance with [3] and [4].

We consider finite dimensional subspaces of the given Sobolev spaces $H^1(\Omega)$, $H^1(\Omega)^3$, $H^1(\Omega)^{3\times 3}$ and $H^{\frac{1}{2}}(\Gamma)$, $H^{-\frac{1}{2}}(\Gamma)$, $H^{\frac{1}{2}}(\Gamma)^3$, $H^{-\frac{1}{2}}(\Gamma)^3$. We denote the chosen bases of these spaces by

$$b_i^{H^1(\Omega)}, \quad b_i^{H^1(\Omega)^3}, \quad b_i^{H^1(\Omega)^{3\times 3}}, \quad b_i^{H^{\frac{1}{2}}(\Gamma)}, \quad b_i^{H^{-\frac{1}{2}}(\Gamma)}, \quad b_i^{H^{\frac{1}{2}}(\Gamma)^3}, \quad b_i^{H^{-\frac{1}{2}}(\Gamma)^3}.$$

In addition, the basis functions $b_i^{H^{-\frac{1}{2}}(\Gamma)}$ and $b_i^{H^{-\frac{1}{2}}(\Gamma)^3}$ are assumed to be piecewise constant, while the others are piecewise linear.

Evaluated in nodal points, the resulting systems are given in matrix-vector notation by

$$\mathbf{M}_0\dot{\mathbf{u}}_{\mathrm{a}} = -\mathbf{D}_0^T\mathbf{v} - \mathbf{C}_0\boldsymbol{\phi}_{\mathrm{a}} + \mathbf{M}_0\mathbf{f}_{\mathrm{a}}, \tag{27}$$
$$\mathbf{M}_1\dot{\mathbf{v}} = \mathbf{D}_0\mathbf{u}_{\mathrm{a}} - \mathbf{C}_1\boldsymbol{\psi}_{\mathrm{a}}, \tag{28}$$
$$\mathbf{B}_{\mathrm{a}}(\partial_t)\begin{pmatrix}\boldsymbol{\phi}_a\\ \boldsymbol{\psi}_a\end{pmatrix} = \begin{pmatrix}\mathbf{C}_0^T\mathbf{u}_{\mathrm{a}}\\ \mathbf{C}_1^T\mathbf{v}\end{pmatrix} \tag{29}$$

for the **acoustic wave equation** and by

$$\rho\mathbf{M}_1\dot{\mathbf{u}}_{\mathrm{e}} = -\mu\mathbf{D}_1^T\mathbf{V} + \lambda\mathbf{D}_0\boldsymbol{\omega} - \tilde{\mathbf{C}}_0\boldsymbol{\phi}_{\mathrm{e}} + \rho\mathbf{M}_1\mathbf{f}_{\mathrm{e}}, \tag{30}$$
$$\mu\mathbf{M}_2\dot{\mathbf{V}} = \mu\mathbf{D}_1\mathbf{u}_{\mathrm{e}} - \mu\tilde{\mathbf{C}}_1\boldsymbol{\psi}_{\mathrm{e}}, \tag{31}$$
$$\lambda\mathbf{M}_0\dot{\boldsymbol{\omega}} = -\lambda\mathbf{D}_0^T\mathbf{u}_{\mathrm{e}} - \lambda\overline{\mathbf{C}}_1\boldsymbol{\psi}_{\mathrm{e}}, \tag{32}$$

$$\mathbf{B}_{\mathrm{e}}(\partial_t)\begin{pmatrix}\boldsymbol{\phi}_{\mathrm{e}}\\ \boldsymbol{\psi}_{\mathrm{e}}\end{pmatrix}=\begin{pmatrix}\tilde{\mathbf{C}}_0^T\mathbf{u}_{\mathrm{e}}\\ \mu\tilde{\mathbf{C}}_1^T\mathbf{V}+\lambda\overline{\mathbf{C}}_1^T\boldsymbol{\omega}\end{pmatrix}, \tag{33}$$

for the **elastic wave equation**, where we use a slightly modified notation as in the cited literature.

Thus, the mass matrices are given by

$$\mathbf{M}_0|_{ij}=\left(b_i^{H^1(\Omega)},b_j^{H^1(\Omega)}\right)_{L^2(\Omega)}, \tag{34}$$

$$\mathbf{M}_1|_{ij}=\left(b_i^{H^1(\Omega)^3},b_j^{H^1(\Omega)^3}\right)_{L^2(\Omega)^3}, \tag{35}$$

$$\mathbf{M}_2|_{ij}=\frac{1}{2}\left(b_i^{H^1(\Omega)^{3\times 3}},b_j^{H^1(\Omega)^{3\times 3}}\right)_{L^2(\Omega)^{3\times 3}}, \tag{36}$$

the matrices involving the differential operators by

$$\mathbf{D}_0|_{ij}=\frac{1}{2}\left(b_i^{H^1(\Omega)^3},\nabla b_j^{H^1(\Omega)}\right)_{L^2(\Omega)^3}-\frac{1}{2}\left(\nabla\cdot b_i^{H^1(\Omega)^3},b_j^{H^1(\Omega)}\right)_{L^2(\Omega)}, \tag{37}$$

$$\mathbf{D}_1|_{ij}=\frac{1}{2}\left(b_i^{H^1(\Omega)^{3\times 3}},\nabla b_j^{H^1(\Omega)^3}\right)_{L^2(\Omega)^{3\times 3}}-\frac{1}{2}\left(\nabla\cdot b_i^{H^1(\Omega)^{3\times 3}},b_j^{H^1(\Omega)^3}\right)_{L^2(\Omega)^3}, \tag{38}$$

and for the boundary, we have

$$\mathbf{C}_0|_{ij}=\tfrac{1}{2}\left\langle b_i^{H^{\frac{1}{2}}(\Gamma)},b_j^{H^{-\frac{1}{2}}(\Gamma)}\right\rangle_{\Gamma},\qquad \tilde{\mathbf{C}}_0|_{ij}=\tfrac{1}{2}\left\langle b_i^{H^{\frac{1}{2}}(\Gamma)^3},b_j^{H^{-\frac{1}{2}}(\Gamma)^3}\right\rangle_{\Gamma}, \tag{39}$$

$$\mathbf{C}_1|_{ij}=-\tfrac{1}{2}\left\langle \gamma_D b_i^{H^1(\Omega)^3}\cdot n,b_j^{H^{\frac{1}{2}}(\Gamma)}\right\rangle_{\Gamma},\qquad \tilde{\mathbf{C}}_1|_{ij}=-\tfrac{1}{2}\left\langle \gamma_D b_i^{H^1(\Omega)^{3\times 3}}n,b_j^{H^{\frac{1}{2}}(\Gamma)^3}\right\rangle_{\Gamma},$$

$$\overline{\mathbf{C}}_1|_{ij}=-\tfrac{1}{2}\left\langle \gamma_D\left(b_i^{H^1(\Omega)}I\right)n,b_j^{H^{\frac{1}{2}}(\Gamma)^3}\right\rangle_{\Gamma}. \tag{40}$$

Further, we define

$$\mathbf{J}_{\mathrm{a}}(s)|_{ij}=\left\langle b_i^{H^{-\frac{1}{2}}(\Gamma)},J_{\mathrm{a}}(s)b_j^{H^{-\frac{1}{2}}(\Gamma)}\right\rangle_{\Gamma},\qquad \mathbf{J}_{\mathrm{e}}(s)|_{ij}=\left\langle b_i^{H^{-\frac{1}{2}}(\Gamma)^3},J_{\mathrm{e}}(s)b_j^{H^{-\frac{1}{2}}(\Gamma)^3}\right\rangle_{\Gamma},$$

$$\mathbf{K}_{\mathrm{a}}(s)|_{ij}=\left\langle b_i^{H^{-\frac{1}{2}}(\Gamma)},K_{\mathrm{a}}(s)b_j^{H^{\frac{1}{2}}(\Gamma)}\right\rangle_{\Gamma},\qquad \mathbf{K}_{\mathrm{e}}(s)|_{ij}=\left\langle b_i^{H^{-\frac{1}{2}}(\Gamma)^3},K_{\mathrm{e}}(s)b_j^{H^{\frac{1}{2}}(\Gamma)^3}\right\rangle_{\Gamma},$$

$$\mathbf{W}_{\mathrm{a}}(s)|_{ij}=\left\langle b_i^{H^{\frac{1}{2}}(\Gamma)},W_{\mathrm{a}}(s)b_j^{H^{\frac{1}{2}}(\Gamma)}\right\rangle_{\Gamma},\qquad \mathbf{W}_{\mathrm{e}}(s)|_{ij}=\left\langle b_i^{H^{\frac{1}{2}}(\Gamma)^3},W_{\mathrm{e}}(s)b_j^{H^{\frac{1}{2}}(\Gamma)^3}\right\rangle_{\Gamma}$$

and thus, represent the matrices $\mathbf{B}(s)$ by

$$\mathbf{B}_{\mathrm{a}}(s) = \begin{pmatrix} s\mathbf{J}_{\mathrm{a}}(s) & \mathbf{K}_{\mathrm{a}}(s) \\ -\mathbf{K}_{\mathrm{a}}^{T}(s) & \frac{1}{s}\mathbf{W}_{\mathrm{a}}(s) \end{pmatrix}, \quad \mathbf{B}_{\mathrm{e}}(s) = \begin{pmatrix} s\mathbf{J}_{\mathrm{e}}(s) & \mathbf{K}_{\mathrm{e}}(s) \\ -\mathbf{K}_{\mathrm{e}}^{T}(s) & \frac{1}{s}\mathbf{W}_{\mathrm{e}}(s) \end{pmatrix}. \tag{41}$$

Remark 4 It should be noted that these representations of the matrices (34)–(41) and involved basis functions emphasize the similar terms $\mathbf{M}_0$, $\mathbf{M}_1$, $\mathbf{D}_0$, $\mathbf{C}_0$, $\mathbf{C}_1$, of the two problems but also point out the differences, e.g., the additional terms for elasticity given as $\mathbf{M}_2$, $\mathbf{D}_1$, $\tilde{\mathbf{C}}_0$, $\tilde{\mathbf{C}}_1$, $\overline{\mathbf{C}}_1$.

In the last step of the discretization, we present the time discretization, where the reader is again referred to the work by Banjai et al. [3] and Eberle [4].

5 Time-Discretization

In order to perform the time discretization, we take a look at the resulting systems of the FEM-BEM coupling (27)–(29) as well as (30)–(33). We want to remark that the interior problem is discretized with a leapfrog method and the exterior via a convolution quadrature method (see, e.g., [7] and [8]). More details are given in [3] for the acoustic case and in [4] for the elastic one.

We define $\triangle t$ as the time-step width and the superscript n indicates the n-th time step. We first present the full discretization for the **acoustic wave equation**

$$\mathbf{M}_1\mathbf{v}^{n+\frac{1}{2}} = \mathbf{M}_1\mathbf{v}^{n} + \frac{1}{2}\triangle t\mathbf{D}_0\mathbf{u}_{\mathrm{a}}^{n} - \frac{1}{2}\triangle t\mathbf{C}_1\boldsymbol{\psi}_{\mathrm{a}}^{n}, \tag{42}$$

$$\mathbf{M}_0\mathbf{u}_{\mathrm{a}}^{n+1} = \mathbf{M}_0\mathbf{u}_{\mathrm{a}}^{n} - \triangle t\mathbf{D}_0^{T}\mathbf{v}^{n+\frac{1}{2}} + \triangle t\mathbf{M}_0\mathbf{f}_{\mathrm{a}}^{n+\frac{1}{2}} - \triangle t\mathbf{C}_0\boldsymbol{\phi}_{\mathrm{a}}^{n+\frac{1}{2}}, \tag{43}$$

$$\mathbf{M}_1\mathbf{v}^{n+1} = \mathbf{M}_1\mathbf{v}^{n+\frac{1}{2}} + \frac{1}{2}\triangle t\mathbf{D}_0\mathbf{u}_{\mathrm{a}}^{n+1} - \frac{1}{2}\triangle t\mathbf{C}_1\boldsymbol{\psi}_{\mathrm{a}}^{n+1}, \tag{44}$$

$$\left[\mathbf{B}_{\mathrm{a}}(\partial_t^{\triangle t})\begin{pmatrix}\boldsymbol{\phi}_{\mathrm{a}} \\ \boldsymbol{\psi}_{\mathrm{a}}\end{pmatrix}\right]^{n+\frac{1}{2}} = \begin{pmatrix} \mathbf{C}_0^{T}\bar{\mathbf{u}}_{\mathrm{a}}^{n+\frac{1}{2}} \\ \mathbf{C}_1^{T}\left(\mathbf{v}^{n+\frac{1}{2}} - \alpha\triangle t^2\mathbf{M}_1^{-1}\mathbf{C}_1\dot{\boldsymbol{\psi}}_{\mathrm{a}}^{n+\frac{1}{2}}\right)\end{pmatrix}, \tag{45}$$

where $\alpha \geq 1$ is a stabilization parameter. The stabilization parameter is important for the stability analysis. As stated in [3], it turns out that the choice $\alpha = 1$ yields a stable scheme under the CFL condition $\triangle t\|\mathbf{M}_1^{-\frac{1}{2}}\mathbf{D}_0\mathbf{M}_0^{-\frac{1}{2}}\| \leq 1$.

Similarly, we introduce the time discrete version of the **elastic wave equation**

$$\mathbf{M}_2\mathbf{V}^{n+\frac{1}{2}} = \mathbf{M}_2\mathbf{V}^{n} + \frac{1}{2}\triangle t\mathbf{D}_1\mathbf{u}_{\mathrm{e}}^{n} - \frac{1}{2}\triangle t\tilde{\mathbf{C}}_1\boldsymbol{\psi}_{\mathrm{e}}^{n}, \tag{46}$$

$$\mathbf{M}_0\boldsymbol{\omega}^{n+\frac{1}{2}} = \mathbf{M}_0\boldsymbol{\omega}^{n} - \frac{1}{2}\Delta t\mathbf{D}_0^T\mathbf{u}_{\mathrm{e}}^{n} - \frac{1}{2}\Delta t\overline{\mathbf{C}}_1\boldsymbol{\phi}_{\mathrm{e}}^{n}, \tag{47}$$

$$\rho\mathbf{M}_1\mathbf{u}_{\mathrm{e}}^{n+1} = \rho\mathbf{M}_1\mathbf{u}_{\mathrm{e}}^{n} - \mu\Delta t\mathbf{D}_1^T\mathbf{V}^{n+\frac{1}{2}} + \lambda\Delta t\mathbf{D}_0\boldsymbol{\omega}^{n+\frac{1}{2}} + \rho\Delta t\mathbf{M}_1\mathbf{f}_{\mathrm{e}}^{n+\frac{1}{2}} - \Delta t\tilde{\mathbf{C}}_0\boldsymbol{\phi}_{\mathrm{e}}^{n+\frac{1}{2}},$$

$$\mathbf{M}_2\mathbf{V}^{n+1} = \mathbf{M}_2\mathbf{V}^{n+\frac{1}{2}} + \frac{1}{2}\Delta t\mathbf{D}_1\mathbf{u}_{\mathrm{e}}^{n+1} - \frac{1}{2}\Delta t\tilde{\mathbf{C}}_1\boldsymbol{\psi}_{\mathrm{e}}^{n+1}, \tag{48}$$

$$\mathbf{M}_0\boldsymbol{\omega}^{n+1} = \mathbf{M}_0\boldsymbol{\omega}^{n+\frac{1}{2}} - \frac{1}{2}\Delta t\mathbf{D}_0^T\mathbf{u}_{\mathrm{e}}^{n+1} - \frac{1}{2}\Delta t\overline{\mathbf{C}}_1\boldsymbol{\psi}_{\mathrm{e}}^{n+1}, \tag{49}$$

$$\left[\mathbf{B}_{\mathrm{e}}(\partial_t^{\Delta t})\begin{pmatrix}\boldsymbol{\phi}_{\mathrm{e}}\\ \boldsymbol{\psi}_{\mathrm{e}}\end{pmatrix}\right]^{n+\frac{1}{2}} = \begin{pmatrix}\tilde{\mathbf{C}}_0^T\bar{\mathbf{u}}_{\mathrm{e}}^{n+\frac{1}{2}}\\ \mu\tilde{\mathbf{C}}_1^T\left(\mathbf{V}^{n+\frac{1}{2}} - \alpha\Delta t^2\mathbf{M_2}^{-1}\tilde{\mathbf{C}}_1\dot{\boldsymbol{\psi}}_{\mathrm{e}}{}^{n+\frac{1}{2}}\right) + \lambda\overline{\mathbf{C}}_1^T\left(\boldsymbol{\omega}^{n+\frac{1}{2}} - \alpha\Delta t^2\mathbf{M_2}^{-1}\overline{\mathbf{C}}_1\dot{\boldsymbol{\psi}}_{\mathrm{e}}{}^{n+\frac{1}{2}}\right)\end{pmatrix}. \tag{50}$$

Remark 5 Comparing the time discretization of the acoustic system (42)–(45) and the elastic system (46)–(50), we see that the application of the two discretization methods (leapfrog method and convolution quadrature method) can be performed for both wave-type equations in a similar way. However, the numerical effort in the elastic case is significantly larger than the acoustic case due to the vector valued setup.

6 Stability Analysis

The stability analysis of the full discretization is based on the stability of the spatial semi-discretization. The estimates for the energy and the boundary functions of the spatial semi-discretization are shown for the acoustic wave equation in [3] and for the elastic wave equation in [4]. The procedure can be summarized by the following steps:

1. Start with the perturbed problem.
2. Take a look at the discrete (field) energy and its bounds.
3. Give the bounds for the boundary function.
4. Determine the error bound for the full discretization.

We want to highlight that the coercivity result (16) from Lemma 1 as well as the estimate (17) from Lemma 2 are essential for the corresponding proofs for the consideration of the discrete energy (step 2.) and the boundary function (step 3.). As such, they build the basis for the entire stability analysis.

In order to point out further connections of the two wave-type equations, we state the discrete energy. For the **acoustic case**, we get

$$E_{\mathrm{a}}^{n} = \frac{1}{2}\left(|\mathbf{u}_{\mathrm{a}}^{n}|^2 + |\mathbf{v}^{n-\frac{1}{2}}|^2 + |\mathbf{v}^{n+\frac{1}{2}}|^2\right)$$

while the energy for the **elastic wave equation** is given by

$$E_{\mathrm{e}}^{n}=\rho\frac{1}{2}|\mathbf{u}_{\mathrm{e}}^{n}|^{2}+\mu\frac{1}{4}\left(|\mathbf{V}^{n+\frac{1}{2}}|^{2}+|\mathbf{V}^{n-\frac{1}{2}}|^{2}\right)+\lambda\frac{1}{2}\left(|\boldsymbol{\omega}^{n+\frac{1}{2}}|^{2}+|\boldsymbol{\omega}^{n-\frac{1}{2}}|^{2}\right)$$

as the discrete versions of (21) and (26).

Next, we take a look at the perturbed version of the systems (42)–(45) and (46)–(50), where $\mathbf{f}_{\mathrm{a}}$, $\mathbf{g}_{\mathrm{a}}$, $\mathbf{f}_{\mathrm{e}}$, $\mathbf{g}_{\mathrm{e}}$, $\mathbf{k}_{\mathrm{e}}$, $\boldsymbol{\theta}_{\mathrm{a}}$, $\boldsymbol{\zeta}_{\mathbf{a}}$, $\boldsymbol{\theta}_{\mathrm{e}}$, and $\boldsymbol{\zeta}_{\mathbf{e}}$ are the corresponding perturbations. Thus, we are led to the perturbed discretization scheme of the **acoustic wave equation**

$$\begin{aligned}
\mathbf{M}_1\mathbf{v}^{n+\frac{1}{2}} &= \mathbf{M}_1\mathbf{v}^{n}+\frac{1}{2}\Delta t\mathbf{D}_0\mathbf{u}_{\mathrm{a}}^{n}-\frac{1}{2}\Delta t\mathbf{C}_1\boldsymbol{\psi}_{\mathrm{a}}^{n}+\frac{1}{2}\Delta t\mathbf{g}_{\mathrm{a}}^{n},\\
\mathbf{M}_0\mathbf{u}_{\mathrm{a}}^{n+1} &= \mathbf{M}_0\mathbf{u}_{\mathrm{a}}^{n}-\Delta t\mathbf{D}_0^T\mathbf{v}^{n+\frac{1}{2}}-\Delta t\mathbf{C}_0\boldsymbol{\phi}_{\mathrm{a}}^{n+\frac{1}{2}}+\Delta t\mathbf{M}_0\mathbf{f}_{\mathrm{a}}^{n+\frac{1}{2}},\\
\mathbf{M}_1\mathbf{v}^{n+1} &= \mathbf{M}_1\mathbf{v}^{n+\frac{1}{2}}+\frac{1}{2}\Delta t\mathbf{D}_0\mathbf{u}_{\mathrm{a}}^{n+1}-\frac{1}{2}\Delta t\mathbf{C}_1\boldsymbol{\psi}_{\mathrm{a}}^{n+1}+\frac{1}{2}\Delta t\mathbf{g}_{\mathrm{a}}^{n+1},
\end{aligned}$$

$$\left[\mathbf{B}_{\mathrm{a}}(\partial_t^{\Delta t})\begin{pmatrix}\boldsymbol{\phi}_{\mathrm{a}}\\ \boldsymbol{\psi}_{\mathrm{a}}\end{pmatrix}\right]^{n+\frac{1}{2}}=\begin{pmatrix}\mathbf{C}_0^T\bar{\mathbf{u}}_{\mathrm{a}}^{n+\frac{1}{2}}\\ \mathbf{C}_1^T\left(\mathbf{v}^{n+\frac{1}{2}}-\alpha\Delta t^2\mathbf{M}_1^{-1}\mathbf{C}_1\dot{\boldsymbol{\psi}}_{\mathrm{a}}^{n+\frac{1}{2}}\right)\end{pmatrix}+\begin{pmatrix}\boldsymbol{\theta}_{\mathrm{a}}^{n+\frac{1}{2}}\\ \boldsymbol{\zeta}_{\mathrm{a}}^{n+\frac{1}{2}}\end{pmatrix}.$$

The perturbed discretization scheme for the **elastic wave equation** reads as

$$\begin{aligned}
\mathbf{M}_2\mathbf{V}^{n+\frac{1}{2}} &= \mathbf{M}_2\mathbf{V}^{n}+\frac{1}{2}\Delta t\mathbf{D}_1\mathbf{u}_{\mathrm{e}}^{n}-\frac{1}{2}\Delta t\tilde{\mathbf{C}}_1\boldsymbol{\psi}_{\mathrm{e}}^{n}+\frac{1}{2}\Delta t\mathbf{g}_{\mathrm{e}}^{n},\\
\mathbf{M}_0\boldsymbol{\omega}^{n+\frac{1}{2}} &= \mathbf{M}_0\boldsymbol{\omega}^{n}-\frac{1}{2}\Delta t\mathbf{D}_0^T\mathbf{u}_{\mathrm{e}}^{n}-\frac{1}{2}\Delta t\overline{\mathbf{C}}_1\boldsymbol{\phi}_{\mathrm{e}}^{n}+\frac{1}{2}\Delta t\mathbf{k}_{\mathrm{e}}^{n},\\
\rho\mathbf{M}_1\mathbf{u}_{\mathrm{e}}^{n+1} &= \rho\mathbf{M}_1\mathbf{u}_{\mathrm{e}}^{n}-\mu\Delta t\mathbf{D}_1^T\mathbf{V}^{n+\frac{1}{2}}+\lambda\Delta t\mathbf{D}_0\boldsymbol{\omega}^{n+\frac{1}{2}}-\Delta t\tilde{\mathbf{C}}_0\boldsymbol{\phi}_{\mathrm{e}}^{n+\frac{1}{2}}+\rho\Delta t\mathbf{M}_1\mathbf{f}_{\mathrm{e}}^{n+\frac{1}{2}},\\
\mathbf{M}_2\mathbf{V}^{n+1} &= \mathbf{M}_2\mathbf{V}^{n+\frac{1}{2}}+\frac{1}{2}\Delta t\mathbf{D}_1\mathbf{u}_{\mathrm{e}}^{n+1}-\frac{1}{2}\Delta t\tilde{\mathbf{C}}_1\boldsymbol{\psi}_{\mathrm{e}}^{n+1}+\frac{1}{2}\Delta t\mathbf{g}_{\mathrm{e}}^{n+1},\\
\mathbf{M}_0\boldsymbol{\omega}^{n+1} &= \mathbf{M}_0\boldsymbol{\omega}^{n+\frac{1}{2}}-\frac{1}{2}\Delta t\mathbf{D}_0^T\mathbf{u}_{\mathrm{e}}^{n+1}-\frac{1}{2}\Delta t\overline{\mathbf{C}}_1\boldsymbol{\psi}_{\mathrm{e}}^{n+1}+\frac{1}{2}\Delta t\mathbf{k}_{\mathrm{e}}^{n+1},
\end{aligned}$$

$$\left[\mathbf{B}_{\mathrm{e}}(\partial_t^{\Delta}t)\begin{pmatrix}\boldsymbol{\phi}_{\mathrm{e}}\\ \boldsymbol{\psi}_{\mathrm{e}}\end{pmatrix}\right]^{n+\frac{1}{2}}=\begin{pmatrix}\tilde{\mathbf{C}}_0^T\bar{\mathbf{u}}_{\mathrm{e}}^{n+\frac{1}{2}}\\ \mu\tilde{\mathbf{C}}_1^T\left(\mathbf{V}^{n+\frac{1}{2}}-\alpha\Delta t^2\mathbf{M}_2{}^{-1}\tilde{\mathbf{C}}_1\dot{\boldsymbol{\psi}}_{\mathrm{e}}^{n+\frac{1}{2}}\right)+\lambda\overline{\mathbf{C}}_1^T\left(\boldsymbol{\omega}^{n+\frac{1}{2}}-\alpha\Delta t^2\mathbf{M}_2{}^{-1}\overline{\mathbf{C}}_1\dot{\boldsymbol{\psi}}_{\mathrm{e}}^{n+\frac{1}{2}}\right)\end{pmatrix}+\begin{pmatrix}\boldsymbol{\theta}_{\mathrm{e}}^{n+\frac{1}{2}}\\ \boldsymbol{\zeta}_{\mathrm{e}}^{n+\frac{1}{2}}\end{pmatrix}.$$

Based on this, we compare the bounds of the energy given in the following lemmas.

Lemma 3 (Lemma 8.1 from [3]) *The discrete energy* E_{a}^n *is bounded at* $t = n\Delta t$ *by*

$$E_{\mathrm{a}}^n \leq C_{\mathrm{a}}\Big(E_{\mathrm{a}}^0 + \frac{t}{2}\Delta t \sum_{j=0}^{n} |\mathbf{f}_{\mathrm{a}}^{j+\frac{1}{2}}|^2 + |\mathbf{g}_{\mathrm{a}}^j|^2$$

$$+ \max(t^2, t^6)\Delta t \sum_{j=0}^{n} \Big(|(\partial_t^{\Delta t})^2 \boldsymbol{\theta}_{\mathrm{a}}^{j+\frac{1}{2}}|^2 + |(\partial_t^{\Delta t})^2 \boldsymbol{\zeta}_{\mathrm{a}}^{j+\frac{1}{2}}|^2 \Big) \Big),$$

where C_{a} *is independent of* h, Δt, *and* n.

Lemma 4 (Lemma 12 from [4]) *The discrete energy* E_{e} *is bounded at* $t = n\Delta t$ *by*

$$E_{\mathrm{e}}^n \leq C_{\mathrm{e}}\Big(E_{\mathrm{e}}^0 + \frac{t}{2}\Delta t \sum_{j=0}^{n} |\rho \mathbf{f}_{\mathrm{e}}^{j+\frac{1}{2}}|^2 + |\mathbf{g}_{\mathrm{e}}^j|^2 + |\mathbf{k}_{\mathrm{e}}^j|^2$$

$$+ \max(t^2, t^6)\Delta t \sum_{j=0}^{n} \Big(|(\partial_t^{\Delta t})^2 \boldsymbol{\theta}_{\mathrm{e}}^{j+\frac{1}{2}}|^2 + |(\partial_t^{\Delta t})^2 \boldsymbol{\zeta}_{\mathrm{e}}^{j+\frac{1}{2}}|^2 \Big) \Big),$$

where C_{e} *is independent of* h, Δt, *and* n.

Remark 6 As a result, we found out that the bounding terms C and $\max(t^2, t^6)\Delta t$ of the discrete energy do not depend on the wave-type equation itself. Thus, the dependence of the estimate on t and Δt is the same for both wave equations.

Summarizing the discrete stability results and the consistency errors from [3] and [4], we finally obtain the error bounds for the full discretization.

Theorem 1 (Theorem 9.1 from [3], Theorem 3 from [4]) *Assume that the initial values and the inhomogeneity of the wave equation have their support in* Ω. *Let the initial values for the semi-discretization be chosen as* $u_{\mathrm{a},h}(0) = P_h u_{\mathrm{a}}(0)$, $v_h(0) = P_h v_h(0)$, $u_{\mathrm{e},h}(0) = P_h u_{\mathrm{e}}(0)$, $V_h(0) = P_h V(0)$, $\omega_h(0) = P_h \omega(0)$, *where* P_h *denotes the* L_2*-orthogonal projection onto the finite element space and* $\phi_{\mathrm{a},h}(0) = P_h \phi_{\mathrm{a}}(0)$, $\phi_{\mathrm{e},h}(0) = P_h \phi_{\mathrm{e}}(0)$, $\psi_{\mathrm{a},h}(0) = P_h \psi_{\mathrm{a}}(0)$, *and* $\psi_{\mathrm{e},h}(0) = P_h \psi_{\mathrm{e}}(0)$ *the corresponding projection onto the boundary element space. If the solution of the wave equation is sufficiently smooth, then the error of the semi-discretization as well as the full discretization under the CFL condition* $\Delta t \|\mathbf{M}_1^{-\frac{1}{2}} \mathbf{D}_0 \mathbf{M}_0^{-\frac{1}{2}}\| \leq 1$ *(acoustic) and* $\frac{\Delta t}{\sqrt{\rho}} \left\| \left(\sqrt{\mu} \mathbf{M}_2^{-\frac{1}{2}} \mathbf{D}_1 \mathbf{M}_1^{-\frac{1}{2}}, -\sqrt{\lambda} \mathbf{M}_0^{-\frac{1}{2}} \mathbf{D}_0^T \mathbf{M}_1^{-\frac{1}{2}} \right)^T \right\| \leq 1$ *(elastic) and the stability* *parameter α is bounded at $t = n\Delta t$ for the* ***acoustic wave equation*** *by*

$$\|u_{\mathrm{a},h}^n - u_{\mathrm{a}}(t)\|_{L_2(\Omega)} + \|v_h^n - v(t)\|_{L_2(\Omega)^3} +$$
$$+ \left(\Delta t \sum_{j=0}^{n-1} \left\| \phi_{\mathrm{a},h}^{j+\frac{1}{2}} - \phi_{\mathrm{a}}\left(t_{j+\frac{1}{2}}\right) \right\|_{H^{-\frac{1}{2}}(\Gamma)}^2 + \left\| \bar{\psi}_{\mathrm{a},h}^{j+\frac{1}{2}} - \psi_{\mathrm{a}}\left(t_{j+\frac{1}{2}}\right) \right\|_{H^{\frac{1}{2}}(\Gamma)}^2 \right)^{\frac{1}{2}}$$
$$\leq C_{\mathrm{a}}(t)(h + (\Delta t)^2),$$

and for the ***elastic wave equation*** *by*

$$\rho\|u_{\mathrm{e},h}^n - u_{\mathrm{e}}(t)(t)\|_{L_2(\Omega)^3} + \frac{1}{2}\mu\|V_h^n - V(t)\|_{L_2(\Omega)^{3\times 3}} + \lambda\|\omega_h^n - \omega(t)\|_{L_2(\Omega)}$$
$$+ \left(\Delta t \sum_{j=0}^{n-1} \left\| \phi_{\mathrm{e},h}^{j+\frac{1}{2}} - \phi_{\mathrm{e}}\left(t_{j+\frac{1}{2}}\right) \right\|_{H^{-\frac{1}{2}}(\Gamma)^3}^2 + \left\| \bar{\psi}_{\mathrm{e},h}^{j+\frac{1}{2}} - \psi_{\mathrm{e}}\left(t_{j+\frac{1}{2}}\right) \right\|_{H^{\frac{1}{2}}(\Gamma)^3}^2 \right)^{\frac{1}{2}}$$
$$\leq C_{\mathrm{e}}(t)(h + (\Delta t)^2),$$

where $C_{\mathrm{a}}(t)$ and $C_{\mathrm{e}}(t)$ grow at most polynomially with t.

Remark 7 All in all, Theorem 1 shows an asymptotically optimal $O\left(h + (\Delta t)^2\right)$ convergence for the full discretized acoustic as well as elastic wave equation. Thus, both problems have the same convergence order.

7 Conclusion

In this survey we gave an overview of the FEM-BEM coupling of a transmission problem and analyzed the connections and differences of the acoustic and elastic wave equation. One of the key issues for the stability analysis was the Calderón operator. In addition, we focused on the space and time discretization of the problems and finally stated the convergence result. In addition, we want to mention [6], where the FEM-BEM coupling for the Maxwell's equations was examined. All in all, these works build the basis for further considerations, e.g., the adoption to other wave-type equations such as the thermoelastic wave equation [2] or the implementation of the FEM-BEM coupling for the elastodynamic wave equation [5].

Acknowledgement The author would like to thank the anonymous reviewer for the helpful remarks and suggestions for the improvement of the paper.

References

1. Abboud, T., Joly, P., Rodriguez, J., Terrasse, I.: Coupling discontinuous Galerkin methods and retarded potentials for transient wave propagation on unbounded domains. J. Comput. Phys. **230**(15), 5877–5907 (2011)
2. Augustin, M., Eberle, S.: FEM-BEM coupling for the thermoelastic wave equation with transparent boundary conditions in 3d. (submitted)
3. Banjai, L., Lubich, Ch., Sayas, F.J.: Stable numerical coupling of exterior and interior problems for the wave equation. Numer. Math. **129**, 611–646 (2015)
4. Eberle, S.: The elastic wave equation and the stable numerical coupling of its interior and exterior problems. Z. Angew. Math. Mech. **98**, 1261–1283 (2018)
5. Eberle, S.: An implementation and numerical experiments of the FEM-BEM coupling for the elastodynamic wave equation in 3d. Z. Angew. Math. Mech. **99**(12), (2019)
6. Kovács, B., Lubich, Ch.: Stable and convergent fully discrete interior-exterior coupling of Maxwell's equations. Numer. Math. **137**, 91–117 (2017)
7. Lubich, Ch.: Convolution quadrature and discretized operational calculus. I. Numer. Math. **52**(2), 129–145 (1988)
8. Lubich, Ch.: Convolution quadrature and discretized operational calculus. II. Numer. Math. **52**(2), 413–425 (1988)

On Hyperbolic Initial-Boundary Value Problems with a Strictly Dissipative Boundary Condition

Matthias Eller

Abstract The well-posedness of the initial-boundary value problems for symmetric hyperbolic systems with strictly dissipative boundary conditions is proved. The regularity assumptions on the coefficients of the differential operator and the boundary condition as well as the boundary itself are quite minimal. Characterizations of strictly dissipative boundary operators are given and the example of Maxwell's equations is discussed.

1 Introduction and Main Result

Consider the symmetric hyperbolic system of first-order

$$Pu := A^0(t,x)\frac{\partial u}{\partial t} + \sum_{j=1}^{d} A^j(t,x)\frac{\partial u}{\partial x_j} + D(t,x)u = f(t,x) \quad \text{in } Q := (0,T)\times\Omega \tag{1}$$

subject to the initial data

$$u(0,x) = u^0(x)\,, \quad x \in \Omega \tag{2}$$

and the boundary condition

$$B(t,x)u(t,x) = g(t,x) \quad \text{in } \Sigma := (0,T)\times\partial\Omega\,. \tag{3}$$

The coefficients A^j, $j = 0, 1, \ldots, d$ are $N \times N$ Hermitian matrix functions and A^0 is uniformly positive definite. The set Ω is an open, bounded, and connected subset of $\mathbb{R}^d$ with a Lipschitz boundary $\Gamma = \partial\Omega$. The elements of the matrices A^j are in $W^{1,\infty}(Q)$ and the elements of D are in $L^\infty(Q)$. Furthermore, B is a

M. Eller (✉)
Georgetown University, Washington, DC, USA
e-mail: mme4@georgetown.edu

W. Dörfler et al. (eds.), *Mathematics of Wave Phenomena*, Trends in Mathematics,
https://doi.org/10.1007/978-3-030-47174-3_8

$p \times N$ matrix with elements in $L^\infty(\Sigma)$ and rank $B \le p \le N$. It is assumed that $g : (t, x) \to \operatorname{Im} B(t, x) \subset \mathbb{C}^p$ almost everywhere and that T is a positive real or infinity. The matrix

$$A(t, x) = \sum_{j=1}^{d} A^j(t, x)\nu_j(x) ,$$

is called the boundary matrix, where ν is the exterior unit normal vector field along $\Gamma = \partial\Omega$. This is a matrix with L^∞ elements and it is assumed to be of uniformly constant signature: The number of positive eigenvalues N_+ and the number of negative eigenvalues N_- is independent of $(t, x) \in \Sigma$ and the non-zero eigenvalues are bounded away from zero. Due to these assumptions, we can always write $A = \Lambda_+^2 - \Lambda_-^2$ where Λ_+ and Λ_- are Hermitian semipositive matrices a.e. $(t, x) \in \Sigma$. If $N_+ = 0$ ($N_- = 0$), then $A = -\Lambda_-^2$ ($A = \Lambda_+^2$).

Definition 1 A boundary operator B is strictly dissipative for the system P, if rank $B = N_-$, $N(A) \subset N(B)$,

$$|Bz| \gtrsim |z| \text{ for all } z \in N(B)^\perp \quad \text{and} \quad \langle Aw, w\rangle \gtrsim |Aw|^2 \text{ for all } w \in N(B) ,$$

a.e. $(t, x) \in \Sigma$.

Here $\langle\cdot, \cdot\rangle$ is the standard scalar product in $\mathbb{C}^N$, $|\cdot|$ is the Euclidean norm in $\mathbb{C}^N$, $N(B)$ is the null space of the matrix B, and we write $a \lesssim b$ whenever there exists a constant $C > 0$ such that $a \le Cb$. Note that a similar definition is given in Definition 3.3 [1]. However, an extra degree of carefulness is required since we allow for characteristic boundaries, that is, the null space of the boundary matrix may not be trivial. The example $B = \Lambda_-$ shows that every symmetric hyperbolic system in a given region Ω admits strictly dissipative boundary operators. Strictly dissipative boundary operators can be characterized by an inequality.

Lemma 1 *A boundary operator B is strictly dissipative for the system P if and only if B has rank N_-, $N(A) \subset N(B)$, and the inequality*

$$\langle Av, v\rangle \gtrsim |Av|^2 - C|Bv|^2 \tag{4}$$

holds for all $v \in \mathbb{C}^N$, a.e. $(t, x) \in \Sigma$ with a positive constant C independent of v, t, x.

Proof Suppose that B is strictly dissipative. Let $v = w + z$ where $z \in N(B)$ and $w \in N(B)^\perp$. Then, using the Cauchy-Schwarz inequality, we compute

$$\begin{aligned}\langle Av, v\rangle &= \langle Az, z\rangle + 2\langle Az, w\rangle + \langle Aw, w\rangle \gtrsim |Az|^2 - \frac{1}{2}|Az|^2 - C_1|w|^2 \\ &= \frac{1}{2}[|Az|^2 + |Aw|^2] - C_2|w|^2 \ge \frac{1}{4}|Av|^2 - C_3|Bw|^2 ,\end{aligned}$$

which results in the desired inequality.

Conversely, if the inequality of the lemma is satisfied, we immediately infer that

$$\langle Az, z\rangle \gtrsim |Az|^2 \quad \text{for all } z \in N(B)\ .$$

Recall that $\dim N(B) = N - N_-$ and that $N(B)^\perp \cap N(A) = \{0\}$. Hence, $w \in N(B)^\perp$ implies $|Aw| \gtrsim |w|$. Furthermore, since $\langle Av, v\rangle = |\Lambda_+ v|^2 - |\Lambda_- v|^2$, we know that $N(B) \subset \{v \in \mathbb{C}^N : |\Lambda_+ v|^2 \geq |\Lambda_- v|^2\}$ which is a conic set. A dimensional argument gives $N(B)^\perp \subset \{v \in \mathbb{C}^N : |\Lambda_+ v|^2 < |\Lambda_- v|^2\} \cup \{0\}$ and thus $\langle Aw, w\rangle \lesssim -|w|^2$ for all $w \in N(B)^\perp$. The inequality $|Bw| \gtrsim |w|$ for all $w \in N(B)^\perp$ follows then from (4). □

The main result of this note is the following theorem.

Theorem 1 *The initial-boundary value problem* $Pu = f \in L^2(Q, \mathbb{C}^N)$, $u^0 = u \in L^2(\Omega, \mathbb{C}^N)$, $Bu = g \in L^2(\Sigma, \mathbb{C}^p)$, *where* B *is strictly dissipative, is well-posed in* L^2. *More precisely, there exists a unique weak solution* $u \in C([0, T], L^2(\Omega, \mathbb{C}^N))$ *such that* $Au\Big|_\Sigma \in L^2(\Sigma, \mathbb{C}^N)$ *and there exists a constant* γ_0 *such that for* $\gamma \geq \gamma_0$

$$\|e^{-\gamma T}u(T)\|^2_\Omega + \gamma\|e^{-\gamma t}u\|^2_Q + \|e^{-\gamma t}Au\|^2_\Sigma \lesssim \|u^0\|^2_\Omega + \frac{1}{\gamma}\|e^{-\gamma t}f\|^2_Q + \|e^{-\gamma t}g\|^2_\Sigma\ . \tag{5}$$

All norms are in L^2, the lower index indicates the set of integration. To the best of our knowledge, this theorem has not been proved anywhere in the literature. The work by Lax and Phillips [9] comes already rather close, but there is no discussion of hyperbolic systems and the boundary condition is only homogeneous. Still, the notion of semi-strong solutions is taken from this work [9, Section 2]. Even the treatise of first-order systems by Benzonia-Gavage and Serre [1] does not contain this result. The result is stated in the context of Maxwell's equations in [2, Proposition 1.1] without a proof. Lately there has been renewed interest in this result, mainly due to works on quasi-linear Maxwell equations, see for example [14, Section 3].

The theorem above is somehow special since it puts rather minimal assumptions on the regularity of the coefficients of the operator P, the boundary operator B, and the regularity of the boundary Γ. Note that the Lipschitz boundary allows for corners and the essential boundedness of the boundary operator gives a lot of flexibility which may be useful in applied problems.

Most of the literature on hyperbolic boundary problems work with strictly hyperbolic systems and a boundary condition which satisfies the Kreiss-Sakamoto condition. This condition is from the theoretical point very satisfying, since it is also necessary for well-posedness. Furthermore, Kreiss-Sakamoto conditions can be non-local. The proof requires the use of pseudo-differential operators. Hence, the operator, the boundary condition, and the boundary have to have more regularity. The classical work which explains all this in the smooth C^∞-case is the book by Charazain and Piriou [3, Chapter 7]. Later Métivier relaxed the assumptions on the coefficients and the boundary operator to $W^{1,\infty}$, at least in the case of zero initial data [11]. For a detailed study of the initial-boundary value problem and the most recent results in the context of strictly hyperbolic or constantly hyperbolic system we refer to [12].

However some important systems are neither strictly hyperbolic nor constantly hyperbolic. Among them are Maxwell's equations and the equations of elastodynamics, at least in the anisotropic cases. Furthermore, rather recently Métivier has constructed an example of a symmetric hyperbolic system with variable coefficients and a boundary operator which satisfies the Kreiss-Sakamoto condition where L^2 well-posedness fails [12]. This example is somewhat surprising since in the case of constant coefficients L^2 well-posedness always holds for symmetric hyperbolic systems with a Kreiss-Sakamoto boundary operator [7]. Hence, we believe that strictly dissipative boundary conditions are of interest due to the fact that L^2 well-posedness always holds and that it can be established under minimal regularity assumptions.

The proof of Theorem 1 will follow the approach pioneered by Friedrichs [6] for the pure Cauchy problem with the adjustment to boundary problems made by Lax and Phillips [9]. The existence of the weak solution follows by a duality argument from the unique solvability of a dual problem. Furthermore, Friedrichs proved that each weak solution is also a strong solution, that is, the limit of a sequence of functions of higher regularity. This is done by regularizing the weak solution by means of mollifiers. In the case of an initial-boundary value problem the regularization can be performed only in directions tangential to the boundary. If the boundary matrix a has non-zero determinant, then also the derivatives in direction normal to the boundary will improve. However, if the boundary is characteristic, that is the boundary matrix has zero eigenvalues, then the notion of strong solutions has to be adjusted to semi-strong solutions [9].

Since a duality argument is used to establish the weak solution for the initial-boundary value problem (1)–(3), we need to show that there exists an adjoint problem with good properties. The following proposition guarantees that corresponding to the strictly dissipative boundary condition B there exists always a strictly dissipative boundary condition F for the operator

$$P^* = -A^0\frac{\partial}{\partial t} - \sum_{j=1}^{d} A^j\frac{\partial}{\partial x_j} + D^H - \frac{\partial A^0}{\partial t} - \sum_{j=1}^{d}\frac{\partial A^j}{\partial x_j},$$

the formal adjoint of P. Here and henceforth X^H denotes the Hermitian transpose of the matrix X.

Proposition 1 *Suppose that $B \in L^\infty(\Sigma, \mathbb{C}^{p\times N})$ where $p \geq N_-$ and B is strictly dissipative in the sense of Definition 1. Then there exist matrices $E \in L^\infty(\Sigma, \mathbb{C}^{p\times N})$ and $F, G \in L^\infty(\Sigma, \mathbb{C}^{N_+\times N})$ such that $A = -B^H E + G^H F$ and $N(E), N(G), N(F)$ all contain $N(A)$, a.e. $(t,x) \in \Sigma$. Furthermore, the matrix F satisfies*

$$|Fz| \gtrsim |z| \text{ for all } z \in N(F)^\perp \quad \text{and} \quad -\langle Az, z\rangle \gtrsim |Az|^2 \text{ for all } z \in N(F),$$

a.e. $(t,x) \in \Sigma$.

The proof is relegated to the appendix. With this result in hand we can give a precise definition of a weak solution.

Definition 2 A function $u \in L^2(Q, \mathbb{C}^N)$ is a weak solution to the initial-boundary value problem (1)–(3) if for all $v \in H^1(Q, \mathbb{C}^N)$ satisfying $Fv = 0$ on Σ and $v(T, x) = 0$ for all $x \in \Omega$ the identity

$$\int_Q \langle u, P^* v\rangle \, dxdt = \int_Q \langle f, v\rangle \, dtdx + \int_\Omega \langle A^0 u^0, v\rangle\Big|_{t=0} dx + \int_\Sigma \langle g, Ev\rangle \, d\Sigma$$

holds. Here F is the adjoint boundary operator introduced in the proposition above where the matrix E has been introduced as well.

Using integration by parts one can show that any weak solution of regularity C^1 is in fact a classical solution to the initial-boundary value problem.

The Proof of Theorem 1 will be discussed in the next section. Section 3 considers an example and in Sect. 4 we give another equivalent formulation of strictly dissipative boundary conditions.

2 Proof of Theorem 1

1. The a priori Estimate Let $u \in H^1(Q, \mathbb{C}^N)$ and set $v = e^{-\gamma t}$ for some $\gamma > 0$. Then

$$\frac{\partial}{\partial t}(e^{-\gamma t} u) = -\gamma v + \frac{\partial v}{\partial t} \quad \text{and} \quad \gamma \langle A^0 v, v\rangle + \langle Pv, v\rangle = \langle e^{-\gamma t} Pu, v\rangle$$

and using the symmetry of the coefficient matrices A^j for $j = 1, 2, \ldots, N$ and integration by parts gives

$$\begin{aligned}
\int_Q \langle Pv, v\rangle \, dtdx =& \frac{1}{2}\int_Q \frac{\partial}{\partial t}\langle A^0 v, v\rangle \, dtdx - \frac{1}{2}\int_Q \langle (\partial_t A^0) v, v\rangle \, dtdx \\
&+ \frac{1}{2}\int_Q \sum_{j=1}^d \frac{\partial}{\partial x_j}\langle A^j v, v\rangle \, dxdt \\
&- \frac{1}{2}\sum_{j=1}^d \int_Q \langle (\partial_j A^j) v, v\rangle \, dxdt + \int_Q \langle Dv, v\rangle \, dxdt \\
=& \frac{1}{2}\int_\Omega \langle A^0 v, v\rangle\Big|_{t=T} dx - \frac{1}{2}\int_\Omega \langle A^0 v, v\rangle\Big|_{t=0} dx \\
&+ \frac{1}{2}\int_\Sigma \langle Av, v\rangle \, d\Sigma + \int_Q \langle \tilde{D} v, v\rangle \, dxdt \, ,
\end{aligned}$$

where

$$\tilde{D} = D - \frac{1}{2}\partial_t A^0 - \frac{1}{2}\sum_{j=1}^{d} \partial_j A^j \quad \text{and} \quad \partial_j = \frac{\partial}{\partial x_j} .$$

Note that via the Sobolev embedding theorem $v \in C([0, T], L^2(\Omega, \mathbb{C}^N))$. Hence, we have $v(t, \cdot) \in L^2(\Omega, \mathbb{C}^N)$ for all $t \in [0, T]$ and the integrals over Ω are well defined. Combining this formula with the above formula for $\langle e^{-\gamma t} Pu, v\rangle$ gives

$$\begin{aligned} &\frac{1}{2}\int_\Omega \langle A^0 v, v\rangle(T)\, dx + \gamma \int_Q \langle A^0 v, v\rangle\, dtdx + \frac{1}{2}\int_\Sigma \langle Av, v\rangle\, d\Sigma \\ &\qquad = \frac{1}{2}\int_\Omega \langle A^0 v, v\rangle(0)\, dx - \int_Q \langle \tilde{D} v, v\rangle\, dxdt + \int_Q \langle e^{-\gamma t} Pu, v\rangle\, dtdx \end{aligned}$$

This identity will be turned into an inequality if we make the following estimates

$$\Re \int_Q \langle \tilde{D} v, v\rangle\, dxdt \lesssim \|v\|_Q^2 , \qquad \int_Q \langle A^0 v, v\rangle\, d\Sigma \gtrsim \|v\|_Q^2 ,$$

$$\int_\Omega \langle A^0 v, v\rangle(T)\, dx \gtrsim \|v(T)\|_\Omega^2 , \qquad \int_\Omega \langle A^0 v, v\rangle(0)\, dx \lesssim \|v(0)\|_\Omega^2 ,$$

and

$$\Re \int_Q \langle e^{-\gamma t} Pu, v\rangle\, dtdx \le \alpha\gamma \|v\|_Q^2 + \frac{1}{4\alpha\gamma}\|e^{-\gamma t} Pu\|_Q^2 \qquad \text{for all } \alpha > 0 .$$

This leads to

$$\|v(T)\|_\Omega^2 + \gamma\|v\|_Q^2 + \int_\Sigma \langle Av, v\rangle\, d\Sigma \lesssim \|v(0)\|_\Omega^2 + \frac{1}{\gamma}\|e^{-\gamma t} Pu\|_Q^2 ,$$

for γ sufficiently large. Now the boundary operator comes into play. Applying the inequality of Lemma 1 to the last term of the left-hand side leads to

$$\|e^{-\gamma T} u(T)\|_\Omega^2 + \gamma\|e^{-\gamma t} u\|_Q^2 + \|e^{-\gamma t} Au\|_\Sigma^2 \lesssim \|u(0)\|_\Omega^2 + \frac{1}{\gamma}\|e^{-\gamma t} Pu\|_Q^2 + \|e^{-\gamma t} Bu\|_\Sigma^2 . \tag{6}$$

It should be pointed out that the assumption $v \in H^1(Q, \mathbb{C}^N)$ can be slightly relaxed to functions $v \in L^2(Q, \mathbb{C}^N)$ satisfying $\partial v/\partial t$, $A^j \partial v/\partial x_j \in L^2(Q, \mathbb{C}^N)$ for $j = 1, 2, \ldots, d$.

2. Existence of a Weak Solution with Zero Vanishing Initial Data The existence of the solution is now proved by means of duality: The uniqueness of the solution of the dual problem will imply the existence of a solution to the primal problem.

Let $P_\gamma = P + \gamma A^0$. Then $e^{-\gamma t} Pu = P_\gamma v$ and the formally adjoint operator is

$$P_\gamma^* = -A^0 \frac{\partial}{\partial t} + \gamma A^0 - \frac{\partial A^0}{\partial t} - \sum_{j=1}^{d} \frac{\partial A^j}{\partial x_j} + D^H .$$

The estimate (6) can be written in the form of an a priori estimate for the operator P_γ, that is

$$\|v(T)\|_\Omega^2 + \gamma \|v\|_Q^2 + \|Av\|_\Sigma^2 \lesssim \|v(0)\|_\Omega^2 + \frac{1}{\gamma} \|P_\gamma v\|_Q^2 + \|Bv\|_\Sigma^2 \,, \text{ for all } v \in H^1(Q, \mathbb{C}^N) .$$

By Proposition 1 there exists a strictly dissipative boundary operator F for the adjoint operator P_γ^* going backwards in time. Hence, as in the first part of the proof one establishes an a priori estimate for P_γ^*,

$$\|w(0)\|_\Omega^2 + \gamma \|w\|_Q^2 + \|Aw\|_\Sigma^2 \lesssim \|w(T)\|_\Omega^2 + \frac{1}{\gamma} \|P_\gamma^* w\|_Q^2 + \|Fw\|_\Sigma^2 \,, \text{ for all } w \in H^1(Q, \mathbb{C}^N) . \tag{7}$$

Define now the function spaces

$$Y = \{w \in H^1(Q, \mathbb{C}^N) \,:\, w(T) = 0, \; Fw\Big|_\Sigma = 0\} \quad \text{and} \quad Z = \{P_\gamma^* w \,:\, w \in Y\} ,$$

where one notes that Z is a subspace of $L^2(Q, \mathbb{C}^N)$. From Proposition 1 we infer the estimate $\|Ew\|_\Sigma \lesssim \|Aw\|_\Sigma$ for all $w \in Y$. For given $f \in L^2(Q, \mathbb{C}^N)$ and $g \in L^2(\Sigma, \mathbb{C}^p)$ the anti-linear functional on Y

$$\int_Q \langle e^{-\gamma t} f, w\rangle \, dt dx + \int_\Sigma \langle e^{-\gamma t} g, Ew\rangle \, d\Sigma ,$$

satisfies because of (7) the estimate

$$\left| \int_Q \langle e^{-\gamma t} f, w\rangle \, dt dx + \int_\Sigma \langle e^{-\gamma t} g, Ew\rangle \, d\Sigma \right| \\ \leq \|e^{-\gamma t} f\|_Q \|w\|_Q + \|e^{-\gamma t} g\|_\Sigma \|Ew\|_\Sigma \lesssim \|P_\gamma^* w\|_Q .$$

Hence, it is a bounded anti-linear functional on Z. By the Hahn-Banach theorem this functional can be extended to a bounded anti-linear functional on $L^2(Q, \mathbb{C}^N)$

and by the Riesz representation theorem, there exists a unique $v \in L^2(Q, \mathbb{C}^N)$ such that

$$\int_Q \langle v, P_\gamma^* w\rangle \, dtdx = \int_Q \langle e^{-\gamma t} f, w\rangle \, dtdx + \int_\Sigma \langle e^{-\gamma t} g, Ew\rangle \, d\Sigma \qquad \text{for all } w \in Y . \tag{8}$$

This proves that v is a weak solution to the initial-boundary value problem $P_\gamma v = e^{-\gamma t} f$ in Q, $v(0, \cdot) = 0$, $Bv = e^{-\gamma t} g$ in the sense of Definition 2.

The vanishing initial data allow us to extend the initial-boundary value problem as a pure boundary problem to the infinite space-time cylinder $(-\infty, T) \times \Omega$ by setting $v \equiv 0$, $f \equiv 0$, $g \equiv 0$ for $t < 0$, and $A^j(t, x) = A^j(0, x)$, $D \equiv 0$ for $t < 0$, $B(t, x) = \Lambda_-(0, x)$ for $t < 0$. Abusing notation we will denote all these extensions with the same letters.

As a matter of fact, this whole construction works also in the case $T = \infty$. Furthermore, in the case of finite T, one can easily extend the problem to the unbounded time interval by setting the data f and g zero for $t > T$. Similarly, the operator P_γ and the boundary operator can be defined for all $t > T$ by setting $A^j(t, x) = A^j(T, x)$, $D(t, x) \equiv 0$, $B(t, x) = \Lambda_-(T, x)$ for $t > T$.

3. Every Weak Solution Constructed Above is a Semi-strong Solution Here we show that the weak solution constructed above is actually the strong limit of solutions of higher regularity. This is done using Friedrichs's mollifiers and local coordinates. We will assume that the problem has been extended to all $t \in \mathbb{R}$ as outlined above.

There exists a finite cover of Ω by open sets $K_0, K_1, \ldots, K_m$ and a partition of unity $\{\theta_l\}_{l=0}^m$ subordinated to this partition of unity, that is $\operatorname{supp} \theta_l \subset K_l$ and $\sum_{l=0}^m \theta_l(x) = 1$ for all $x \in \Omega$. Furthermore, $K_0 \cap \partial\Omega = \emptyset$, $K_l \cap \partial\Omega \neq \emptyset$ for $l = 1, 2, \ldots, m$, and in each set K_l one can choose local coordinates (denoted for convenience also by $x_1, x_2, \ldots, x_d$) in a way such that $K_l \cap \Omega \cap \{x_d > 0\}$ and $K_l \cap \partial\Omega \subset \{x_d = 0\}$ for $l = 1, 2, \ldots, d$. No change of coordinates is needed in K_0.

Changing the coordinates will also effect the coefficient matrices of the symmetric hyperbolic system. However, symmetry of the coefficient matrices is preserved. In each set K_l we are dealing with the operator

$$A_l^0 \frac{\partial}{\partial t} + \gamma A_l^0 + \sum_{j=1}^d A_l^j \frac{\partial}{\partial x_j} + D_l \,, \qquad l = 1, 2, \ldots, m \,.$$

Note that the boundary matrix is $A_l = A_l^d$. Furthermore, the sets $K_1, \ldots, K_m$ are chosen sufficiently small such that in each K_l the boundary matrix A can be block diagonalized by a unitary matrix X_l with entries in $W^{1,\infty}$, that is

$$X_l^H A_l X_l = \begin{bmatrix} 0 & 0 & 0 \\ 0 & A_{l,+} & 0 \\ 0 & 0 & -A_{l,-} \end{bmatrix} ,$$

where $A_{l,\pm}$ are uniformly positive definite $N_\pm \times N_\pm$ matrices. This block-diagonalization procedure is a non-trivial fact and goes back to a theorem by Kato [8, pp. 32–34], see also Lemma 2.3 [4]. The dependent variable is changed accordingly to $v_l = X_l^H \theta_l v$ for $l = 1, \ldots, m$. Hence, in each coordinate chart $K_1, \ldots, K_m$, the symmetric hyperbolic system has a block-diagonal boundary matrix. In other words, the first $N - N_+ - N_-$ components of the dependent variable are not differentiated in x_d-direction. The localization of the operator P_γ with both, the change of the independent and the change of the dependent variable is denoted by $P_{\gamma,l}$, that is

$$P_{\gamma,l}=\gamma X_l^H A_l^0 X_l+\sum_{j=0}^{d} X_l^H A_l^j X_l \frac{\partial}{\partial x_j}+X^H D X+\sum_{j=0}^{d} X_l^H A_l^j \frac{\partial X_l}{\partial x_j} , \quad l = 1, \ldots, m , \tag{9}$$

where we have set $t = x_0$.

Set $Q_l = (-\infty, \infty) \times (K_l \cap \Omega)$ for $l = 0, 1, \ldots, m$ and $\Sigma_l = (-\infty, \infty) \times (K_l \cap \partial\Omega)$. Then in each set Q_l $(j = 1, 2, \ldots, m)$ the function v_l is a weak solution to the equation

$$P_{\gamma,l} v_l = f_l \text{ where } f_l = X_l^H \left[\theta_l e^{-\gamma t} f + \sum_{j=1}^{d} A^j v \frac{\partial \theta_l}{\partial x_j} \right] \text{ in } Q_l, \ .$$

For brevity set $y = (x_1, \ldots, x_{d-1})$ (the spatially tangential variables) and choose a function $\varphi(t, y) \in C_0^\infty(\mathbb{R}^d)$ such that $\int_{\mathbb{R}^d} \varphi(t, x)\, dt dy = 1$. Set $\varphi^{(k)} = k^d \varphi(kt, ky)$. The regularization of a function or distribution $v(t, y)$ is then given by the convolution $v^{(k)} = \varphi^{(k)} * v$. Recall that for $v \in L^2(\mathbb{R}^d)$ we have $v^{(k)} \to v$ in $L^2(\mathbb{R}^d)$. In each set Q_l, the regularization of the function v_l in all but the x_d-variable will result in $v_l^{(k)} \in L^2(\mathbb{R}_+, C^\infty(\mathbb{R}^d, \mathbb{C}^N))$. (Note the split of variables into the normal variable x_d and the tangential variables (t, y), that is $\mathbb{R}^d \equiv \mathbb{R}_t \times R_y^{d-1}$.)

Two properties of the regularization are of interest, see [9, p. 435–436]. For $a \in W^{1,\infty}(Q_l)$ and $b \in L^\infty(Q_l)$ we have for each $l = 1, \ldots, m$

$$\begin{aligned} &\|(a\partial_j v_l)^{(k)} - a\partial_j v_l^{(k)}\|_{Q_l} \to 0 \text{ as } k \to \infty , \ j = 0, 1, \ldots, d-1 , \\ &\|(b v_l)^{(k)} - b v_l^{(k)}\|_{Q_l} \to 0 \text{ as } k \to \infty . \end{aligned} \tag{10}$$

The first relation of formula (10) is know as the (tangential) Friedrichs Lemma. For the regularizations of the localized solutions v_l one can now show that

$$\begin{aligned} v_l^{(k)} &\to v_l \text{ in } L^2(Q_l, \mathbb{C}^N) , \\ P_{\gamma,l} v_l^{(k)} &\to f_l \text{ in } L^2(Q_l, \mathbb{C}^N) , \end{aligned} \tag{11}$$

for $l = 1, .., m$. The convergence $v_l^{(k)} \to v_l$ in $L^2(Q_l, \mathbb{C}^N)$ follows from properties of the regularization. However, we like to point out that for k sufficiently large, the regularization $v_l^{(k)}$ is compactly supported in $\mathbb{R} \times K_l$. For the second statement, use formula (10) to establish

$$\left\| \tilde{A}_l P_{\gamma,l} v_l^{(k)} - [\tilde{A}_l P_{\gamma,l} v_l]^{(k)} \right\|_{Q_l} \to 0 \quad \text{for } k \to \infty\,, \text{ where } \tilde{A}_l = \begin{bmatrix} I & 0 & 0 \\ 0 & A_{l,+}^{-1} & 0 \\ 0 & 0 & -A_{l,-}^{-1} \end{bmatrix}.$$

The premultiplication with the matrix $\tilde{A}_l$ (where I in the top-left corner is the identity matrix of size $N - N_+ - N_-$), sets the coefficient in front of the last $N_+ + N_-$ components of the normal derivative (the derivative in x_d direction) of v_l to one. Hence, there are no derivatives in the x_d direction in the difference above. Since $\tilde{A}_l P_{\gamma,l} v_l = \tilde{A}_l f_l \in L^2(Q_l, \mathbb{C}^N)$ we have proved that $\tilde{A}_l P_{\gamma,l} v_l^{(k)} \in L^2(Q, \mathbb{C}^N)$ and another application of formula (11) gives that

$$\left\| \tilde{A}_l P_{\gamma,l} v_l^{(k)} - \tilde{A}_l f_l^{(k)} \right\|_{Q_l} \to 0 \quad \text{as } k \to \infty\,.$$

This is equivalent with the second statement in (11). The two statements of (11) are also true for $l = 0$, i.e. in the set K_0. The proof is actually easier, since one can rely on the usual Friedrichs lemma. We like to point out that $\tilde{A}_l P_{\gamma,l} v_l^{(k)} \in L^2(Q_l, \mathbb{C}^N)$ implies that $A_l^j \partial v_l^{(k)} / \partial x_j \in L^2(Q_l, \mathbb{C}^N)$ for $j = 0, 1, \ldots, d$.

By construction the functions $v^{(k)} = \sum_{l=0}^m X_l \theta_l v_l^{(k)}$ are sequences of functions which converge to v in $L^2(Q, \mathbb{C}^N)$. Furthermore, $P_\gamma v^{(k)} \to e^{-\gamma t} f$ in $L^2(Q, \mathbb{C}^N)$. The crucial point is to understand the regularity of the $v^{(k)}$. Clearly, they are square integrable and away from the spatial boundary they have square integrable first partial derivatives, since they inherit this property from v_0. Near the boundary, the functions $A^j \partial v^{(k)} / \partial x_j$ are square integrable ($j = 1, 2, \ldots, d$) since this is true in local coordinates as shown above.

Integration by parts results in the following Green formula

$$\int_Q \langle v^{(k)}, P_\gamma^* w \rangle \, dtdx = \int_Q \langle P_\gamma v^{(k)}, w \rangle \, dtdx - \int_\Sigma \langle A v^{(k)}, w \rangle \, d\Sigma\,,$$

for all $w \in H^1(Q, \mathbb{C}^N)$ satisfying $w(0, x) = w(T, x) = 0$ for all $x \in \Omega$ and $Fw = 0$ in Σ. Using Proposition 1 modifies this formula to

$$\int_Q \langle v^{(k)}, P_\gamma^* w \rangle \, dtdx = \int_Q \langle P_\gamma v^{(k)}, w \rangle \, dtdx + \int_\Sigma \langle B v^{(k)}, E w \rangle \, d\Sigma\,,$$

and subtracting the identity (8) satisfied by v gives

$$\int_Q \langle v^{(k)} - v, P_\gamma^* w\rangle \, dtdx = \int_Q \langle P_\gamma v^{(k)} - e^{-\gamma t} f, v\rangle \, dtdx + \int_\Sigma \langle Bv^{(k)} - e^{-\gamma t} g, Ew\rangle \, d\Sigma \ ,$$

Using the convergence of $v^{(k)} \to v$ and $P_\gamma v^{(k)} \to e^{-\gamma t} f$ in $L^2(Q, \mathbb{C}^N)$ provides

$$\int_\Sigma \langle Bv^{(k)} - e^{-\gamma t} g, Ew\rangle \, d\Sigma \to 0 \quad \text{as } k \to \infty \ .$$

Since the set of traces $Ew \in H^{1/2}(\Sigma, \mathbb{C}^p)$ of functions $w \in H^1(Q, \mathbb{C}^N)$ satisfying $w(0, x) = w(T, x) = 0$ for all $x \in \Omega$ and $Fw = 0$ in Σ is dense in $L^2(\Sigma, \mathbb{C}^p)$, we conclude that $Bv^{(k)} \to e^{-\gamma t} g$ in $L^2(\Sigma, \mathbb{C}^p)$ as $k \to \infty$.

Finally, the a priori estimate (6) can be applied to $v^{(k)}$ on the space time cylinder $(-\delta, t) \times \Omega$ for some $\delta > 0$ and all $-\delta \le t \le T$. The choice of the negative initial time results in $v^{(k)}(\delta, x) = 0$ for large k and

$$\sup_{0 \le t \le T} \|v^{(k)}(t)\|_\Omega^2 + \gamma \|v^{(k)}\|_{Q_\delta}^2 + \|Av^{(k)}\|_{\Sigma_\delta}^2 \lesssim \frac{1}{\gamma} \|P_\gamma v^{(k)}\|_{Q_\delta}^2 + \|Bv^{(k)}\|_{\Sigma_\delta}^2 \ .$$

Here $Q_\delta = (-\delta, T) \times \Omega$ and $\Sigma_\delta = (-\delta, T) \times \Gamma$. Since the right-hand side in this inequality is convergent as $k \to \infty$, the sequences $Av^{(k)}$ and $v^{(k)}$ are Cauchy sequences in $L^2(\Sigma, \mathbb{C}^N)$ and $C([0, T], L^2(\Omega, \mathbb{C}^N))$, respectively. We conclude that the weak solution v with zero initial data, constructed in the second part of the proof, is continuous in time with values in $L^2(\Omega, \mathbb{C}^N)$, has boundary values $Av \in L^2(\Sigma, \mathbb{C}^N)$, and satisfies the estimate

$$\sup_{0 \le t \le T} \|v(t)\|_\Omega^2 + \gamma \|v\|_Q^2 + \|Av\|_\Sigma^2 \lesssim \frac{1}{\gamma} \|e^{-\gamma t} f\|_Q^2 + \|e^{-\gamma t} g\|_\Sigma^2 \ ,$$

which implies the uniqueness of the solution.

4. Non-zero Initial Data For $u^0 \in C_0^\infty(\Omega, \mathbb{C}^N)$ one can construct a solution to the initial-boundary value problem (1)–(3) [13, p. 281]. Due to the finite speed of propagation, the strong solution u_1 to the pure Cauchy problem $Pu_1 = 0$ in $(0, T) \times \mathbb{R}^d$, $u_1(0, \cdot) = u^0$ is zero on $(0, \delta) \times \Gamma$ for some small δ. The consideration of the pure Cauchy problem requires that the coefficients of the operator P are extended to the whole space. However, for $t \in (0, \delta)$ the solution is independent of this extension. Let $\chi \in C_0^\infty(\mathbb{R})$ such that $0 \le \chi(t) \le 1$ and $\chi(t) = 1$ for all $t \le 0$ and $\chi(t) = 0$ for all $t \ge \delta$. Moreover, let u_2 be the semi-strong solution to the initial-boundary value problem

$$Pu_2 = f - \partial_t(A^0 \chi) u_1 \text{ in } Q, \quad Bu_2 = g \text{ in } \Sigma, \quad u_2(0, \cdot) = 0 \ .$$

Then $u = u_1 + u_2$ is a semi-strong solution to the initial-boundary problem (1)–(3) and satisfies the estimate of Theorem 1. For initial data $u^0 \in L^2(\Omega)$ there exists a sequence $u_n^0 \subset C_0^\infty(\Omega, \mathbb{C}^N)$ such that $u_n^0 \to u^0$ in $L^2(\Omega, \mathbb{C}^N)$. The semi-strong solutions u_n to the initial-boundary value problem

$$Pu_n = f \text{ in } Q, \quad Bu_n = g \text{ in } \Sigma, \quad u_n(0, \cdot) = u_n^0$$

satisfy the estimate

$$\|e^{-\gamma T} u_n(T)\|_\Omega^2 + \gamma \|e^{-\gamma t} u_n\|_Q^2 + \|e^{-\gamma t} A u_n\|_\Sigma^2 \lesssim \|u_n^0\|_\Omega^2 + \frac{1}{\gamma} \|e^{-\gamma t} f\|_Q^2 + \|e^{-\gamma t} g\|_\Sigma^2 .$$

We conclude that $u_n \to u$ in $L^2(Q, \mathbb{C}^N)$ and in $C([0,T], L^2(\Omega, \mathbb{C}^N))$ and that $Au_n \to Au \in L^2(\Sigma, \mathbb{C}^N)$. Furthermore, u is a weak solution to the initial-boundary value problem (1)–(3). Using a diagonal sequence $u_k^{(k)}$ of regularizations, one observes that u is a semi-strong solution as well.

3 Maxwell's Equation

Maxwell's equations

$$\begin{aligned} \partial_t(\varepsilon e) - \nabla \times h + \sigma e &= -j \\ \partial_t(\mu h) + \nabla \times h &= 0 \end{aligned} \qquad \text{in } Q = (0, T) \times \Omega \tag{12}$$

are probably the best example of a symmetric hyperbolic system where the spatial boundary Σ is characteristic. Here $d = 3$, $N = 6$, and the vector fields $e = e(t, x)$ and $h = h(t, x)$ denote the electric field and the magnetic field density, respectively. The coefficients in this system are the electric permittivity $\varepsilon(t, x)$, the magnetic permeability $\mu(t, x)$, and the conductivity $\sigma(t, x)$. While the electric permeability and the magnetic permittivity are assumed to be Hermitian uniformly positive definite matrices, the conductivity is Hermitian and positive semidefinite. The function $j(t, x)$ on the right-hand side is the electric current density in Ω. Note that the 6×6 boundary matrix

$$A = \begin{bmatrix} 0 & -\nu\times \\ \nu\times & 0 \end{bmatrix}$$

has the eigenvalues $1, -1, 0$, all with multiplicity 2. Here

$$\nu\times = \begin{bmatrix} 0 & -\nu_3 & \nu_2 \\ \nu_3 & 0 & -\nu_1 \\ -\nu_2 & \nu_1 & 0 \end{bmatrix}$$

and $N(A) = \operatorname{span}\left(\begin{bmatrix}\nu\\0\end{bmatrix}, \begin{bmatrix}0\\\nu\end{bmatrix}\right)$. A strictly dissipative boundary operator is given by

$$B = \begin{bmatrix}\nu\times & -\alpha(I_3 - \nu\nu^\top)\end{bmatrix} , \tag{13}$$

where I_3 is the 3×3 identity matrix and $\alpha \in L^\infty(\Sigma)$ is strictly positive, that is $\alpha \gtrsim 1$ for almost all $(t, x) \in \Sigma$. Note that B is a 3×6 matrix of rank 2 and that

$$B\begin{bmatrix}e\\h\end{bmatrix} = \nu \times e - \alpha h_\tau ,$$

where $h_\tau = (\nu \times h) \times \nu$ denotes tangential component of h along Γ. One verifies that $\begin{bmatrix}e\\h\end{bmatrix} \in N(B)$ if and only if $\nu \times e = \alpha h_\tau$. Then, compute

$$\left\langle A\begin{bmatrix}e\\h\end{bmatrix}, \begin{bmatrix}e\\h\end{bmatrix}\right\rangle = 2\Re\langle \nu \times e, h\rangle$$

and for $\begin{bmatrix}e\\h\end{bmatrix} \in N(B)$ one obtains

$$\left\langle A\begin{bmatrix}e\\h\end{bmatrix}, \begin{bmatrix}e\\h\end{bmatrix}\right\rangle = \alpha^{-1}|\nu \times e|^2 + \alpha|h_\tau|^2 \gtrsim \left|A\begin{bmatrix}e\\h\end{bmatrix}\right|^2 .$$

Next let $\begin{bmatrix}e\\h\end{bmatrix} \in N(B)^\perp$. Then $\langle \nu, e\rangle = 0$, $\nu \times e = -\alpha^{-1}h_\tau$, and hence, for $\begin{bmatrix}e\\h\end{bmatrix} \in N(B)^\perp$ one has

$$B\begin{bmatrix}e\\h\end{bmatrix} = \nu \times e + \alpha^2(\nu \times e) ,$$

These last two displayed formulas verify that this boundary operator is strictly dissipative in the sense of Definition 1.

Next we will show that there is an strictly dissipative boundary condition for the adjoint operator. We will not construct a boundary operator F as a 2×6 matrix as stipulated by Proposition 1. Rather we will present a boundary operator

$$F = \begin{bmatrix}\nu\times & \alpha(I_3 - \nu\nu^\top)\end{bmatrix} .$$

which is a 3×6 matrix of rank 2 and very similar to B. Setting

$$E = \frac{1}{2\alpha}B \quad \text{and} \quad G = \frac{1}{2\alpha}F$$

we have

$$G^H F - B^H C = \frac{1}{2\alpha}(F^H F - B^H B) = \frac{1}{2}\begin{bmatrix} \alpha^{-1}(I_3 - \nu\nu^\top) & -\nu\times \\ \nu\times & \alpha(I_3 - \nu\nu^\top) \end{bmatrix}$$
$$- \frac{1}{2}\begin{bmatrix} \alpha^{-1}(I_3 - \nu\nu^\top) & \nu\times \\ -\nu\times & \alpha(I_3 - \nu\nu^\top) \end{bmatrix} = A\,,$$

using the formulas

$$(\nu\times)^2 = \nu\nu^\top - I_3, \quad (\nu\times)(I - \nu\nu^\top) = (I - \nu\nu^\top)(\nu\times) = \nu\times,$$
$$\text{and} \quad (I - \nu\nu^\top)^2 = I - \nu\nu^\top\,.$$

Remark 1 In the case $\alpha = 0$ the boundary operator B is not strictly dissipative. One verifies that $\langle Az, z\rangle = 0$ for all $z \in N(B)$. The boundary operator B is then conservative and the initial-boundary value problem is still uniquely solvable, but it is not well-posed in L^2. There is a loss of derivatives on the boundary [4, 5].

4 More on Strictly Dissipative Boundary Operators

At every point $(t, x) \in \Sigma$, the vector space $\mathbb{C}^N$ admits the orthogonal decomposition into a direct sum of invariant subspaces of the boundary matrix A, that is $\mathbb{C}^N = N(A) \oplus S_- \oplus S_+$. Here $S_\pm$ is the invariant subspace of A spanned by all eigenvectors with positive (negative) eigenvalues. Correspondingly, there is the decomposition $u = u_0 + u_- + u_+$ for any $u \in \mathbb{C}^N$. Furthermore, we recall that the boundary matrix can be written as a difference $A = \Lambda_+^2 - \Lambda_-^2$ where $\Lambda_\pm$ are two Hermitian semipositive definite matrix functions.

Proposition 2 *A boundary operator of the form $B = \Lambda_- + O\Lambda_+$ is strictly dissipative for any matrix $O \in L^\infty(\Sigma, \mathbb{C}^{N\times N})$ with range equal to a subspace of S_- and spectral norm uniformly less than 1, that is $|Ov| \le c|v|$ for all $v \in \mathbb{C}^N$, a.e. $(t, x) \in \Sigma$ and $0 < c < 1$.*

Conversely, any strictly dissipative boundary operator $B \in L^\infty(\Sigma, \mathbb{C}^{N\times N})$ is of the form $R(\Lambda_- + O\Lambda_+)$ where $R, O \in L^\infty(\Sigma, \mathbb{C}^{N\times N})$ and R is invertible and O is subject to the restrictions given above.

Remark 2 Because of the matrix product $O\Lambda_+$ in the boundary operator, we can restrict ourselves to matrices O with $S_- \oplus N(A) \subset N(O)$. Hence, O represents a linear mapping from S_+ to S_- with spectral norm less than one.

Proof Let $B = \Lambda_- + O\Lambda_+$, where O satisfies the conditions listed in the proposition above. We will show that B must be strictly dissipative in the sense of Definition 1. Choose $\delta > 0$. Then using the Cauchy-Schwarz inequality one computes

$$\begin{aligned}\langle Av, v\rangle =&|\Lambda_+ v_+|^2 - |\Lambda_- v_-|^2 = |\Lambda_+ v_+|^2 + \delta|\Lambda_- v_-|^2 - (1+\delta)|Bv - O\Lambda_+ v_+|^2 \\ \geq&|\Lambda_+ v_+|^2 + \delta|\Lambda_- v_-|^2 - (1+\delta)|Bv|^2 - (1+\delta)|O\Lambda_+ v_+|^2 \\ &- \frac{1+\delta}{\delta}|Bv|^2 - (1+\delta)\delta|O\Lambda_+ v_+|^2 \\ \geq&|\Lambda_+ v_+|^2 + \delta|\Lambda_- v_-|^2 - \frac{(1+\delta)^2}{\delta}|Bv|^2 - (1+\delta)^2|O\Lambda_+ v_+|^2 .\end{aligned}$$

Since $|O\Lambda_+ v_+| \leq c|\Lambda_+ v_+|$, one obtains by choosing $\delta = \frac{1}{c}\sqrt{\frac{1+c^2}{2}} - 1$ that

$$\langle Av, v\rangle \gtrsim |Av|^2 - C|Bv|^2 .$$

Furthermore, the condition on the range of O guarantees that the rank of B is less or equal than N_- and the inequality just proved insures that the rank of B is at least equal to N_-. Strict dissipativity follows now from Lemma 1.

Suppose now that $B \in L^\infty(\Sigma, \mathbb{C}^{N\times N})$ is a strictly dissipative boundary operator. Since rank of B is equal to N_-, there exists an invertible matrix $R_1 \in L^\infty(\Sigma, \mathbb{C}^{N\times N})$ such that the range of $R_1 B$ is the subspace S_-. Since $N(A) \subset N(B)$, there is the decomposition

$$R_1 Bv = B_+ v_+ + B_- v_-$$

and the matrix B_- represents is an isomorphism on S_-. Indeed, if $B_- v_- = 0$ for some $v_- \in S_- \setminus \{0\}$, then $Bv_- = 0$ and hence $\langle Av_-, v_-\rangle = -|\Lambda_- v_-|^2 < 0$ contradicts the fact that B is strictly dissipative. Hence, there exists another invertible matrix R_2 such that

$$R_2 R_1 Bv = R_2 B_+ v_+ + \Lambda_- v_- .$$

Furthermore, since Λ_+ represents an isomorphism on S_+, we can write with $R = R_2 R_1$

$$RBv = O\Lambda_+ v_+ + \Lambda_- v_-$$

for some matrix $O \in L^\infty(\Sigma, \mathbb{C}^{N\times N})$ with range equal to a subspace of S_-. Note that O can be chosen with $S_- \oplus N(A) \subset N(O)$. Now let $z \in N(B)$. Then, by strict dissipativity, $c|\Lambda_+ z_+| \geq |\Lambda_- z_-|$ for some positive $c < 1$ and since $|O\Lambda_+ z_+| = |\Lambda_- z_-|$ we conclude that

$$|O\Lambda_+ v| \leq c|\Lambda_+ v| , \qquad \text{for all } z \in N(B) .$$

Finally, the matrix Λ_+ facilitates a linear mapping from $N(B)$ into S_+. The kernel of this mapping is $N(A)$ and because of the strict dissipativity of B, we see that this linear mapping is surjective. Hence, the inequality above proves that O has spectral norm uniformly less than one, a.e. $(t, x) \in \Sigma$. □

If the boundary operator is given in the form $B = \Lambda_- + O\Lambda_+$, then an adjoint boundary condition is always given by $F = \Lambda_+ + O^H\Lambda_-$. In view of Remark 2 the matrix O^H represents a linear operator from S_- to S_+ with spectral norm uniformly less than one. With $E = \Lambda_-$ and $G = \Lambda_+$ one verifies that

$$\begin{aligned} -B^H E + G^H F &= -(\Lambda_- + \Lambda_+ O^H)\Lambda_- + \Lambda_+(\Lambda_+ + O^H\Lambda_-) \\ &= -\Lambda_-^2 - \Lambda_+ O^H \Lambda_- + \Lambda_+^2 + \Lambda_+ O^H \Lambda_- = A \,. \end{aligned}$$

The proof of the previous proposition shows that F is a strictly dissipative boundary operator for P^*.

This result can be used to obtain a larger set of dissipative boundary operators for Maxwell's equations (12). If $u = \begin{bmatrix} e \\ h \end{bmatrix}$, then the orthogonal decomposition into invariant subspaces of the matrix A is given by

$$u_0 = \begin{bmatrix} \langle e, \nu\rangle\nu \\ \langle h, \nu\rangle\nu \end{bmatrix}, \quad u_- = \begin{bmatrix} e_\tau + \nu \times h \\ -\nu \times e + h_\tau \end{bmatrix}, \quad u_+ = \begin{bmatrix} e_\tau - \nu \times h \\ \nu \times e + h_\tau \end{bmatrix},$$

and the decomposition of the boundary matrix into the difference of two semi-positive definite matrices is

$$A = \begin{bmatrix} I_3 - \nu\nu^\top & -\nu\times \\ \nu\times & I_3 - \nu\nu^\top \end{bmatrix} - \begin{bmatrix} I_3 - \nu\nu^\top & \nu\times \\ -\nu\times & I_3 - \nu\nu^\top \end{bmatrix}.$$

Since $u_\pm = \Lambda_\pm u$ we have

$$\begin{aligned} Bu = u_- + Ou_+ &= \begin{bmatrix} e_\tau + \nu \times h \\ -\nu \times e + h_\tau \end{bmatrix} + O \begin{bmatrix} e_\tau - \nu \times h \\ \nu \times e + h_\tau \end{bmatrix} \\ &= \begin{bmatrix} e_\tau + \nu \times h \\ -\nu \times e + h_\tau \end{bmatrix} + \begin{bmatrix} O_{11} & O_{12} \\ O_{21} & O_{22} \end{bmatrix} \begin{bmatrix} e_\tau - \nu \times h \\ \nu \times e + h_\tau \end{bmatrix}. \end{aligned}$$

One convenient choice is $-O_{22} = \mathscr{O}$, $O_{12} = O_{21} = 0$, and $O_{11} = -(\nu\times)\mathscr{O}(\nu\times)$, where $\mathscr{O} \in L^\infty(\Sigma, \mathbb{C}^{3\times 3})$ has spectral norm less than one and $\mathscr{O}^H\nu = 0$. Noting the redundancies between the first three rows and the last three rows of Bu, we can just ignore the first three rows of Bu and simplify the boundary operator to

$$B' \begin{bmatrix} e \\ h \end{bmatrix} = \nu \times e - h_\tau + \mathscr{O}(\nu \times e + h_\tau)\,. \tag{14}$$

This is a natural generalization of the boundary operator (13), discussed in the previous section which can be presented in the from

$$B\begin{bmatrix} e \\ h \end{bmatrix} = \nu \times e - h_\tau + \beta(\nu \times e + h_\tau) ,$$

for some function $\beta \in L^\infty(\Sigma)$ satisfying $|\beta| \leq c$ for some positive constant $c < 1$. In fact $\alpha = \frac{1-\beta}{1+\beta}$.

In certain cases it is possible to enlarge the class of boundary operators which provide well-posedness in L^2. We demonstrate this with a simple example. Consider the 2×2 system

$$\partial_t u_1 + \partial_x u_1 = f_1$$

$$\partial_t u_2 - \partial_x u_2 = f_2$$

on the half line $\{x > 0\}$. In this case $A = \begin{bmatrix} -1 & 0 \\ 0 & 1 \end{bmatrix}$ and the strictly dissipative boundary operators are of the form $Bu = u_1 + \beta u_2$ for all $|\beta| < 1$. Since the system consists of two scalar equations we can multiply the second equation by $\alpha > 1$ and the first one by $1/\alpha$ without destroying the symmetry. For this modified system the boundary matrix is $A = \begin{bmatrix} -1/\alpha & 0 \\ 0 & \alpha \end{bmatrix}$ and the strictly dissipative boundary operators are of the form $Bu = u_1 + \alpha\beta u_2$ for all $|\beta| < 1$. However, since α can be chosen arbitrarily large, one sees that the modified system (and hence the original one) is well-posed in L^2 with boundary operators of the form $Bu = u_1 + \beta' u_2$ for any real number β'. These boundary operators are exactly the (local) boundary operators which satisfy the Kreiss-Sakamoto condition.

Majda and Osher [10, Section 2] show that Maxwell's equations with $\varepsilon = \mu = I_3$ and $\Omega = \{x_3 > 0\}$ can be decoupled into two 3×3 systems by means of a unitary tangential Fourier multiplier. With some effort the system can be decoupled (by a unitary pseudo-differential operator) in a region with a sufficiently smooth boundary ($C^{1,1}$ should suffice) as long as ε and μ are scalar functions. In other words the system has to be isotropic. This decoupling gives extra flexibility in the analysis of the system which can be used to prove well-posedness in the sense of Theorem 1 as long as the matrix function $\mathscr{O}$ in (14) has spectral radius uniformly less than one. In the quoted paper by Majda and Osher, it is proved that these are exactly the local boundary conditions of Kreiss-Sakamoto type.

If ε and μ are not scalar multiples of I_3 (the anisotropic case), then the system cannot be decoupled, at least not in a generic case. We conjecture that in this case strictly dissipative boundary operators and local boundary operators of Kreiss-Sakamoto type coincide.

Acknowledgments The author wishes to thank J. Naumann (Berlin) for asking the author about the complete proof of Proposition 1.1 in [2] and for reading and commenting on earlier versions of the manuscript. Thanks are also due to the anonymous referee of this paper for a number of helpful comments and suggestions.

Appendix

The singular value decomposition of B is $B = UTV^H$ where $V \in L^\infty(\Sigma, \mathbb{C}^{N\times N})$ and $U \in L^\infty(\Sigma, \mathbb{C}^{p\times p})$ are unitary matrices and $T \in L^\infty(\Sigma, \mathbb{R}^{p\times N})$ is of the form

$$T = \begin{bmatrix} S & 0 \\ 0 & 0 \end{bmatrix}\begin{matrix} \}N_- \\ \}p-N_- \end{matrix} \qquad \text{where} \qquad S = \begin{bmatrix} s_1 & 0 & \cdots & 0 \\ 0 & s_2 & \cdots & 0 \\ \vdots & \vdots & \ddots & 0 \\ 0 & 0 & \cdots & s_{N_-} \end{bmatrix},$$

since B is strictly dissipative. The singular values $s_1, \ldots, s_{N_-}$ are uniformly positive a.e. $(t, x) \in \Sigma$. Define now $E = -UT^\# V^H A$ where

$$T^\# = \begin{bmatrix} S^{-1} & 0 \\ 0 & 0 \end{bmatrix}.$$

The unitary matrix $V = [v_1 \ \ldots \ v_N]$ is structured as follows. The first N_- columns span $N(B)^\perp$ and the last $N - N_+ - N_-$ columns span $N(A)$. This can be done since $N(A) \subset N(B)$. Introduce two $N_+ \times N$ matrices G and F by

$$G = \begin{bmatrix} v_{N_-+1}^H \\ \vdots \\ v_{N_-+N_+}^H \end{bmatrix} \qquad \text{and} \qquad F = GA\,,$$

and observe that $N(G) = N(B)^\perp \oplus N(A)$. One computes

$$G^H F = G^H GA = V \begin{bmatrix} 0 & 0 & 0 \\ 0 & I_{N_+} & 0 \\ 0 & 0 & 0 \end{bmatrix} V^H A$$

and

$$-B^H E + G^H F = VT^H U^H UT^\# V^H A + G^H GA = VT^H T^\# V^H A + G^H GA$$

$$= V \begin{bmatrix} I_{N_-} & 0 & 0 \\ 0 & 0 & 0 \\ 0 & 0 & 0 \end{bmatrix} V^H A + V \begin{bmatrix} 0 & 0 & 0 \\ 0 & I_{N_+} & 0 \\ 0 & 0 & 0 \end{bmatrix} V^H A = A\,,$$

since the first $N_- + N_-$ columns of V are an orthonormal basis of $N(A)^{\perp} = \operatorname{Im} A$. This proves the first statement.

For the second statement we start by proving $N(F)^{\perp} = AN(B)$ and $N(F) = [AN(B)]^{\perp}$. Let $y \in N(F)$. Then $GAy = 0$, that is $Ay \in N(G)$ and hence, $\langle y, Az\rangle = \langle Ay, z\rangle = 0$ for all $z \in N(B)$. This shows $N(F) \subset [AN(B)]^{\perp}$. Now let $y \in [AN(B)]^{\perp}$. Then $0 = \langle y, Az\rangle$ for all $z \in N(B)$, which implies $Ay \in N(G)$. But this means $y \in N(F)$. Note that $\dim N(F)^{\perp} = N_+$.

Let $z \in N(B) \cap N(F)$. Then $\langle Az, z\rangle = 0$ and by the strictly dissipativity of B we have $\langle Az, z\rangle \gtrsim |Az|^2$. Hence $Az = 0$ and we have established $N(A) = N(B) \cap N(F)$. Let now $w_1, w_2, .., w_{N_-}, v_{N_+ + N_- + 1}, \ldots, v_N$ be a basis for $N(F)$. Then $w_1, w_2, .., w_{N_-}, v_{N_- + 1}, \ldots, v_N$ is a basis for $\mathbb{C}^N$ and the square matrix W, whose columns are the vectors of this basis, block-diagonalizes A, that is

$$W^H A W = \begin{bmatrix} A_- & 0 & 0 \\ 0 & A_+ & 0 \\ 0 & 0 & 0 \end{bmatrix}.$$

By the strict dissipativity of B the matrix A_+ is uniformly positive definite. Hence, since the matrix A is assumed to have constant signature independent of $(t, x) \in \Sigma$ almost everywhere, we infer that A_- must be uniformly negative definite. Thus $\langle Az, z\rangle \lesssim -|Az|^2$ for all $z \in N(F)$.

Finally, observe that $F^H F = AG^H GA$ where the matrix $G^H G$ is the orthogonal projection onto $N(B) \ominus N(A)$. By the strict dissipativity A is uniformly positive definite on $N(B) \ominus N(A)$. Thus, the matrix $F^H F$ has N_+ positive eigenvalues bounded away from zero, a.e. $(t, x) \in \Sigma$.

References

1. Benzoni-Gavage, S., Serre, D.: Multidimensional Hyperbolic Partial Differential Equations. Oxford Mathematical Monographs. The Clarendon Press Oxford University Press, Oxford (2007). First-order systems and applications
2. Cagnol, J., Eller, M.: Boundary regularity for Maxwell's equations with applications to shape optimization. J. Differ. Eq. **250**(2), 1114–1136 (2011)
3. Chazarain, J., Piriou, A.: Introduction to the theory of linear partial differential equations. In: Studies in Mathematics and its Applications, vol. 14. North-Holland, Amsterdam (1982). Translated from the French
4. Eller, M.: On symmetric hyperbolic boundary problems with nonhomogeneous conservative boundary conditions. SIAM J. Math. Anal. **4**(1), 1925–1949 (2012)
5. Eller, M.: Loss of derivatives for hyperbolic boundary problems with constant coefficients. Discrete Contin. Dyn. Syst. Ser. B **23**(3), 1347–1361 (2018)
6. Friedrichs, K.O.: Symmetric hyperbolic linear differential equations. Comm. Pure Appl. Math. **7**, 345–392 (1954)
7. Guès, O., Métivier, G., Williams, M., Zumbrun, K.: Uniform stability estimates for constant-coefficient symmetric hyperbolic boundary value problems. Comm. Partial Differ. Eq. **32**(4–6), 579–590 (2007)

8. Kato, T.: Perturbation Theory for Linear Operators, 2nd edn. Springer, Berlin (1976). Grundlehren der Mathematischen Wissenschaften, Band 132
9. Lax, P.D., Phillips, R.S.: Local boundary conditions for dissipative symmetric linear differential operators. Comm. Pure Appl. Math. **13**, 427–455 (1960)
10. Majda, A., Osher, S.: Initial-boundary value problems for hyperbolic equations with uniformly characteristic boundary. Comm. Pure Appl. Math. **28**(5), 607–675 (1975)
11. Métivier, G.: Stability of multidimensional shocks. In: Advances in the Theory of Shock Waves. Progress in Nonlinear Differential Equations and Their Applications, vol. 47, pp. 25–103. Birkhäuser, Boston (2001)
12. Métivier, G.: On the L^2 well posedness of hyperbolic initial boundary value problems. Ann. Inst. Fourier (Grenoble) **67**(5), 1809–1863 (2017)
13. Rauch, J.: $\mathcal{L}_2$ is a continuable initial condition for Kreiss' mixed problems. Comm. Pure Appl. Math. **25**, 265–285 (1972)
14. Schnaubelt, R., Spitz, M.: Local wellposedness of quasilinear Maxwell equations with absorbing boundary conditions. Evol. Equ. Control Theory (2018). arXiv.org 1812.03803

On the Spectral Stability of Standing Waves of Nonlocal $\mathcal{PT}$ Symmetric Systems

Wen Feng and Milena Stanislavova

Abstract We consider standing wave solutions of the nonlocal NLS and the nonlocal Klein–Gordon Equations. Using a variety of different techniques such as energy estimates, direct spectral calculations and index count theorems, together with the spectral properties of operators, we prove spectral stability of these waves as solutions of five different nonlocal models.

1 Introduction

The nonlinear Schrödinger equation (NLS)

$$iq_t(x,t) = q_{xx}(x,t) - 2\sigma\bar{q}(-x,t)q^2(x,t)$$

is the generic model for the evolution of slowly varying wave packets in nonlinear wave systems such as those describing the water waves, Bose–Einstein condensates and in nonlinear optics. The discovery in 1972 by Zakharov and Shabat that the NLS equation possesses a Lax pair (see [18]) and can be solved via the inverse scattering method generated tremendous excitement and activity in the field, see [9], [14]. This was soon followed by the general framework to find integrable systems solvable by the Inverse Scattering Transform (ITS) developed by Ablowitz, Kaup, Newell and Segur (AKNS) [1–3]. The idea is to consider a linear spectral problem called the scattering problem, whose compatibility conditions lead to a system that

Milena Stanislavova is partially supported by NSF-DMS, # 1516245.

W. Feng
Department of Mathematics, Niagara University, NY, USA
e-mail: wfeng@niagara.edu

M. Stanislavova (✉)
Department of Mathematics, University of Kansas, Lawrence, KS, USA
e-mail: stanis@ku.edu

W. Dörfler et al. (eds.), *Mathematics of Wave Phenomena*, Trends in Mathematics,
https://doi.org/10.1007/978-3-030-47174-3_9

is equivalent with the original equation via a symmetry reduction. For the classical NLS equation, this is done using the standard symmetry reduction

$$r(x,t) = \sigma \bar{q}(x,t), \quad \sigma = \pm 1.$$

Recently, Ablowitz and Musslimani [4, 5] used the symmetry reduction

$$r(x,t) = \sigma \bar{q}(-x,t), \quad \sigma = \pm 1$$

of the general AKNS scattering problems where the nonlocality appears in both space and time or time alone and found a new nonlocal NLS equation that can be written as

$$iq_t(x,t) = q_{xx}(x,t) - 2\sigma \bar{q}(-x,t) q^2(x,t)$$

they also proved the integrability of this new equation, and introduced new reverse space-time and reverse time NLS equations.

The term nonlocal equation refers to the fact that the nonlinearity depends on the value of the unknown function q at the point $-x$, rather than the usual x variable.

This equation is applied in optics [8, 15–17] for both focusing and defocusing nonlinearities and is also shown [7] to be gauge equivalent to an unconventional system of coupled Landau–Lifshitz equations.

Recently, there has been renewed interest in these systems, because the nonlocal equation can be viewed as a linear Schrödinger equation with a self-induced potential

$$V[q,x,t] = -2\sigma q(x,t) \bar{q}(-x,t)$$

satisfying the parity-time ($\mathcal{PT}$) symmetry condition

$$V[q,x,t] = \bar{V}[q,-x,t].$$

It has to be noted that $\mathcal{PT}$ symmetric systems have attracted considerable attention in recent years, see the papers [6, 13] and the numerous references therein.

2 Nonlocal NLS Models

As discussed in the introduction, we consider the nonlocal NLS equation

$$iq_t(x,t) = q_{xx}(x,t) + 2q^2(x,t)\bar{q}(-x,t), \quad (t,x) \in \mathbf{R} \times \mathbf{R}. \tag{1}$$

In addition, we shall be interested in the reverse time nonlocal NLS equation

$$iq_t(x,t) = q_{xx}(x,t) + 2q^2(x,t)q(x,-t), (t,x) \in \mathbf{R} \times \mathbf{R}, \tag{2}$$

and reverse space-time nonlocal NLS equation

$$iq_t(x,t) = q_{xx}(x,t) + 2q^2(x,t)q(-x,-t), (t,x) \in \mathbf{R} \times \mathbf{R} \tag{3}$$

These nonlocal equations were introduced by Ablowitz and Musslimani [4, 5]. In this paper, we set out to describe the stability of their standing waves. Since these systems are Hamiltonian in nature, the spectrum is symmetric with respect to the real and imaginary axis, which necessitates the following definition.

Definition 1 We say that the wave is spectrally stable if the spectrum of the linearized operator is on the imaginary axis.

Our main results for the nonlocal NLS type equations are as follows.

Theorem 1 *The standing wave solutions $e^{-i\omega t}\sqrt{\omega}\,\mathrm{sech}(\sqrt{\omega}x)$ of the nonlocal NLS equation (1) are spectrally stable.*

The standing wave solutions $e^{i\omega t}\sqrt{\frac{\omega}{2}}\tanh(\sqrt{\frac{\omega}{2}}x)$ of the reverse time nonlocal NLS equation (2) are spectrally stable.

Finally, the standing wave solutions $e^{i\omega t}\sqrt{\frac{\omega}{2}}\tanh(\sqrt{\frac{\omega}{2}}x)$ of the reverse space - time nonlocal NLS equation (3) are spectrally stable as well.

2.1 Nonlocal Space NLS Model

We will be interested in the stability properties of special pulse solutions for either (1), (2) or (3). That is, we are looking solution in the form $e^{-i\omega t}\phi(x)$, where $\phi(x) = \phi(-x)$ and ϕ vanishes at both infinities. For each of these equations, these reduce to the elliptic equation

$$-\phi'' + \omega\phi - 2\phi^3 = 0. \tag{4}$$

This last equation can be multiplied on both sides by ϕ' to get $\phi''\phi' = \omega\phi\phi' - 2\phi^3\phi'$. Integrate once to get $\phi' = \pm\phi(\omega-\phi^2)^{\frac{1}{2}}$. It follows that $\int \frac{d\phi}{\phi(\omega-\phi^2)^{1/2}} = \pm\int dx$. We have obtained the explicit form of the wave $\phi = \sqrt{\omega}\,\mathrm{sech}(\sqrt{\omega}x)$.

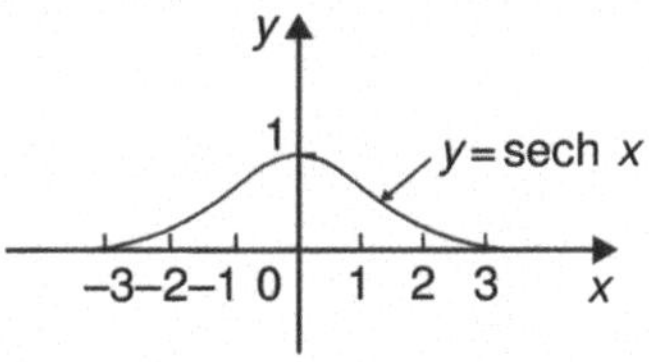

2.1.1 The Eigenvalue Problem for the Nonlocal Space NLS Model

We first linearize around the standing wave $q(x,t) = e^{-i\omega t}(\phi(x) + u(x,t))$ in (1), which leads to

$$wu + iu_t - u_{xx} - 2\phi^2\bar{u}(-x,t) - 4\phi^2 u = 0.$$

Taking the real and imaginary parts leads us to

$$w(u_1+iu_2)+i(u_1+iu_2)_t-(u_1+iu_2)_{xx}-2\phi^2(u_1(-x,t)-iu_2(-x,t))-4\phi^2(u_1+iu_2) = 0.$$

The resulting 2×2 system looks like:

$$\begin{cases} wu_1 - (u_2)_t - (u_1)_{xx} - 4\phi^2 u_1 - 2\phi^2 u_1(-x,t) = 0 \\ wu_2 + (u_1)_t - (u_2)_{xx} - 4\phi^2 u_2 + 2\phi^2 u_2(-x,t) = 0 \end{cases} \tag{5}$$

This eigenvalue problem is not in a standard form, as we see the unknown functions depending on the usual in this context non-local variable $-x$. We introduce new variables in order to obtain an equivalent system, which is local in nature. Namely, consider variables U_1, V_1, U_2 and V_2 as follows:

$$U_1(x,t) = \frac{u_1(x,t)+u_1(-x,t)}{2}, \quad \text{even in } x,$$

$$V_1(x,t) = \frac{u_1(x,t)-u_1(-x,t)}{2}, \quad \text{odd in } x,$$

$$U_2(x,t) = \frac{u_2(x,t)+u_2(-x,t)}{2}, \quad \text{even in } x,$$

$$V_2(x,t) = \frac{u_2(x,t)-u_2(-x,t)}{2}, \quad \text{odd in } x.$$

Further, $u_1(x,t) = U_1 + V_1$, $u_2(x,t) = U_2 + V_2$, $u_1(-x,t) = U_1 - V_1$ and $u_2(-x,t) = U_2 - V_2$. The system becomes:

$$\begin{cases} -(U_2+V_2)_t + w(U_1+V_1) - (U_1+V_1)_{xx} - 4\phi^2(U_1+V_1) - 2\phi^2(U_1-V_1) = 0 \\ (U_1+V_1)_t + w(U_2+V_2) - (U_2+V_2)_{xx} - 4\phi^2(U_2+V_2) + 2\phi^2(U_2-V_2) = 0 \end{cases}$$

Since U_1, U_2 are even and V_1, V_2 are odd, this can be written as a system of four equations:

$$\begin{cases} (U_2)_t = -(U_1)_{xx} + wU_1 - 6\phi^2 U_1 \\ (V_2)_t = -(V_1)_{xx} + wV_1 - 2\phi^2 V_1 \\ (U_1)_t = (U_2)_{xx} - wU_2 + 2\phi^2 U_2 \\ (V_1)_t = (V_2)_{xx} - wV_2 + 6\phi^2 V_2 \end{cases}$$

for $(U_1, V_1, U_2, V_2) \in L^2_{\text{even}} \times L^2_{\text{odd}} \times L^2_{\text{even}} \times L^2_{\text{odd}}$.
Introduce the operators $L_+ = -\partial_{xx} + w - 6\phi^2$, $L_- = -\partial_{xx} + w - 2\phi^2$ acting on $H^2_{\text{even}}(\mathbf{R})$ or $H^2_{\text{odd}}(\mathbf{R})$.

Then the system can be written in the form

$$\begin{pmatrix} U_1 \\ V_1 \\ U_2 \\ V_2 \end{pmatrix}_t = JL \begin{pmatrix} U_1 \\ V_1 \\ U_2 \\ V_2 \end{pmatrix},$$

since

$$\begin{pmatrix} U_1 \\ V_1 \\ U_2 \\ V_2 \end{pmatrix}_t = \begin{pmatrix} 0 & 0 & -1 & 0 \\ 0 & 0 & 0 & -1 \\ 1 & 0 & 0 & 0 \\ 0 & 1 & 0 & 0 \end{pmatrix} \begin{pmatrix} L_+ & 0 & 0 & 0 \\ 0 & L_- & 0 & 0 \\ 0 & 0 & L_- & 0 \\ 0 & 0 & 0 & L_+ \end{pmatrix} \begin{pmatrix} U_1 \\ V_1 \\ U_2 \\ V_2 \end{pmatrix}$$

In here we use $J = \begin{pmatrix} 0 & 0 & -1 & 0 \\ 0 & 0 & 0 & -1 \\ 1 & 0 & 0 & 0 \\ 0 & 1 & 0 & 0 \end{pmatrix}$ and $L = \begin{pmatrix} L_+ & 0 & 0 & 0 \\ 0 & L_- & 0 & 0 \\ 0 & 0 & L_- & 0 \\ 0 & 0 & 0 & L_+ \end{pmatrix}$.

We introduce first the instability index counting theory, which will be the main theoretical tool for us. Then, we apply the instability index count to the standing waves of the nonlocal space NLS for the spectral stability.

2.1.2 Stability Analysis of the Waves

We start by outlining the instability index count theory, as developed in [10–12]. We consider the eigenvalue problem in the form

$$\mathcal{J}\mathcal{L}f = \lambda f, \tag{6}$$

where $\mathcal{J}$ is assumed to be bounded, invertible and skew-symmetric ($\mathcal{J}^* = -\mathcal{J}$), while $(\mathcal{L}, D(\mathcal{L}))$ is self-adjoint($\mathcal{L}^* = \mathcal{L}$) and not necessarily bounded, with finite dimensional kernel $Ker[\mathcal{L}]$. In addition, we assume that $\mathcal{L}$ has a finite Morse index finite number of negative eigenvalues, that is $n(\mathcal{L})$, and $\mathcal{J}^{-1} : Ker[\mathcal{L}] \to Ker[\mathcal{L}]^{\perp}$. Here, the orthogonality is understood with respect to the dot product of the underlying Hilbert space $H : D(\mathcal{L}) \subset H$.

Let k_r denote the number of positive eigenvalues of (6), k_c be the number of quadruplets of eigenvalues with non-zero real and imaginary parts, and k_i^-, the number of pairs of purely imaginary eigenvalues with negative Krein-signature.[1] Introduce the matrix D as follows. Let $Ker[\mathcal{L}] = \{\phi_1, \ldots, \phi_n\}$, then

$$D_{ij} := \langle \mathcal{L}^{-1}[\mathcal{J}^{-1}\phi_i], \mathcal{J}^{-1}\phi_j \rangle. \tag{7}$$

Note that the last formula makes sense, since $\mathcal{J}^{-1}\phi_i \in Ker[\mathcal{L}]^{\perp}$. Thus $\mathcal{L}^{-1}[\mathcal{J}^{-1}\phi_i]$ is well-defined. The index counting theorem, see Theorem 1, [11] states that if $det(D) \neq 0$, then[2]

$$k_r + 2k_c + 2k_i^- = n(\mathcal{L}) - n(D), \tag{8}$$

For the nonlocal space NLS equation (1), $L_-\phi = 0$. Also $\phi = \sqrt{\omega}\,\mathrm{sech}(\sqrt{\omega}x)$ does not have zeros. Using Sturm–Liouville theory, we deduce that $L_- \geq 0$.

Again using (4) and differenting with respect to ω on both sides, it follows that $L_+(-\frac{d\phi}{d\omega}) = \phi$. Thus

$$\left\langle L_+^{-1}\phi, \phi \right\rangle = -\frac{1}{2}\frac{d}{d\omega}\|\phi_\omega\|^2 = -\frac{1}{2}\frac{d}{d\omega}\int_{-\infty}^{\infty} \omega\,\mathrm{sech}^2(\sqrt{\omega}x)\,dx = -\frac{1}{2\sqrt{\omega}} < 0.$$

Further $\phi'' = \omega\phi - 2\phi^3$, by taking derivative on both sides, we have $L_+\phi' = 0$.

$\phi' = -\omega\tanh(\sqrt{\omega}x)\,\mathrm{sech}(\sqrt{w}x)$, $\phi' = 0$ as $x = 0$, and ϕ' changes sign once. $L_+\phi = -4\phi^3$, $\langle L_+\phi, \phi\rangle = -4\int_{-\infty}^{\infty}\phi^4\,dx < 0$.

Using Sturm–Liouville theory again, L_+ has a simple negative eigenvalue.

$$Ker(L) = Ker\begin{pmatrix} L_+ & 0 & 0 & 0 \\ 0 & L_- & 0 & 0 \\ 0 & 0 & L_- & 0 \\ 0 & 0 & 0 & L_+ \end{pmatrix} = \left\{ \begin{pmatrix} 0 \\ 0 \\ 0 \\ \phi' \end{pmatrix}, \begin{pmatrix} \phi' \\ 0 \\ 0 \\ 0 \end{pmatrix}, \begin{pmatrix} 0 \\ \phi \\ 0 \\ 0 \end{pmatrix}, \begin{pmatrix} 0 \\ 0 \\ \phi \\ 0 \end{pmatrix} \right\}$$

[1]The precise definition of those is provided in many references, for example in [10]. This will be irrelevant for us, since in our applications $k_i^- = 0$.

[2]Where we recall that $n(M)$ is the Morse index (i.e. the number of negative eigenvalues) for any self-adjoint operator M.

Then we have

1. L_+ is defined on $L^2(\mathbb{R})$ with domain $H^2(\mathbb{R})$, has a unique, simple negative eigenvalue whose eigenfunction is even; zero is simple with associated eigenfunction ϕ', and the essential spectrum is $[\omega, \infty)$.
2. L_- is defined on $L^2(\mathbb{R})$ with domain $H^2(\mathbb{R})$, has no negative eigenvalue ; zero is simple with associated eigenfunction ϕ, and the essential spectrum is $[\omega, \infty)$.
3. J is bounded, invertible and skew-symmetric ($J^* = -J$). In addition, J^{-1} : $Ker[L] \to Ker[L]^{\perp}$.

Then we need to compute $n(D)$.

$$D = \begin{pmatrix} \langle L^{-1}J^{-1}\phi_1, J^{-1}\phi_1\rangle & \langle L^{-1}J^{-1}\phi_1, J^{-1}\phi_2\rangle & \langle L^{-1}J^{-1}\phi_1, J^{-1}\phi_3\rangle & \langle L^{-1}J^{-1}\phi_1, J^{-1}\phi_4\rangle \\ \langle L^{-1}J^{-1}\phi_2, J^{-1}\phi_1\rangle & \langle L^{-1}J^{-1}\phi_2, J^{-1}\phi_2\rangle & \langle L^{-1}J^{-1}\phi_2, J^{-1}\phi_3\rangle & \langle L^{-1}J^{-1}\phi_2, J^{-1}\phi_4\rangle \\ \langle L^{-1}J^{-1}\phi_3, J^{-1}\phi_2\rangle & \langle L^{-1}J^{-1}\phi_3, J^{-1}\phi_2\rangle & \langle L^{-1}J^{-1}\phi_3, J^{-1}\phi_3\rangle & \langle L^{-1}J^{-1}\phi_3, J^{-1}\phi_4\rangle \\ \langle L^{-1}J^{-1}\phi_4, J^{-1}\phi_1\rangle & \langle L^{-1}J^{-1}\phi_4, J^{-1}\phi_1\rangle & \langle L^{-1}J^{-1}\phi_4, J^{-1}\phi_3\rangle & \langle L^{-1}J^{-1}\phi_4, J^{-1}\phi_4\rangle \end{pmatrix}$$

where $\phi_1 = \begin{pmatrix} 0 \\ 0 \\ 0 \\ \phi' \end{pmatrix}$, $\phi_2 = \begin{pmatrix} \phi' \\ 0 \\ 0 \\ 0 \end{pmatrix}$, $\phi_3 = \begin{pmatrix} 0 \\ \phi \\ 0 \\ 0 \end{pmatrix}$, $\phi_4 = \begin{pmatrix} 0 \\ 0 \\ \phi \\ 0 \end{pmatrix}$

$$D_{11} = \left\langle \begin{pmatrix} L_+^{-1} & 0 & 0 & 0 \\ 0 & L_-^{-1} & 0 & 0 \\ 0 & 0 & L_-^{-1} & 0 \\ 0 & 0 & 0 & L_+^{-1} \end{pmatrix} \begin{pmatrix} 0 \\ \phi' \\ 0 \\ 0 \end{pmatrix}, \begin{pmatrix} 0 \\ \phi' \\ 0 \\ 0 \end{pmatrix} \right\rangle = \left\langle L_-^{-1}\phi', \phi' \right\rangle > 0.$$

Similarly,

$$D_{22} = \left\langle L_-^{-1}\phi', \phi' \right\rangle > 0. D_{33} = \left\langle L_+^{-1}\phi, \phi \right\rangle < 0. D_{44} = \left\langle L_+^{-1}\phi, \phi \right\rangle < 0.$$

And $D_{ij} = 0$, where $i \neq j$, and $i, j \in \{1, 2, 3, 4\}$.

Thus $n(D) = 2$ and $n(L) = 2$, $n_{\text{unstable}}(JL) = 0$, thus the standing wave solutions $e^{-i\omega t}\sqrt{\omega}\,\text{sech}(\sqrt{\omega}x)$ of the nonlocal NLS equation (1) are spectrally stable.

2.2 *Nonlocal Time NLS and Nonlocal in Space and Time NLS Models*

We again consider solutions in the form $q(x, t) = e^{i\omega t}\phi(x)$, $\phi(-x) = -\phi(x)$ of (2), which yields the profile equation

$$\phi'' + \omega\phi - 2\phi^3 = 0 \tag{9}$$

Then, one can check that we have explicit front solutions in the form

$$\phi(x) = \sqrt{\frac{\omega}{2}} \tanh(\sqrt{\frac{\omega}{2}} x).$$

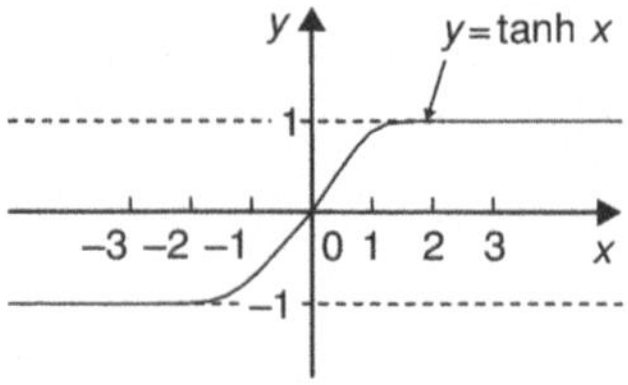

We are interested in the solutions in the form $q(x,t) = e^{i\omega t}\phi(x)$ of $\phi(-x) = -\phi(x)$ of (3), which yields the same profile equation as (9).

2.2.1 The Eigenvalue Problem for Nonlocal Time NLS Model

For the reverse time nonlocal NLS model, (2), we linearize at the solution $q = e^{i\omega t}(\phi(x) + u(x,t))$.
Ignoring all second and higher order terms leads us to the eigenvalue problem $-we^{iwt}(\phi+u)+ie^{iwt}u_t - e^{iwt}(\phi''+u_{xx})+2e^{iwt}(\phi^2+2\phi u)(\phi(x)+u(x,-t)) = 0$ and we obtain the equation for a complex solution $u(x,t)$

$$-wu + iu_t - u_{xx} + 2\phi^2 u(x,-t) + 4\phi^2 u = 0.$$

Separating the real and imaginary parts leads to

$$-w(u_1+iu_2)+i(u_1+iu_2)_t-(u_1+iu_2)_{xx}+2\phi^2(u_1(x,-t)+iu_2(x,-t))+4\phi^2(u_1+iu_2) = 0,$$

The resulting 2×2 system looks like:

$$\begin{cases} -wu_1 - (u_2)_t - (u_1)_{xx} + 4\phi^2 u_1 + 2\phi^2 u_1(x,-t) = 0 \\ -wu_2 + (u_1)_t - (u_2)_{xx} + 4\phi^2 u_2 + 2\phi^2 u_2(x,-t) = 0 \end{cases}$$

We introduce new variables to formally get rid of the nonlocality of the system, Letting U_1, V_1, U_2 and V_2 as follows:

$$U_1 = \frac{u_1(x,t) + u_1(x,-t)}{2}, \text{ even in } t,$$

$$V_1 = \frac{u_1(x,t) - u_1(x,-t)}{2}, \text{ odd in } t$$

$$U_2 = \frac{u_2(x,t) + u_2(x,-t)}{2}, \quad \text{even in } t$$

$$V_2 = \frac{u_2(x,t) - u_2(x,-t)}{2}, \quad \text{odd in } t,$$

allows us to rewrite the eigenvalue problem in the following system

$$\begin{cases} (U_2+V_2)_t + w(U_1+V_1) + (U_1+V_1)_{xx} - 4\phi^2(U_1+V_1) - 2\phi^2(U_1-V_1) = 0 \\ (U_1+V_1)_t - w(U_2+V_2) - (U_2+V_2)_{xx} + 4\phi^2(U_2+V_2) + 2\phi^2(U_2-V_2) = 0 \end{cases}$$

Since U_1, U_2 are even in time and V_1, V_2 are odd in time. This can be written as a system of four equations

$$\begin{cases} (U_2)_t = -(U_1)_{xx} - wU_1 + 6\phi^2 U_1 \\ (V_2)_t = -(V_1)_{xx} - wV_1 + 2\phi^2 V_1 \\ (U_1)_t = (U_2)_{xx} + wU_2 - 6\phi^2 U_2 \\ (V_1)_t = (V_2)_{xx} + wV_2 - 2\phi^2 V_2 \end{cases}$$

for $(U_1, V_1, U_2, V_2) \in L^2_{\text{even}}(\mathbf{R}) \times L^2_{\text{odd}}(\mathbf{R}) \times L^2_{\text{even}}(\mathbf{R}) \times L^2_{\text{odd}}(\mathbf{R})$.
Introduce the operators $L_+ = -\partial_{xx} - w + 6\phi^2$, $L_- = -\partial_{xx} - w + 2\phi^2$ acting on H^2_{even} or H^2_{odd}.

Then the system can be written in the form

$$\begin{pmatrix} U_1 \\ V_1 \\ U_2 \\ V_2 \end{pmatrix}_t = \begin{pmatrix} -L_+U_2 \\ -L_-V_2 \\ L_+U_1 \\ L_-V_1 \end{pmatrix} \text{ after transformations } \begin{pmatrix} U_1 \\ V_1 \\ U_2 \\ V_2 \end{pmatrix} \to e^{\lambda t} \begin{pmatrix} U_1 \\ V_1 \\ U_2 \\ V_2 \end{pmatrix}$$

We have the eigenvalue problem

$$\lambda \begin{pmatrix} U_1 \\ V_1 \end{pmatrix} = \begin{pmatrix} -L_+ & 0 \\ 0 & -L_- \end{pmatrix} \begin{pmatrix} U_2 \\ V_2 \end{pmatrix}, \quad \lambda \begin{pmatrix} U_2 \\ V_2 \end{pmatrix} = \begin{pmatrix} L_+ & 0 \\ 0 & L_- \end{pmatrix} \begin{pmatrix} U_1 \\ V_1 \end{pmatrix} \tag{10}$$

2.2.2 The Eigenvalue Problem of the Nonlocal in Space and Time NLS Models

For the reverse space-time nonlocal NLS model, (3), we linearize at the solution $q = e^{i\omega t}(\phi(x) + u(x,t))$.

The resulting 2×2 system looks like:

$$\begin{cases} -wu_1 - (u_2)_t - (u_1)_{xx} - 4\phi^2 u_1 + 2\phi^2 u_1(-x,-t) = 0 \\ -wu_2 + (u_1)_t - (u_2)_{xx} - 4\phi^2 u_2 - 2\phi^2 u_2(-x,-t) = 0 \end{cases}$$

We introduce new variables to formally get rid of the nonlocality of the system, Letting U_1, V_1, U_2 and V_2 as follows:

$$U_1 = \frac{u_1(x,t) + u_1(-x,-t)}{2}, \quad \text{even in } x, t,$$

$$V_1 = \frac{u_1(x,t) - u_1(-x,-t)}{2}, \quad \text{odd in } x, t$$

$$U_2 = \frac{u_2(x,t) + u_2(-x,-t)}{2}, \quad \text{even in } x, t$$

$$V_2 = \frac{u_2(x,t) - u_2(-x,-t)}{2}, \quad \text{odd in } x, t,$$

allows us to rewrite the eigenvalue problem in the following system

$$\begin{cases} (U_2)_t = -(U_1)_{xx} - wU_1 - 6\phi^2 U_1 \\ (V_2)_t = -(V_1)_{xx} - wV_1 - 2\phi^2 V_1 \\ (U_1)_t = (U_2)_{xx} + wU_2 + 6\phi^2 U_2 \\ (V_1)_t = (V_2)_{xx} + wV_2 + 2\phi^2 V_2 \end{cases}$$

for $(U_1, V_1, U_2, V_2) \in L^2_{\text{even}} \times L^2_{\text{odd}} \times L^2_{\text{even}} \times L^2_{\text{odd}}$.
Introduce the operators $L_+ = -\partial_{xx} - w - 6\phi^2$, $L_- = -\partial_{xx} - w - 2\phi^2$ acting on H^2_{even} or H^2_{odd}.

Then the system can be written in the form

$$\begin{pmatrix} U_1 \\ V_1 \\ U_2 \\ V_2 \end{pmatrix}_t = \begin{pmatrix} -L_+U_2 \\ -L_-V_2 \\ L_+U_1 \\ L_-V_1 \end{pmatrix} \text{ after transformations } \begin{pmatrix} U_1 \\ V_1 \\ U_2 \\ V_2 \end{pmatrix} \to e^{\lambda t} \begin{pmatrix} U_1 \\ V_1 \\ U_2 \\ V_2 \end{pmatrix}$$

We have the same eigenvalue problem (10).
From (10), we obtain

$$\lambda^2 \begin{pmatrix} U_1 \\ V_1 \end{pmatrix} = \begin{pmatrix} -L_+^2 & 0 \\ 0 & -L_-^2 \end{pmatrix} \begin{pmatrix} U_1 \\ V_1 \end{pmatrix}$$

Then $L_+^2 U_1 = -\lambda^2 U_1$ and $L_+^2 V_1 = -\lambda^2 V_1$, it follows that λ is pure imaginary. So there are no eigenvalues such that $\Re\lambda > 0$. It follows that the standing wave solutions $e^{i\omega t}\sqrt{\frac{\omega}{2}}\tanh(\sqrt{\frac{\omega}{2}}x)$ of Eq. (2) are stable in the sense that the eigenvalue is pure imaginary. Additionally, the standing wave solutions $e^{i\omega t}\sqrt{\frac{\omega}{2}}\tanh(\sqrt{\frac{\omega}{2}}x)$ of Eq. (3) are stable in the sense that the eigenvalue is pure imaginary.

3 Klein–Gordon Models

In this section, we consider the following two nonlocal Klein–Gordon (KG) equations. The nonlocal time KG equation is

$$q_{tt}(x,t) - q_{xx}(x,t) + q(x,t) = 2q^2(x,t)q(-x,t), \quad (t,x) \in \mathbf{R} \times \mathbf{R}. \tag{11}$$

In addition, we shall be interested in the reverse space-time nonlocal KG equation

$$q_{tt}(x,t) - q_{xx}(x,t) + q(x,t) = 2q^2(x,t)q(-x,-t), \quad (t,x) \in \mathbf{R} \times \mathbf{R}. \tag{12}$$

The stability results for the nonlocal KG models are as below.

Theorem 2 *The standing wave solutions $e^{i\omega t}\sqrt{1-\omega^2}\,\mathrm{sech}(\sqrt{1-\omega^2}x)$ of the reverse time nonlocal KG equation (11) are spectrally stable for $\omega^2 < 1$.*

The standing wave solutions $e^{i\omega t}\sqrt{\frac{\omega^2-1}{2}}\tanh(\sqrt{\frac{\omega^2-1}{2}}x)$ of the reverse space-time nonlocal KG equation (12) are spectrally stable, for $\omega^2 > 1$.

3.1 Reverse Time Nonlocal KG Equation

For the reverse time nonlocal KG equation, let us look at the standing waves $q(x,t) = e^{i\omega t}\phi(x)$, which yields the profile equation

$$\phi'' + (1-\omega^2)\phi - 2\phi^3 = 0 \tag{13}$$

Then we obtain the explicit form of the wave $\phi(x) = \sqrt{1-\omega^2}\,\mathrm{sech}(\sqrt{1-\omega^2}x)$.

We linearize at the solution $q = e^{i\omega t}(\phi(x) + u(x,t))$. Ignoring all second and higher order terms and separating the real and imaginary parts leads to a 2×2 system:

$$\begin{cases} 2\omega(u_1)_t - \omega^2 u_2 + (u_2)_{tt} - (u_2)_{xx} + u_2 - 2\phi^2 u_2(x,-t) - 4\phi^2 u_2 = 0 \\ -2\omega(u_2)_t - \omega^2 u_1 + (u_1)_{tt} - (u_1)_{xx} + u_1 - 2\phi^2 u_1(x,-t) - 4\phi^2 u_1 = 0 \end{cases}$$

We introduce new variables to formally get rid of the nonlocality of the system, Letting $U_1,\ V_1,\ U_2$ and V_2 as follows:

$$U_1 = \frac{u_1(x,t)+u_1(x,-t)}{2}, \text{ even in } t,$$

$$V_1 = \frac{u_1(x,t)-u_1(x,-t)}{2}, \text{ odd in } t$$

$$U_2 = \frac{u_2(x,t)+u_2(x,-t)}{2}, \text{ even in } t$$

$$V_2 = \frac{u_2(x,t)-u_2(x,-t)}{2}, \text{ odd in } t,$$

allows us to rewrite the eigenvalue problem in the following system

$$\begin{cases} (U_1)_{tt} - 2\omega(U_2)_t - \omega^2 U_1 - (U_1)_{xx} + U_1 - 6\phi^2 U_1 = 0 \\ (V_1)_{tt} - 2\omega(V_2)_t - \omega^2 V_1 - (V_1)_{xx} + V_1 - 2\phi^2 V_1 = 0 \\ (U_2)_{tt} + 2\omega(U_1)_t - \omega^2 U_2 - (U_2)_{xx} + U_2 - 6\phi^2 U_2 = 0 \\ (V_2)_{tt} + 2\omega(V_1)_t - \omega^2 V_2 - (V_2)_{xx} + V_2 - 2\phi^2 V_2 = 0 \end{cases}$$

for $(U_1, V_1, U_2, V_2) \in L^2_{\text{even}}(\mathbf{R}) \times L^2_{\text{odd}}(\mathbf{R}) \times L^2_{\text{even}}(\mathbf{R}) \times L^2_{\text{odd}}(\mathbf{R})$.

Then the system can be written in the form

$$\begin{pmatrix} U_1 \\ V_1 \\ U_2 \\ V_2 \end{pmatrix}_{tt} + \begin{pmatrix} 0 & 0 & -2\omega & 0 \\ 0 & 0 & 0 & -2\omega \\ 2\omega & 0 & 0 & 0 \\ 0 & 2\omega & 0 & 0 \end{pmatrix} \begin{pmatrix} U_1 \\ V_1 \\ U_2 \\ V_2 \end{pmatrix}_t + \begin{pmatrix} L_+ & 0 & 0 & 0 \\ 0 & L_- & 0 & 0 \\ 0 & 0 & L_+ & 0 \\ 0 & 0 & 0 & L_- \end{pmatrix} \begin{pmatrix} U_1 \\ V_1 \\ U_2 \\ V_2 \end{pmatrix} = \begin{pmatrix} 0 \\ 0 \\ 0 \\ 0 \end{pmatrix}$$

Equivalently, it can be written as a first-order system

$$\begin{pmatrix} U_1 \\ V_1 \\ U_2 \\ V_2 \\ (U_1)_t \\ (V_1)_t \\ (U_2)_t \\ (V_2)_t \end{pmatrix}_t = \begin{pmatrix} 0 & 0 & 0 & 0 & 1 & 0 & 0 & 0 \\ 0 & 0 & 0 & 0 & 0 & 1 & 0 & 0 \\ 0 & 0 & 0 & 0 & 0 & 0 & 1 & 0 \\ 0 & 0 & 0 & 0 & 0 & 0 & 0 & 1 \\ -L_+ & 0 & 0 & 0 & 0 & 0 & 2\omega & 0 \\ 0 & -L_- & 0 & 0 & 0 & 0 & 0 & 2\omega \\ 0 & 0 & -L_+ & 0 & -2\omega & 0 & 0 & 0 \\ 0 & 0 & 0 & -L_- & 0 & -2\omega & 0 & 0 \end{pmatrix} \begin{pmatrix} U_1 \\ V_1 \\ U_2 \\ V_2 \\ (U_1)_t \\ (V_1)_t \\ (U_2)_t \\ (V_2)_t \end{pmatrix},$$

where the operators $L_+ = -\partial_{xx}+(1-\omega^2)-6\phi^2$, $L_- = -\partial_{xx}+(1-\omega^2)-2\phi^2$ acting on H^2_{even} or H^2_{odd}. We know that $L_-\phi = 0$. Also $\phi = \sqrt{1-\omega^2}\,\text{sech}(\sqrt{1-\omega^2}x)$ does not have zeros. Using Sturm–Liouville theory, we deduce that $L_- \geq 0$. Further, we have $L_+\phi' = 0$, where ϕ' changes sign once. Finally,

$$\langle L_+\phi, \phi\rangle = -4\int_{-\infty}^{\infty} \phi^4\, dx < 0.$$

By Sturm–Liouville theory again, L_+ has a simple negative eigenvalue. One might want to use the index count theory to prove the stability. However, since J^{-1} does not map $Ker[\mathcal{L}]$ into $Ker[\mathcal{L}^{\perp}]$ as required in the index counting theory, the index count theory does not apply.

Instead, we will study the spectral problem directly. The corresponding eigenvalue problem is

$$\lambda \begin{pmatrix} U_1 \\ V_1 \\ U_2 \\ V_2 \\ (U_1)_t \\ (V_1)_t \\ (U_2)_t \\ (V_2)_t \end{pmatrix} = \begin{pmatrix} 0 & 0 & 0 & 0 & 1 & 0 & 0 & 0 \\ 0 & 0 & 0 & 0 & 0 & 1 & 0 & 0 \\ 0 & 0 & 0 & 0 & 0 & 0 & 1 & 0 \\ 0 & 0 & 0 & 0 & 0 & 0 & 0 & 1 \\ -L_+ & 0 & 0 & 0 & 0 & 0 & 2\omega & 0 \\ 0 & -L_- & 0 & 0 & 0 & 0 & 0 & 2\omega \\ 0 & 0 & -L_+ & 0 & -2\omega & 0 & 0 & 0 \\ 0 & 0 & 0 & -L_- & 0 & -2\omega & 0 & 0 \end{pmatrix} \begin{pmatrix} U_1 \\ V_1 \\ U_2 \\ V_2 \\ (U_1)_t \\ (V_1)_t \\ (U_2)_t \\ (V_2)_t \end{pmatrix},$$

We will show by contradiction that λ can not have positive real part. Consider the following subsystem, which is closed

$$\begin{pmatrix} V_1 \\ V_2 \end{pmatrix}_{tt} + 2\omega \begin{pmatrix} 0 & -1 \\ 1 & 0 \end{pmatrix} \begin{pmatrix} V_1 \\ V_2 \end{pmatrix}_t + \begin{pmatrix} L_- & 0 \\ 0 & L_- \end{pmatrix} \begin{pmatrix} V_1 \\ V_2 \end{pmatrix} = \begin{pmatrix} 0 \\ 0 \end{pmatrix} \tag{14}$$

Since it is enough to show λ cannot have positive real part for the subsystem. In order to consider the subsystem, we need to set up an auxiliary theorem. Let us go back to second-order in time equations in the general form

$$V_{tt} + 2\omega J V_t + LV = 0, \tag{15}$$

where L is a self-adjoint operator with domain $D(L) = H^2(\mathbf{R})$ and J is skew-symmetric that sends real-valued functions into real-valued functions.

The following theorem provides a general stability criteria—this is somewhat similar (and can be used) to the stability of traveling fronts for the Klein–Gordon problem.

Theorem 3 *Let L be a self-adjoint operator, $L \geq 0$, $Ker(L)$ is finite dimensional and $L|_{Ker(L)^\perp} \geq \delta > 0$. Assume that J is skew-symmetric and maps real-valued into real-valued functions. Then, the problem* (15) *is spectrally stable.*

Before we proceed with the proof of Theorem 3, let us see how it implies the stability for (14). We just apply it for the skew-symmetric $J = \begin{pmatrix} 0 & -1 \\ 1 & 0 \end{pmatrix}$ and the non-negative operator $L = \begin{pmatrix} L_- & 0 \\ 0 & L_- \end{pmatrix}$.

Proof We start with a simple energy estimate argument[3] for the problem (15). Assume, without loss of generality that V is real-valued. Then, taking a dot product with $2V_t$ yields

$$\partial_t \left[\|V_t(t,.)\|^2 + \langle LV(t), V(t)\rangle \right] = 0$$

Taking spectral decomposition $V(t,\cdot) = a(t) + z(t)$, where $a(t) \in Ker(L)$, $z(t) \perp Ker(L)$ yields

$$\|a'(t)\|^2 + \|z_t(t)\|^2 + \langle Lz(t), z(t)\rangle = \|a'(0)\|^2 + \|z_t(0)\|^2 + \langle Lz(0), z(0)\rangle.$$

Thus,

$$\sup_{t>0}[\|a'(t)\|^2 + \|z_t(t)\|^2 + \|L^{1/2}z(t)\|^2] \leq \|a'(0)\|^2 + \|z_t(0)\|^2 + \|L^{1/2}z(0)\|^2. \tag{16}$$

Observe that since $Ker(L) \subset H^2$ is finite dimensional, it is spanned by a finite number of L^2 normalized functions, $\{\varphi_j\}$, $j = 1, \ldots, N$. Thus, for any function $f = \sum_{j=1}^N \lambda_j \varphi_j$, we have

$$\|f\|_{H^1} \leq \sum_{j=1}^N |\lambda_j| \|\varphi_j\|_{H^1} \leq \sqrt{N} \left(\sum_{j=1}^N |\lambda_j|^2 \right)^{1/2} \max_{1\leq j\leq N} \|\varphi_j\|_{H^1} =$$
$$= \sqrt{N} \max_{1\leq j\leq N} \|\varphi_j\|_{H^1} \|f\|_{L^2}.$$

Thus, we have that there exists an universal constant C, so that $\|f\|_{L^2} \leq \|f\|_{H^1} \leq C\|f\|_{L^2}$ for every $f \in Ker(L)$. So, for every function $V \in H^1(\mathbf{R}) : V = a+z, a \in Ker(L), z \perp Ker(L)$,

$$\|V\|_{H^1(\mathbf{R})} \sim \|a\|_{L^2} + \|z\|_{H^1} \sim \|a\|_{L^2} + \|z\|_{L^2} + \|L^{1/2}z\|_{L^2}. \tag{17}$$

[3] Note that all the norms that appear $\|\cdot\|$ denote the standard L^2 norm, $\|\cdot\|_{L^2}$.

In the last equivalence, we have used the fact that since $Id + L : Ker(L)^{\perp} \to Ker(L)^{\perp}$ is a positive operator and $D(L) = H^2$, the quadratic form $Q(z) = \langle (Id + L)z, z\rangle$ defines a quantity equivalent to $\|z\|^2_{H^1 \cap Ker(L)^{\perp}}$.

With these preparations, we claim that (16) implies the stability. Before addressing that, let us reformulate (15) in the equivalent form

$$u_t = Hu \tag{18}$$

where $u = \begin{pmatrix} V \\ W \end{pmatrix}$, with $W = V_t$ and $H = \begin{pmatrix} 0 & 1 \\ -L & -2\omega J \end{pmatrix}$.

Indeed, assume instability for (15), or equivalently for (18). That is, there exists an eigenvalue in the form $\alpha + i\beta$, $\alpha > 0$, for H, with an eigenfunction $f = \begin{pmatrix} V \\ (\alpha + i\beta)V \end{pmatrix} \in D(\mathcal{H})$. Then, the equation $u_t = Hu$ has a solution in the form $e^{t(\alpha+i\beta)} f$, which clearly grows exponentially in time, $\|e^{t(\alpha+i\beta)} f\|_{L^1} \geq ce^{t\alpha}$.

Since $V = a(t) + z(t)$, we have by (17) and then (16),

$$\begin{aligned}\|V\|_{H^1} &\leq \|a(t)\| + \|z(t)\|_{H^1} \leq \|a(t)\| + C_L(\|L^{1/2}z(t)\|_{L^2} + \|z(t)\|_{L^2}) \\ &\leq \|a(t)\| + C[\|z(t)\|_{L^2} + |a'(0)| + \|z_t(0)\| + \|L^{1/2}z(0)\|].\end{aligned}$$

Next, we have two more bounds from (16).

$$\begin{aligned}\|a(t)\| &\leq \|a(0)\| + \int_0^t \|a'(\tau)\| d\tau \leq \|a(0)\| + t[\|a'(0)\| + \|z_t(0)\| + \|L^{1/2}z(0)\|] \\ \|z(t,\cdot)\|_{L^2} &\leq \|z(0)\|_{L^2} + \int_0^t \|z_t(\tau,\cdot)\|_{L^2} d\tau \\ &\leq \|z(0)\|_{H^1} + t[\|a'(0)\| + \|z_t(0)\| + \|L^{1/2}z(0)\|].\end{aligned}$$

Putting all this together implies

$$\|V(t)\|_{H^1} \leq C_H(1+t)[\|a(0)\| + \|a'(0)\| + \|z(0)\|_{H^1} + \|z_t(0)\|_{H^1} + \|L^{1/2}z(0)\|]$$

All in all,

$$\begin{aligned}ce^{t\alpha} &\leq \|e^{t(\alpha+i\beta)} f\|_{H^1} \leq \\ &\leq C(1+t)[\|a(0)\| + \|a'(0)\| + \|z(0)\|_{H^2} + \|z_t(0)\|_{H^2} + \|L^{1/2}z(0)\|].\end{aligned}$$

This leads to a contradiction as $t \to \infty$, as one has linear growth on the right-hand side and exponential growth on the left-hand side. □

3.2 Reverse Space-Time Nonlocal KG Equation

For the reverse space-time nonlocal KG equation, consider the standing waves $q(x,t) = e^{i\omega t}\phi(x)$, with $\phi(-x) = -\phi(x)$ and we obtain the profile equation

$$\phi'' + (\omega^2 - 1)\phi - 2\phi^3 = 0 \tag{19}$$

Then we obtain the explicit form of the wave $\phi(x) = \sqrt{\frac{\omega^2-1}{2}}\tanh(\sqrt{\frac{\omega^2-1}{2}}x)$.

We linearize at the solution $q = e^{i\omega t}(\phi(x) + u(x,t))$. Ignoring all the higher order terms and separating the real and imaginary parts leads to a 2×2 system :

$$\begin{cases} 2\omega(u_1)_t - \omega^2 u_2 + (u_2)_{tt} - (u_2)_{xx} + u_2 - 2\phi^2 u_2(-x,-t) - 4\phi^2 u_2 = 0 \\ -2\omega(u_2)_t - \omega^2 u_1 + (u_1)_{tt} - (u_1)_{xx} + u_1 - 2\phi^2 u_1(-x,-t) - 4\phi^2 u_1 = 0 \end{cases}$$

Letting $U_1,\ V_1,\ U_2$ and V_2 as follows:

$$U_1 = \frac{u_1(x,t) + u_1(-x,-t)}{2}, \text{ even in } x, t,$$

$$V_1 = \frac{u_1(x,t) - u_1(-x,-t)}{2}, \text{ odd in } x, t$$

$$U_2 = \frac{u_2(x,t) + u_2(-x,-t)}{2}, \text{ even in } x, t$$

$$V_2 = \frac{u_2(x,t) - u_2(-x,-t)}{2}, \text{ odd in } x, t,$$

allows us to rewrite the eigenvalue problem as below

$$\begin{cases} (U_1)_{tt} - 2\omega(U_2)_t - \omega^2 U_1 - (U_1)_{xx} + U_1 + 2\phi^2 U_1 = 0 \\ (V_1)_{tt} - 2\omega(V_2)_t - \omega^2 V_1 - (V_1)_{xx} + V_1 + 6\phi^2 V_1 = 0 \\ (U_2)_{tt} + 2\omega(U_1)_t - \omega^2 U_2 - (U_2)_{xx} + U_2 + 2\phi^2 U_2 = 0 \\ (V_2)_{tt} + 2\omega(V_1)_t - \omega^2 V_2 - (V_2)_{xx} + V_2 + 6\phi^2 V_2 = 0 \end{cases}$$

for $(U_1, V_1, U_2, V_2) \in L^2_{\text{even}}(\mathbf{R}) \times L^2_{\text{odd}}(\mathbf{R}) \times L^2_{\text{even}}(\mathbf{R}) \times L^2_{\text{odd}}(\mathbf{R})$.

Then the system will be

$$\begin{pmatrix} U_1 \\ V_1 \\ U_2 \\ V_2 \end{pmatrix}_{tt} + \begin{pmatrix} 0 & 0 & -2\omega & 0 \\ 0 & 0 & 0 & -2\omega \\ 2\omega & 0 & 0 & 0 \\ 0 & 2\omega & 0 & 0 \end{pmatrix} \begin{pmatrix} U_1 \\ V_1 \\ U_2 \\ V_2 \end{pmatrix}_t + \begin{pmatrix} L_+ & 0 & 0 & 0 \\ 0 & L_- & 0 & 0 \\ 0 & 0 & L_+ & 0 \\ 0 & 0 & 0 & L_- \end{pmatrix} \begin{pmatrix} U_1 \\ V_1 \\ U_2 \\ V_2 \end{pmatrix} = \begin{pmatrix} 0 \\ 0 \\ 0 \\ 0 \end{pmatrix},$$

where the operators $L_+ = -\partial_{xx} + (1-\omega^2) + 2\phi^2$, $L_- = -\partial_{xx} + (1-\omega^2) + 6\phi^2$ acting on H^2_{even} or H^2_{odd}. We can see that it shares the same system for the reverse time nonlocal KG equation. Then let us look at the spectral properties for L_+ and L_-.

We know that $L_-\phi' = 0$. Also $\phi' = \frac{\omega^2-1}{2}\operatorname{sech}^2(\sqrt{\frac{\omega^2-1}{2}}x)$ does not have zeros. Using Sturm–Liouville theory, we deduce that $L_- \geq 0$. Further we have $L_+\phi = 0$ and ϕ changes sign once. By Sturm–Liouville theory again, L_+ has a simple negative eigenvalue. Since L_+ and L_- have the same spectral properties as the reverse time nonlocal. We could also show by contradiction that λ can not have positive real part for the same closed subsystem. Using Theorem 3, we obtain the stability result for the standing wave solutions $e^{i\omega t}\sqrt{\frac{\omega^2-1}{2}}\tanh(\sqrt{\frac{\omega^2-1}{2}}x)$ of the reverse space-time nonlocal KG equation.

References

1. Ablowitz, M.J., Kaup, D.J., Newell, A.C., Segur, H.: Nonlinear-evolution equations of physical significance. Phys. Rev. Lett. **31**, 125– 127 (1973)
2. Ablowitz, M.J., Kaup, D.J., Newell, A.C., Segur, H.: Method for solving the sine-Gordon equation. Phys. Rev. Lett. **30**, 1262–1264 (1973)
3. Ablowitz, M.J., Kaup, D.J., Newell, A.C., Segur, H.: The inverse scattering transform—Fourier analysis for nonlinear problems. Stud. Appl. Math. **53**, 249– 315 (1974)
4. Ablowitz, M.J., Musslimani, Z.H.: Integrable nonlocal nonlinear equations. Stud. Appl. Math. **139**(1), 7–59 (2017)
5. Ablowitz, M.J., Musslimani, Z.H.: Integrable nonlocal nonlinear Schrödinger equation. Phys. Rev. Lett **110**, 064105 (2013)
6. Bender, C.M., Boettcher, S.: Real spectra in non-Hermitian Hamiltonians having $\mathcal{PT}$ symmetry. Phys. Rev. Lett. **80**, 5243–5246 (1998)
7. Gadzhimuradov, T.A., Agalarov, A.M.: Towards a gauge-equivalent magnetic structure of the nonlocal nonlinear Schrödinger equation. Phys. Rev. A **93**, 062124 (2016)
8. El-Ganainy, R., Makris, K.G., Christodoulides, D.N., Musslimani, Z.H.: Opt. Lett. **32**, 2632 (2007)
9. Gardner, C.S., Greene, J.M., Kruskal, M.D., Miura, R.M.: Method for solving the Korteweg-deVries equation. Phys. Rev. Lett. **19**, 1095–1098 (1967)
10. Kapitula, T.M., Kevrekidis, P.G., Sandstede, B.: Counting eigenvalues via Krein signature in infinite-dimensional Hamitonial systems. Phys. D **3–4**, 263–282 (2004)
11. Kapitula, T., Kevrekidis, P.G., Sandstede, B.: Addendum: "counting eigenvalues via the Krein signature in infinite-dimensional Hamiltonian systems. Phys. D **195**(3–4), 263–282 (2004); Phys. D **201**(1–2), 199–201 (2005)
12. Kapitula, T., Promislow, K.: Spectral and Dynamical Stability of Nonlinear Waves. Applied Mathematical Sciences, vol. 185. Springer, New York (2013)
13. Konotop, V.V., Yang, J., Zezyulin, D.A.: Nonlinear waves in $\mathcal{PT}$-symmetric systems. Rev. Mod. Phys. **88**, 035002 (2016)
14. Lax, P.D.: Integrals of nonlinear equations of evolution and solitary waves. Commun. Pure Appl. Math. **21**, 467–490 (1968)
15. Makris, K.G., El-Ganainy, R., Christodoulides, D.N., Musslimani, Z. H.: Phys. Rev. Lett. **100**, 103904 (2008)

16. Ruter, C.E., Makris, K.G., El-Ganainy, R., Christodoulides, D.N., Segev, M., Kip, D.: Nat. Phys. **6**, 192 (2010)
17. Regensburger, A., Bersch, C., Miri, M.-A., Onishchukov, G., Christodoulides, D.N., Peschel, U.: Nature (London) **488**, 167 (2012)
18. Zakharov, V.E., Shabat, A.B.: Exact theory of two-dimensional self-focusing and one-dimensional self-modulation of waves in nonlinear media. J. Exp. Theor. Phys. **34**, 62–69 (1972)

Sparse Regularization of Inverse Problems by Operator-Adapted Frame Thresholding

Jürgen Frikel and Markus Haltmeier

Abstract We analyze sparse frame based regularization of inverse problems by means of a diagonal frame decomposition (DFD) for the forward operator, which generalizes the SVD. The DFD allows to define a non-iterative (direct) operator-adapted frame thresholding approach which we show to provide a convergent regularization method with linear convergence rates. These results will be compared to the well-known analysis and synthesis variants of sparse ℓ^1-regularization which are usually implemented thorough iterative schemes. If the frame is a basis (non-redundant case), the three versions of sparse regularization, namely synthesis and analysis variants of ℓ^1-regularization as well as the DFD thresholding are equivalent. However, in the redundant case, those three approaches are pairwise different.

1 Introduction

This paper is concerned with inverse problems of the form

$$y^\delta = \mathbf{A}\boldsymbol{x} + z\,, \tag{1}$$

where $\mathbf{A}\colon \mathbb{D}(\mathbf{A}) \subseteq \mathbb{X} \to \mathbb{Y}$ is a linear operator between Hilbert spaces, and z denotes the data distortion (noise). We allow unbounded operators and assume that $\mathbb{D}(\mathbf{A})$ is dense. Moreover, we assume that the unknown object $\boldsymbol{x}$ is an element of a closed subspace space $\mathbb{X}_0 \subseteq \mathbb{X}$ on which $\mathbf{A}$ is bounded. We are particularly interested in problems, where (1) is ill-posed in which case the solution of (1) (if existent) is either not unique or the solution operator is not continuous (hence, the solution process is unstable with respect to data perturbations). In order to stabilize

J. Frikel
Department of Computer Science and Mathematics, OTH Regensburg, Regensburg, Germany
e-mail: juergen.frikel@oth-regensburg.de

M. Haltmeier (✉)
Department of Mathematics, University of Innsbruck, Innsbruck, Austria
e-mail: markus.haltmeier@uibk.ac.at

W. Dörfler et al. (eds.), *Mathematics of Wave Phenomena*, Trends in Mathematics,
https://doi.org/10.1007/978-3-030-47174-3_10

the inversion of (1) one has to apply regularization methods, cf. [10, 22]. The basic idea of regularization is to include a-priori information about the unknown object into the solution process.

In this paper, we use sparsity based regularization, where the a-priori assumption on the unknown object is sparsity of $\boldsymbol{x}$ with respect to a frame $(u_\lambda)_{\lambda\in\Lambda}$ of $\mathbb{X}$, cf. [2, 6, 13, 16, 20, 22]. That is, we regularize the recovery of $\boldsymbol{x}$ from measurements (1) by enforcing sparsity of $\boldsymbol{x}$ with respect to a suitably chosen frame of $\mathbb{X}$. Sparse regularization is well investigated and has been applied to many different imaging problems, and by now there are many algorithms available that implement sparse regularization. However, when dealing with frames, there are at least two fundamentally different concepts implementing sparsity, namely the synthesis and the analysis variant. The reason for this lies in the fact that expansions of $\boldsymbol{x} \in \mathbb{X}$ with respect to frames are not unique (which is in contrast to basis expansions). In the synthesis variant, it is assumed that the unknown is a sparse linear combination of frame elements, whereas, in the analysis variant, it is required that the inner products $\langle u_\lambda, \boldsymbol{x}\rangle$ with respect to a given frame are sparse. The difference between these approaches has been pointed out clearly in [9].

Sparse regularization is widely used in inverse problems as it provides good regularization results and is able to preserve or emphasize features (e.g., edges) in the reconstruction. However, this often comes at the price of speed, since most of the algorithms implementing sparse regularization are based on variational formulations that are solved by iterative schemes.

In the present paper, we investigate a third variant of sparse regularization that is based on an operator adapted diagonal frame decomposition (DFD) of the unknown object, cf. [3, 5, 8] and which allows to define a direct (non-iterative) sparse regularization method. In the noise-free case ($z = 0$), explicit reproducing formulas for the unknown object can be derived from the DFD, where the frame coefficients of $\boldsymbol{x}$ are calculated directly from the data $\boldsymbol{y} = \mathbf{A}\boldsymbol{x}$. In the presence of noise ($z \neq 0$), regularized versions of those formulas are obtained by applying component-wise soft-thresholding to the calculated coefficients, where the soft-thresholding operator is defined as follows:

Definition 1 (Soft-Thresholding) Let Λ be some index set.

- For $\eta, d \in \mathbb{K}$ let $\operatorname{soft}(\eta, d) := \operatorname{sign}(\eta) \max\{0, |\eta| - |d|\}$.
- For $\boldsymbol{\eta}, \boldsymbol{d} \in \mathbb{K}^\Lambda$ we define the component-wise soft-thresholding by

$$\mathbb{S}_{\boldsymbol{d}}(\boldsymbol{\eta}) := (\operatorname{soft}(\eta_\lambda, d_\lambda))_{\lambda\in\Lambda} \,. \tag{2}$$

Here and below we define $\operatorname{sign}(\eta) := \eta/|\eta|$ for $\eta \in \mathbb{K}\setminus\{0\}$ and $\operatorname{sign}(0) := 0$. The advantage of the DFD-variant of sparse regularization lies in the fact that it admits a non-iterative (direct) and fast implementation of sparse regularization that can be easily implemented for several inverse problems.

We point out, that the three variants of sparse regularization (mentioned above) are equivalent if orthonormal bases are used instead of frames, but they are fundamentally different in the redundant case.

As the main theoretical results in this paper we show that the third variant of sparse regularization, which we call *DFD-thresholding*, defines a convergent regularization method and we derive linear convergence rates for sparse solutions. For the basis case, the same results follow from existing results for ℓ^1-regularization [6, 13, 14]. In the redundant case, such results follow from [14] for the synthesis approach and from [15] for the analysis approach. In case of DFD-thresholding, we are not aware of any results concerning convergence analysis or convergence rates.

Outline

This paper is organized as follows. In Sect. 2 we define the diagonal frame decompositions of operators and give several examples of diagonal frame expansions for various operators using wavelet, curvelet, and shearlet frames. In Sect. 3 we review the convergence theory of ℓ^1-regularization and the convergence rates. In Sect. 4, we show that DFD-thresholding is a convergent regularization method and derive its convergence rates.

2 Diagonal Frame Decomposition

In this section, we introduce the concept of operator adapted diagonal frame decompositions (DFD) and discuss some classical examples of such DFDs in the case of the classical 2D Radon transform and the forward operator of photoacoustic tomography with a flat observation surface.

2.1 *Formal Definition*

We define the operator adapted diagonal frame decomposition as a generalization of the wavelet vaguelette decomposition and the diagonal curvelet or shearlet decompositions to general frames, cf. [3, 5, 8].

Definition 2 (Diagonal Frame Decomposition (DFD)) Let $(u_\lambda)_{\lambda\in\Lambda} \in \mathbb{X}^\Lambda$, $(v_\lambda)_{\lambda\in\Lambda} \in \mathbb{Y}^\Lambda$ and let $(\kappa_\lambda)_{\lambda\in\Lambda}$ be a family of positive numbers. For a linear operator $\mathbf{A}\colon \mathbb{D}(\mathbf{A}) \subseteq \mathbb{X} \to \mathbb{Y}$, we call $(u_\lambda, v_\lambda, \kappa_\lambda)_{\lambda\in\Lambda}$ a *diagonal frame decomposition (DFD) for* $\mathbf{A}$, if the following conditions hold:

(D1) $(u_\lambda)_\lambda$ is a frame of $\mathbb{X}$,
(D2) $(v_\lambda)_\lambda$ is a frame of $\overline{\operatorname{ran}(\mathbf{A})} = \overline{\mathbf{A}(\mathbb{X})}$,
(D3) $\forall \lambda \in \Lambda\colon \kappa_\lambda \neq 0 \wedge \mathbf{A}^* v_\lambda = \kappa_\lambda u_\lambda$.

Remark 1 The DFD generalizes the singular value decomposition (SVD) and the wavelet-vaguelette decomposition (WVD) (cf. [8]) as it allows the systems $(u_\lambda)_\lambda$ and $(v_\lambda)_\lambda$ to be non-orthogonal and redundant. Note that by (D2) and (D3), the frame $(u_\lambda)_\lambda$ satisfies $u_\lambda \in \mathbf{A}^*(\overline{\operatorname{ran}(\mathbf{A})}) = \operatorname{ran}(\mathbf{A}^*)$, where we have made use of the identity $\operatorname{ran}(\mathbf{A})^{\perp} = \ker(\mathbf{A}^*)$. For typical inverse problems this yields a notable smoothness assumption on the involved elements of the frame $(u_\lambda)_\lambda$.

Although the SVD has proven itself to be a useful tool for analyzing and solving inverse problems it has the following drawbacks: First, ONBs that are provided by the SVD (though optimally adapted to the operator in consideration), in many cases, don't provide sparse representations of signals of interest and, hence, are not suitable for the use in sparse regularization. In particular, frames that provide sparse representation of signals (such as wavelets or wavelet-like systems) are often not part of the SVD. Second, SVD is often very hard to compute and not known analytically for many practical applications.

To overcome some of those difficulties, wavelet-vaguelette decompositions were introduced. Nevertheless, this concept builds upon expansions of signals with respect to orthogonal wavelet-systems, which may not provide an optimal sparse representation of signals of interest, e.g., signals with sharp edges. Thus, by allowing general frames, the BCD offers great flexibility in the choice of a suitable function system for sparse regularization while retaining the advantages.

Definition 3 For a frame $(u_\lambda)_{\lambda\in\Lambda}$ of $\mathbb{X}$, the *synthesis operator* is defined as $\mathbf{U}\colon \ell^2(\Lambda) \to \mathbb{Y}\colon \boldsymbol{\xi} \mapsto \sum_{\lambda\in\Lambda} \xi_\lambda u_\lambda$ and the corresponding *analysis operator* is defined as its adjoint, $\mathbf{U}^*\colon \mathbb{X} \to \ell^2(\Lambda)\colon \boldsymbol{x} \mapsto (\langle \boldsymbol{x}, u_\lambda\rangle)_{\lambda\in\Lambda}$.

In what follows, the synthesis operator of a frame will be always denoted with the corresponding upper case letter, e.g., if $(v_\lambda)_{\lambda\in\Lambda}$ is a frame, then $\mathbf{V}$ denotes the corresponding synthesis and $\mathbf{V}^*$ the analysis operator.

In order to simplify the notation, we will also refer to $(\mathbf{U}, \mathbf{V}, \boldsymbol{\kappa})$ as DFD instead of using the full notation $(u_\lambda, v_\lambda, \kappa_\lambda)_{\lambda\in\Lambda}$.

If a DFD exists for an operator $\mathbf{A}$, it immediately gives rise to a *reproducing formula*

$$\boldsymbol{x} = \sum_{\lambda\in\Lambda} \langle \boldsymbol{x}, u_\lambda\rangle \, \bar{u}_\lambda = \sum_{\lambda\in\Lambda} \kappa_\lambda^{-1} \langle \mathbf{A}\boldsymbol{x}, v_\lambda\rangle \, \bar{u}_\lambda = \bar{\mathbf{U}} \circ \mathbf{M}_{\boldsymbol{\kappa}}^{+} \circ \mathbf{V}^*(\mathbf{A}\boldsymbol{x}), \tag{3}$$

where $(\bar{u}_\lambda)_{\lambda\in\Lambda}$ is a dual frame to $(u_\lambda)_{\lambda\in\Lambda}$ (cf. [4]) and $\bar{\mathbf{U}}$ the corresponding synthesis operator. Moreover, $\mathbf{M}_{\boldsymbol{\kappa}}^{+}$ denotes the Moore–Penrose inverse of $\mathbf{M}_{\boldsymbol{\kappa}}$ and performs point-wise division with $\boldsymbol{\kappa}$, i.e.

$$(\mathbf{M}_{\boldsymbol{\kappa}}^{+}(\boldsymbol{\eta}))_\lambda := \begin{cases} (\eta_\lambda/\kappa_\lambda)_{\lambda\in\Lambda} & \text{if } \kappa_\lambda \neq 0 \\ 0 & \text{otherwise}. \end{cases} \tag{4}$$

Hence, from given (clean) data $\boldsymbol{y}$, one can calculate the frame coefficient of $\boldsymbol{x}$ and obtain a reconstruction via (3). The key to the practical use of this reproducing formulas is the efficient implementation of the analysis and synthesis operators $\mathbf{V}^*$ and $\bar{\mathbf{U}}$, respectively. For particular cases, we will provide efficient and easy to implement algorithms for the evaluation of $\mathbf{V}^*$ and $\bar{\mathbf{U}}$.

Note that, the reproducing formula (3) (similarly to the SVD) reveals the ill-posedness of the operator equation through the decay of the *quasi-singular values* κ_λ. A regularized version of the reproducing formula (3) can be obtained by incorporating soft-thresholding of the frame coefficients. In Sect. 4, we present a complete analysis of this approach as we are not aware of any results for the general DFD-thresholding in the context of regularization theory.

We now provide several examples of DFDs, including the wavelet vaguelette decomposition and the diagonal curvelet decomposition [3] for the Radon transform as well as a DFD for the forward operator of photoacoustic tomography with flat observation surface.

2.2 Radon Transform

Definition 4 (Radon Transform) The Radon transform $\mathbf{R}\colon L^2(B_1(0)) \to L^2(\mathbb{S}^1 \times \mathbb{R})$ is defined by

$$\forall(\theta, s) \in \mathbb{S}^1 \times \mathbb{R}: \quad \mathbf{R}f(\theta, s) = \int_{\mathbb{R}} f(s\theta + t\theta^\perp)\, \mathrm{d}s\,. \tag{5}$$

It is well known that the Radon transform is bounded on $L^2(B_1(0))$, see [19]. Let $\mathbf{F}g(\theta, \omega) = \int_{\mathbb{R}} g(\theta, s)e^{-i\omega s}$ be the Fourier transform with respect to the first component and consider the Riesz potential [19]

$$(\mathbf{I}^{-\alpha} g)(\theta, \omega) := \frac{1}{2\pi} \int_{\mathbb{R}} |\omega|^\alpha\, (\mathbf{F}g)(\theta, \omega)e^{i\omega s}\, \mathrm{d}s \tag{6}$$

for $\alpha > -1$. The following hold:

(R1) The commutation relation $(\mathbf{I}^{-\alpha} \circ \mathbf{R})f = (\mathbf{R} \circ (-\Delta)^{\alpha/2})f$.
(R2) The filtered backprojection formula $f = (4\pi)^{-1}\mathbf{R}^*(\mathbf{I}^{-1} \circ \mathbf{R})f =: \mathbf{R}^\sharp\mathbf{R}f$.
(R3) Isometry property $4\pi\, \langle f_1, f_2\rangle_{L^2} = \langle \mathbf{I}^{-1} \circ \mathbf{R}f_1, \mathbf{R}f_2\rangle_{L^2}$.

Using these ingredients, one can obtain a DFD for the Radon transform as follows:

DFD for the Radon Transform
Let $(u_\lambda)_{\lambda\in\Lambda}$ be either an orthonormal basis of wavelets with compact support, a (band-limited) curvelet or a shearlet tight frame with $\lambda = (j, k, \beta) \in \Lambda$ where $j \geq 0$ is the scale index. Then $(\mathbf{U}, \mathbf{V}, \kappa)$ is a DFD with

$$v_\lambda := 2^{-j/2}\,(4\pi)^{-1}(\mathbf{I}^{-1} \circ \mathbf{R})u_\lambda \tag{7}$$

(continued)

$$\kappa_\lambda := 2^{-j/2}. \tag{8}$$

These results have been obtained in [8] for wavelet bases, in [3] for curvelet systems and in [5] for the shearlet frame. All cases are shown in similar manner and basically follow from (R1), (R2) and the fact that that $2^{-j/2}(-\Delta)^{1/4}(4\pi)^{-1}u_\lambda \simeq u_\lambda$ for any of the considered systems. The limited data case has been studied in [11].

Equation (7) implies

$$\langle g, v_\lambda \rangle = 2^{-j/2}(4\pi)^{-1}\left\langle g, \mathbf{I}^{-1} \circ \mathbf{R}u_\lambda \right\rangle = 2^{-j/2}(4\pi)^{-1}\left\langle \mathbf{R}^* \circ \mathbf{I}^{-1}g, u_\lambda \right\rangle. \tag{9}$$

This gives an efficient numerical algorithm for the evaluation of $\mathbf{V}^*$ provided that $\mathbf{U}^*$ is associated with an efficient algorithm. This is in particular the case for the wavelet, shearlet and curvelet frames as above.

Remark 2 We would like to note that in order to define a DFD for the case of curvelets or shearlets one needs to consider the Radon transform on subspaces of $L^2(\mathbb{R}^2)$ consisting of functions that are defined on unbounded domains (since band-limited curvelets or shearlets have non-compact support). However, since the Radon transform is an unbounded operator on $L^2(\mathbb{R}^2)$, the reproducing formula (3) will not hold any more in general. The reproducing formulas are at least available for the case that the object $\boldsymbol{x}$ can be represented as a finite linear combination of curvelets or shearlets (cf. [3] and [5]).

Another possibility would be to consider projections of curvelet or shearlet frames onto the space $L^2(B_1(0))$, which would yield a frame for this space (cf. [4]) and then define the DFD in the same way as above. Because the Radon transform is continuous on $L^2(B_1(0))$, the reproducing formula (3) will hold for general linear combinations.

Computing DFD Coefficients for the Radon Transform

Let $\mathbf{U}$ be a wavelet, shearlet of curvelet frame and define $\mathbf{V}$ by (7).

(a) Input: $g \in L^2(\mathbb{S}^1 \times \mathbb{R})$.
(b) Compute $f_{\text{FBP}} := (4\pi)^{-1}\mathbf{R}^* \circ \mathbf{I}^{-1}g$.
(c) Compute $\boldsymbol{\eta} := \mathbf{U}^* f_{\text{FBP}}$ via wavelet, curvelet or shearlet transform.
(d) Apply rescaling $\boldsymbol{\eta} \leftarrow (2^{-j/2}\eta_\lambda)_{\lambda\in\Lambda}$.
(e) Output: Coefficients $\boldsymbol{\eta}$.

2.3 Inversion of the Wave Equation

We consider a planar geometry, which has been considered in our previous work [12]. Let $C_0^\infty(\mathbb{H}_+)$ denote the space of compactly supported functions $h\colon \mathbb{R}^2 \to \mathbb{R}$ that are supported in the half space $\mathbb{H}_+ := \mathbb{R} \times (0, \infty)$. For $f \in C_0^\infty(\mathbb{H}_+)$ consider the initial value problem

$$\begin{aligned} (\partial_t^2 - \Delta u(x, y, t) &= 0, && (x, y, t) \in \mathbb{R}^2 \times \mathbb{R} \\ u(x, y, 0) &= y^{1/2} f(x, y) && (x, y) \in \mathbb{R}^2 \\ \partial_t u(x, y, 0) &= 0 && (x, y) \in \mathbb{R}^2 . \end{aligned} \tag{10}$$

The trace map $\mathbf{A}\colon f \mapsto t^{-1/2} g$ where $g(x, t) := u(x, y = 0, t)\chi\{t \geq 0\}$ for $(x, t) \in \mathbb{R}^2$ is known to be an isometry from $L^2(\mathbb{H}_+)$ to $L^2(\mathbb{H}_+)$, see [1, 12, 18]. In particular, the operator $\mathbf{A}$ is continuous.

Definition 5 (Forward Operator for the Wave Equation) We define $\mathbf{A}\colon L^2(\mathbb{H}_+) \to L^2(\mathbb{H}_+)$ by $\mathbf{A} f := t^{-1/2} u$, for $f \in C_0^\infty(\mathbb{H}_+)$, where u is the solution of (10), and extending it by continuity to $L^2(\mathbb{H}_+)$.

The isometry property implies that any frame gives a DFD $(\mathbf{U}, \mathbf{V}, \boldsymbol{\kappa})$ by setting $v_\lambda = \mathbf{A} u_\lambda$ and $\kappa_\lambda = 1$. This, in particular, yields a wavelet vaguelette decomposition and a diagonal curvelet decomposition for the wave equation.

DFD for the Wave Equation
Let $(u_\lambda)_{\lambda\in\Lambda}$ be either a wavelet frame, a curvelet frame or a shearlet frame with $\lambda = (j, k, \beta) \in \Lambda$ where $j \geq 0$ is the scale index. Then $(\mathbf{U}, \mathbf{V}, \boldsymbol{\kappa})$ is a DFD with

$$v_\lambda := \mathbf{A} u_\lambda \tag{11}$$

$$\kappa_\lambda := 1 . \tag{12}$$

As noted in [12] this result directly follows from the isometry property and the associated inversion formula $f = \mathbf{A}^* \mathbf{A} f$.

The isometry property also gives an efficient numerical algorithm for computing analysis coefficients with respect to the frame $\mathbf{V}$ in the case that $\mathbf{U}$ is associated with an efficient algorithm.

Computing DFD Coefficients for the Wave Equation
Let $\mathbf{U}$ be the curvelet frame and define $\mathbf{V}$ by (11).

(a) Input: $g \in L^2(\mathbb{H}_+)$.
(b) Compute $f_{\mathrm{FBP}} := \mathbf{A}^* g$.
(c) Compute $\boldsymbol{\eta} := \mathbf{U}^* f_{\mathrm{FBP}}$ via wavelet, curvelet or shearlet transform.
(d) Output: Coefficients $\boldsymbol{\eta}$.

The algorithm described above can be used for any problem where the forward operator $\mathbf{A}$ is an isometry. In the case of the wavelet transform this simple procedure has been previously used in [12].

3 Sparse ℓ^1-Regularization

There are two fundamentally different and well-studied instances of sparse frame based regularization, namely ℓ^1-analysis regularization and ℓ^1-synthesis regularization. They are defined by

$$\mathbf{B}_\alpha^{\mathrm{ANA}}(\boldsymbol{y}_\delta) := \operatorname*{arg\,min}_{\boldsymbol{x} \in \mathbb{X}} \left\{ \frac{1}{2} \left\| \mathbf{A}\boldsymbol{x} - \boldsymbol{y}_\delta \right\|^2 + \alpha \sum_\lambda d_\lambda \left| \langle u_\lambda, \boldsymbol{x} \rangle \right| \right\} \tag{13}$$

$$\mathbf{B}_\alpha^{\mathrm{SYN}}(\boldsymbol{y}_\delta) := \mathbf{W} \left(\operatorname*{arg\,min}_{\boldsymbol{\xi} \in \ell^2(\Lambda)} \left\{ \frac{1}{2} \left\| \mathbf{A}\mathbf{W}\boldsymbol{\xi} - \boldsymbol{y}_\delta \right\|^2 + \alpha \sum_\lambda d_\lambda \left| \xi_\lambda \right| \right\} \right), \tag{14}$$

respectively, with weights $d_\lambda > 0$.

Definition 6 We call $\boldsymbol{\xi} \in \ell^2(\Lambda)$ sparse if the set $\{\lambda \in \Lambda \mid \xi_\lambda \neq 0\}$ is finite.

If $\boldsymbol{\xi} = (\xi_\lambda)_{\lambda \in \Lambda} \in \ell^2(\Lambda)$ is sparse, we write $\mathrm{Sign}(\boldsymbol{\xi}) := \{\boldsymbol{z} = (z_\lambda)_{\lambda \in \Lambda} \in \ell^2(\Lambda) \mid z_\lambda \in \mathrm{Sign}(\xi_\lambda)\}$ where $\mathrm{Sign}(\,\cdot\,)\colon \mathbb{R} \to \mathbb{R}$ is the multi-valued signum function defined by $\mathrm{Sign}(0) = [-1, 1]$ and $\mathrm{Sign}(x) = \{x / |x|\}$ for $x \neq 0$. We will use the notation

$$\|\cdot\|_{\boldsymbol{d},1} : \ell^2(\Lambda) \to \mathbb{R} \cup \{\infty\} \colon \boldsymbol{\xi} \mapsto \begin{cases} \sum_\lambda d_\lambda \left|\xi_\lambda\right| & \text{if } (d_\lambda \xi_\lambda)_{\lambda \in \Lambda} \in \ell^1(\Lambda) \\ \infty & \text{otherwise}\,. \end{cases} \tag{15}$$

Any element in the set $\operatorname{arg\,min}\{\|\mathbf{U}^*(\boldsymbol{x})\|_{\boldsymbol{d},1} \mid \mathbf{A}\boldsymbol{x} = \boldsymbol{y}\}$ is called $\|\mathbf{U}^*(\,\cdot\,)\|_{\boldsymbol{d},1}$-minimizing solution of $\mathbf{A}(\boldsymbol{x}) = \boldsymbol{y}$. Note that $\|\mathbf{U}^*(\,\cdot\,)\|_{\boldsymbol{d},1}$-minimizing solutions exists whenever there is any solution $\boldsymbol{x}$ with $\|\mathbf{U}^*(\boldsymbol{x})\|_{\boldsymbol{d},1} < \infty$, as follows from [22, Theorem 3.25].

Below we recall well-posedness and convergence results for both variants. These results hold under the following quite week assumptions:

(A1) $\mathbf{A}\colon \mathbb{X} \to \mathbb{Y}$ is bounded linear;
(A2) $\mathbf{U}, \mathbf{W}$ are synthesis operators of frames $(u_\lambda)_{\lambda\in\Lambda}$, $(w_\lambda)_{\lambda\in\Lambda}$ of $\mathbb{X}$;
(A3) $\boldsymbol{d} = (d_\lambda)_{\lambda\in\Lambda} \in \mathbb{R}^\Lambda$ satisfies $\inf\{d_\lambda \mid \lambda \in \Lambda\} > 0$.

For certain sparse elements we will state linear error estimates which have been derived in [13, 15]. See [2, 6, 16, 20, 22] for some further works on sparse ℓ^1-regularization, and [7, 17, 21] for wavelet regularization methods.

3.1 ℓ^1-Analysis Regularization

Let us define the ℓ^1-analysis Tikhonov functional by

$$\mathcal{A}_{\alpha,\boldsymbol{y}}\colon \mathbb{X} \to \mathbb{R}\cup\{\infty\}\colon \boldsymbol{x} \mapsto \frac{1}{2}\|\mathbf{A}\boldsymbol{x} - \boldsymbol{y}\|^2 + \alpha\left\|\mathbf{U}^*\boldsymbol{x}\right\|_{\boldsymbol{d},1}. \tag{16}$$

Then we have $\mathbf{B}_\alpha^{\mathrm{ANA}}(\boldsymbol{y}) = \arg\min \mathcal{A}_{\alpha,\boldsymbol{y}}$.

Proposition 1 (Convergence of Analysis Regularization) *Let (A1)–(A3) be satisfied, suppose $\boldsymbol{y} \in \mathbb{Y}$, $\alpha > 0$, $(\boldsymbol{y}^k)_{k\in\mathbb{N}} \in \mathbb{Y}^\mathbb{N}$ with $\boldsymbol{y}^k \to \boldsymbol{y}$, and choose $\boldsymbol{x}^k \in \arg\min \mathcal{A}_{\alpha,\boldsymbol{y}^k}$.*

- *Existence: The functional $\mathcal{A}_{\alpha,\boldsymbol{y}}$ has at least one minimizer.*
- *Stability: There exists a subsequence $(\boldsymbol{x}^{k(\ell)})_{\ell\in\mathbb{N}}$ of $(\boldsymbol{x}^k)_{k\in\mathbb{N}}$ and a minimizer $\boldsymbol{x}_\alpha \in \arg\min \mathcal{A}_{\alpha,\boldsymbol{y}}$ such that $\|\boldsymbol{x}^{k(\ell)} - \boldsymbol{x}_\alpha\| \to 0$. If the minimizer $\boldsymbol{x}_\alpha$ of $\mathcal{A}_{\alpha,\boldsymbol{y}}$ is unique, then $\|\boldsymbol{x}^k - \boldsymbol{x}_\alpha\| \to 0$.*
- *Convergence: Assume $\boldsymbol{y} = \mathbf{A}\boldsymbol{x}$ for $\boldsymbol{x} \in \mathbb{X}$ with $\|\mathbf{U}^*\boldsymbol{x}\|_{\boldsymbol{d},1} < \infty$ and suppose $\|\boldsymbol{y}^k - \boldsymbol{y}\| \le \delta_k$ with $(\delta_k)_{k\in\mathbb{N}} \to 0$. Consider a parameter choice $(\alpha_k)_k \in (0,\infty)^\mathbb{N}$ such that $\lim_{k\to\infty}\alpha_k = \lim_{k\to\infty}\delta_k^2/\alpha_k = 0$. Then there is an $\|\mathbf{U}^*(\,\cdot\,)\|_{\boldsymbol{d},1}$-minimizing solution $\boldsymbol{x}^+$ of $\mathbf{A}(\boldsymbol{x}) = \boldsymbol{y}$ and a subsequence $(\boldsymbol{x}^{k(\ell)})_{\ell\in\mathbb{N}}$ with $\|\boldsymbol{x}^{k(\ell)} - \boldsymbol{x}^+\| \to 0$. If the $\|\mathbf{U}^*(\,\cdot\,)\|_{\boldsymbol{d},1}$-minimizing solution is unique, then $\|\boldsymbol{x}^k - \boldsymbol{x}^+\| \to 0$.*

Proof See [13, Propositions 5, 6 and 7]. □

In order to derive convergence rates, one has to make additional assumptions on the exact solution $\boldsymbol{x}^+$ to be recovered. Besides the sparsity this requires a certain interplay between $\boldsymbol{x}^+$ and the forward operator $\mathbf{A}$.

(A4) $\mathbf{U}^*\boldsymbol{x}^+$ is sparse;
(A5) $\exists z \in \operatorname{Sign}(\mathbf{U}^*\boldsymbol{x}^+)\colon \mathbf{U}z \in \operatorname{ran}(\mathbf{A}^*)$;
(A6) $\exists t \in (0,1)\colon \mathbf{A}$ is injective on $\operatorname{span}\{u_\lambda\colon |z_\lambda| > t\}$.

Assumption (A5) is a so-called source condition and the main restrictive assumption. It requires that there exists an element $z \in \operatorname{Sign}(\mathbf{U}^*\boldsymbol{x})$ that satisfies the smoothness assumption $\mathbf{U}z \in \operatorname{ran}(\mathbf{A}^*)$. Because $z \in \ell^2(\Lambda)$, the space $\operatorname{span}\{u_\lambda \mid |z_\lambda| > t\}$ is finite dimensional. Therefore, Condition (A6) requires injectivity on a certain finite dimensional subspace.

Proposition 2 (Convergence Rates for Analysis Regularization) *Suppose (A1)–(A6) hold. Then, for a parameter choice $\alpha = C\delta$, there is a constant $c_+ \in (0, \infty)$ such that for all $\boldsymbol{y}_\delta \in \mathbb{Y}$ with $\|\mathbf{A}\boldsymbol{x} - \boldsymbol{y}_\delta\| \leq \delta$ and every minimizer $\boldsymbol{x}^{\alpha,\delta} \in \arg\min \mathcal{A}_{\alpha,\boldsymbol{y}_\delta}$ we have $\|\boldsymbol{x}^{\alpha,\delta} - \boldsymbol{x}^+\| \leq c_+\delta$.*

Proof See [15, Theorem III.8]. □

3.2 Synthesis Regularization

Let use denote the ℓ^1-synthesis Tikhonov functional by

$$\mathcal{S}_{\alpha,\boldsymbol{y}} \colon \ell^2(\Lambda) \to \mathbb{R} \cup \{\infty\} \colon \boldsymbol{x} \mapsto \frac{1}{2}\|\mathbf{A}\mathbf{W}\boldsymbol{\xi} - \boldsymbol{y}\|^2 + \alpha\|\boldsymbol{\xi}\|_{\boldsymbol{d},1}\,. \tag{17}$$

Then it holds $\mathbf{B}_\alpha^{\mathrm{SYN}}(\boldsymbol{y}) = \mathbf{W}(\arg\min \mathcal{S}_{\alpha,\boldsymbol{y}})$. Synthesis regularization can be seen as analysis regularization for the coefficient inverse problem $\mathbf{A}\mathbf{W}\boldsymbol{\xi} = \boldsymbol{y}$ and the analysis operator $\mathbf{U}^* = \mathrm{Id}$. Using Proposition 1 we therefore have the following result.

Proposition 3 (Convergence of Synthesis Regularization) *Let (A1)–(A3) be satisfied, suppose $\boldsymbol{y} \in \mathbb{Y}$, $\alpha > 0$, $(\boldsymbol{y}^k)_{k\in\mathbb{N}} \in \mathbb{Y}^{\mathbb{N}}$ with $\boldsymbol{y}^k \to \boldsymbol{y}$ and take $\boldsymbol{\xi}^k \in \arg\min \mathcal{S}_{\alpha,\boldsymbol{y}^k}$.*

- *Existence: The functional $\mathcal{S}_{\alpha,\boldsymbol{y}}$ has at least one minimizer.*
- *Stability: There exists a subsequence $(\boldsymbol{\xi}^{k(\ell)})_{\ell\in\mathbb{N}}$ of $(\boldsymbol{\xi}^k)_{k\in\mathbb{N}}$ and $\boldsymbol{\xi}_\alpha \in \arg\min \mathcal{S}_{\alpha,\boldsymbol{y}}$ such that $(\boldsymbol{\xi}^{k(\ell)})_\ell \to \boldsymbol{\xi}_\alpha$. If the minimizer of $\mathcal{S}_{\alpha,\boldsymbol{y}}$ is unique, then $\|\boldsymbol{\xi}^k - \boldsymbol{\xi}_\alpha\| \to 0$.*
- *Convergence: Assume $\boldsymbol{y} = \mathbf{A}\mathbf{W}\boldsymbol{\xi}$ for $\boldsymbol{\xi} \in \ell^2(\Lambda)$ with $\|\boldsymbol{\xi}\|_{\boldsymbol{d},1} < \infty$ and $\|\boldsymbol{y}^k - \boldsymbol{y}\| \leq \delta_k$ with $(\delta_k)_{k\in\mathbb{N}} \to 0$. Consider a parameter choice $(\alpha_k)_{k\in\mathbb{N}} \in (0,\infty)^{\mathbb{N}}$ with $\lim_{k\to\infty}\alpha_k = \lim_{k\to\infty}\delta_k^2/\alpha_k = 0$. Then there exist an $\|\cdot\|_{\boldsymbol{d},1}$-minimizing solution $\boldsymbol{\xi}^+$ of $(\mathbf{A}\mathbf{W})(\boldsymbol{\xi}) = \boldsymbol{y}$ and a subsequence $(\boldsymbol{\xi}^{k(\ell)})_{\ell\in\mathbb{N}}$ with $\|\boldsymbol{\xi}^{k(\ell)} - \boldsymbol{\xi}^+\| \to 0$. If the $\|\cdot\|_{\boldsymbol{d},1}$-minimizing solution is unique, then $\|\boldsymbol{\xi}^k - \boldsymbol{\xi}^+\| \to 0$.*

Proof Follows from Proposition 1 with $\mathbf{U} = \mathrm{Id}$ and $\mathbf{A}\mathbf{W}$ in place of $\mathbf{A}$. □

We have linear convergence rates under the following additional assumptions on the element to be recovered.

(S4) $\boldsymbol{x}^+ = \mathbf{W}\boldsymbol{\xi}^+$ where $\boldsymbol{\xi}^+ \in \ell^2(\Lambda)$ is sparse;

(S5) $\exists z \in \operatorname{Sign}(\boldsymbol{\xi}^+) \colon z = \operatorname{ran}(\mathbf{W}^*\mathbf{A}^*)$;
(S6) $\exists t \in (0,1) \colon \mathbf{AW}$ is injective on $\operatorname{span}\{e_\lambda \colon |z_\alpha| > t\}$.

Proposition 4 (Convergence Rates for Synthesis Regularization) *Suppose that (A1)–(A3) and (S4)–(S6) hold. Then, for a parameter choice $\alpha = C\delta$, there is a constant $c_+ \in (0,\infty)$ such that for all $\boldsymbol{y}_\delta \in \mathbb{Y}$ with $\|\mathbf{A}\boldsymbol{x} - \boldsymbol{y}_\delta\| \leq \delta$, every minimizer $\boldsymbol{\xi}^{\alpha,\delta} \in \arg\min \mathcal{S}_{\alpha,\boldsymbol{y}_\delta}$ we have $\|\boldsymbol{\xi}^{\alpha,\delta} - \boldsymbol{\xi}^+\| \leq c_+\delta$.*

Proof Follows from Proposition 2. □

Because $\mathbf{W}$ is bounded, the above convergence results can be transferred to convergence in the signal space $\mathbb{X}$. In particular, we have stability $\|\mathbf{W}(\boldsymbol{\xi}^{k(\ell)}) - \mathbf{W}(\boldsymbol{\xi}_\alpha)\| \to 0$ and convergence $\|\mathbf{W}\boldsymbol{\xi}^{k(\ell)} - \mathbf{W}\boldsymbol{\xi}^+\| \to 0$ under the assumptions made in Proposition 3, and linear convergence rates $\|\mathbf{W}\boldsymbol{\xi}^{\alpha,\delta} - \boldsymbol{x}^+\| \leq \tilde{c}_+\delta$ under the assumptions made in Proposition 4.

3.3 Sparse Regularization Using an SVD

In the special case that $\mathbf{U}$ is part of an SVD, then analysis and synthesis regularization are equivalent and can be computed explicitly by soft-thresholding of the expansion coefficients.

Theorem 1 (Equivalence in the SVD Case) *Let* $(\mathbf{U}, \mathbf{V}, \boldsymbol{\kappa})$ *be an SVD for* $\mathbf{A}$*, let* $\boldsymbol{y}_\delta \in \mathbb{Y}$ *and consider* (13) *and* (14) *with* $\mathbf{W} = \mathbf{U}$*. Then*

$$\mathbf{B}_\alpha^{\mathrm{ANA}}(\boldsymbol{y}_\delta) = \mathbf{B}_\alpha^{\mathrm{SYN}}(\boldsymbol{y}_\delta) = \{(\mathbf{U} \circ \mathbf{M}_{\boldsymbol{\kappa}}^+ \circ \mathbb{S}_{\alpha\boldsymbol{d}/\boldsymbol{\kappa}} \circ \mathbf{V}^*)(\boldsymbol{y}_\delta)\},$$

equals the soft-thresholding estimator in the SVD system.

Proof Because $\mathbf{U}$ is an orthonormal basis of $\mathbb{X}$, we have $\boldsymbol{x} = \mathbf{U}\boldsymbol{\xi} \Leftrightarrow \xi = \mathbf{U}^*\boldsymbol{x}$ which implies that $\mathbf{B}_\alpha^{\mathrm{ANA}}(\boldsymbol{y}_\delta) = \mathbf{B}_\alpha^{\mathrm{SYN}}(\boldsymbol{y}_\delta)$. Now let $\boldsymbol{x}_\alpha \in \mathbf{B}_\alpha^{\mathrm{SYN}}(\boldsymbol{y}_\delta)$ be any minimizer of the ℓ^1-analysis Tikhonov functional $\mathcal{A}_{\alpha,\boldsymbol{y}}$. Let $\mathbf{P}_{\operatorname{ran}(\mathbf{A})^\perp}$ denote the orthogonal projection on $\operatorname{ran}(\mathbf{A})^\perp$. We have

$$\begin{aligned}
\mathcal{A}_{\alpha,\boldsymbol{y}}(\boldsymbol{x}) &= \frac{1}{2}\|\mathbf{A}\boldsymbol{x} - \boldsymbol{y}\|^2 + \alpha\left\|\mathbf{U}^*\boldsymbol{x}\right\|_{\boldsymbol{d},1} \\
&= \left\|\mathbf{P}_{\operatorname{ran}(\mathbf{A})^\perp}(\boldsymbol{y})\right\|^2 + \sum_{\lambda\in\Lambda}\frac{1}{2}\left|\langle\mathbf{A}\boldsymbol{x} - \boldsymbol{y}, v_\lambda\rangle\right|^2 + \sum_{\lambda\in\Lambda}\alpha d_\lambda\left|\langle\boldsymbol{x}, u_\lambda\rangle\right| \\
&= \left\|\mathbf{P}_{\operatorname{ran}(\mathbf{A})^\perp}(\boldsymbol{y})\right\|^2 + \sum_{\lambda\in\Lambda}\frac{1}{2}\left|\kappa_\lambda\langle\boldsymbol{x}, u_\lambda\rangle - \langle\boldsymbol{y}, v_\lambda\rangle\right|^2 + \alpha d_\lambda\left|\langle\boldsymbol{x}, u_\lambda\rangle\right|.
\end{aligned}$$

The latter sum is minimized by componentwise soft-thresholding. This shows $\boldsymbol{x}_{\alpha,\delta} = (\mathbf{U} \circ \mathbb{S}_{\alpha\boldsymbol{d}/\boldsymbol{\kappa}^2} \circ \mathbf{M}_{\boldsymbol{\kappa}}^+ \circ \mathbf{V}^*)(\boldsymbol{y}_\delta) = (\mathbf{U} \circ \mathbf{M}_{\boldsymbol{\kappa}}^+ \circ \mathbb{S}_{\alpha\boldsymbol{d}/\boldsymbol{\kappa}} \circ \mathbf{V}^*)(\boldsymbol{y}_\delta)$ and concludes the proof. □

In the case that $(\mathbf{U}, \mathbf{V}, \boldsymbol{\kappa})$ is a redundant DFD expansion and not an SVD, then (13), (14), and the soft-thresholding estimator

$$(\mathbf{U} \circ \mathbf{M}_{\boldsymbol{\kappa}}^{+} \circ \mathbb{S}_{\alpha \boldsymbol{d}/\boldsymbol{\kappa}} \circ \mathbf{V}^{*})(\boldsymbol{y}_{\delta}) \tag{18}$$

are all non-equivalent. Further, in this case, (13) and (14) have to be computed by iterative minimization algorithms. This requires repeated application of the forward and adjoint problem and therefore is time consuming. In the following section, we study DFD thresholding which is the analog of (18) for redundant systems. Despite the non-equivalence to ℓ^1-regularization, we are able to derive the same type of convergence results and linear convergence rates as for the analysis and synthesis variants of ℓ^1-regularization.

4 Regularization via DFD Thresholding

Throughout this section we fix the following assumptions

(B1) $\mathbf{A}\colon \mathbb{X} \to \mathbb{Y}$ is bounded linear.
(B2) $(\mathbf{U}, \mathbf{V}, \boldsymbol{\kappa})$ is s DFD for $\mathbf{A}$.
(B3) $\boldsymbol{d} = (d_\lambda)_{\lambda \in \Lambda} \in \mathbb{R}^{\Lambda}$ satisfies $\inf\{d_\lambda \mid \lambda \in \Lambda\} > 0$.

In this section we show well-posedness, convergence and convergence rates for DFD soft-thresholding. We denote by $(\bar{u}_\lambda)_{\lambda \in \Lambda}$ a dual frame to $(u_\lambda)_{\lambda \in \Lambda}$ and $\bar{\mathbf{U}}$ the corresponding synthesis operator.

4.1 *DFD Soft-Thresholding*

Any DFD gives an explicit inversion formula $\boldsymbol{x} = (\bar{\mathbf{U}} \circ \mathbf{M}_{\boldsymbol{\kappa}}^{+} \circ \mathbf{V}^{*})(\mathbf{A}\boldsymbol{x})$ where $\mathbf{M}_{\boldsymbol{\kappa}}^{+}$ is defined by (4). For ill-posed problems, $\kappa_\lambda \to 0$ and therefore the above reproducing formula is unstable when applied to noisy data $\boldsymbol{y}_\delta$ instead of $\mathbf{A}\boldsymbol{x}$. Below we stabilize the inversion by including the soft-thresholding operation.

Definition 7 (DFD Soft-Thresholding) Let $(\mathbf{U}, \mathbf{V}, \boldsymbol{\kappa})$ be a DFD for $\mathbf{A}$. We define the nonlinear DFD soft-thresholding estimator by

$$\mathbf{B}_{\alpha}^{\mathrm{DFD}}\colon \mathbb{Y} \to \mathbb{X}\colon \boldsymbol{y} \mapsto (\bar{\mathbf{U}} \circ \mathbf{M}_{\boldsymbol{\kappa}}^{+} \circ \mathbb{S}_{\alpha \boldsymbol{d}/\boldsymbol{\kappa}} \circ \mathbf{V}^{*})(\boldsymbol{y})\,. \tag{19}$$

If $(u_\lambda)_\lambda$, $(v_\lambda)_\lambda$ are ONBs, then Theorem 1 shows that (13) and (14) are equivalent to (19). In the case of general frames, $\mathbf{B}_{\alpha}^{\mathrm{ANA}}$, $\mathbf{B}_{\alpha}^{\mathrm{SYN}}$ and $\mathbf{B}_{\alpha}^{\mathrm{DFD}}$ are all different.

As the main result in this paper we show that DFD soft-thresholding yields the same theoretical results as ℓ^1-regularization. Assuming efficient implementations for $\bar{\mathbf{U}}$ and $\mathbf{V}^*$, the DFD estimator has the advantage that it can be calculated non-iteratively and is therefore much faster than $\mathbf{B}_{\alpha}^{\mathrm{SYN}}$ and $\mathbf{B}_{\alpha}^{\mathrm{DFD}}$.

Consider the ℓ^1-Tikhonov functional for the multiplication operator $\mathbf{M}_\kappa$,

$$\mathcal{M}_{\alpha,\eta} \colon \ell^2(\Lambda) \to \mathbb{R} \cup \{\infty\} \colon \boldsymbol{\xi} \mapsto \frac{1}{2}\|\mathbf{M}_\kappa \boldsymbol{\xi} - \boldsymbol{\eta}\|^2 + \alpha \|\boldsymbol{\xi}\|_{\boldsymbol{d},1} \,. \tag{20}$$

The proof strategy used in this paper is based on the following Lemma.

Lemma 1 (ℓ^1-Minimization for Multiplication Operators)

(a) $\forall \alpha \in \mathbb{R}_{>0} \forall \boldsymbol{\eta} \in \ell^2(\Lambda) \colon$ $\mathcal{M}_{\alpha,\eta}$ *has a unique minimizer.*
(b) $\forall \alpha \in \mathbb{R}_{>0} \forall \boldsymbol{\eta} \in \ell^2(\Lambda) \colon$ $(\mathbf{M}_\kappa^+ \circ \mathbb{S}_{\alpha \boldsymbol{d}/\kappa})(\boldsymbol{\eta}) = \arg\min \mathcal{M}_{\alpha,\eta}$.
(c) $\mathbf{M}_\kappa^+ \circ \mathbb{S}_{\alpha \boldsymbol{d}/\kappa} \colon \ell^2(\Lambda) \to \ell^2(\Lambda)$ *is continuous.*
(d) $\mathbf{V}^* \mathbf{A} = \mathbf{M}_\kappa \mathbf{U}^*$.

Proof Because (Id, Id, κ) is an SVD for $\mathbf{M}_\kappa$, Items (a), (b) follow from Theorem 1, the equivalence of ℓ^1-regularization and soft-thresholding in the SVD case. Item (c) follows from Proposition 3. Moreover, the equality $(\mathbf{V}^* \mathbf{A} \boldsymbol{x})_\lambda = \langle v_\lambda, \mathbf{A} \boldsymbol{x} \rangle = \langle \mathbf{A}^* v_\lambda, \boldsymbol{x} \rangle = \kappa_\lambda \langle u_\lambda, \boldsymbol{x} \rangle = (\mathbf{M}_\kappa \mathbf{U}^* \boldsymbol{x})_\lambda$ shows Item (d). □

Note that the continuity of $\mathbf{M}_\kappa^+ \circ \mathbb{S}_{\alpha \boldsymbol{d}/\kappa}$ (see Item (b) in the above lemma) is not obvious as it is the composition of the soft thresholding $\mathbb{S}_{\alpha \boldsymbol{d}/\kappa}$ with the discontinuous operator in $\mathbf{M}_\kappa^+$. The characterization in Item (b) of $\mathbf{M}_\kappa^+ \circ \mathbb{S}_{\alpha \boldsymbol{d}/\kappa}$ as minimizer of the ℓ^1-Tikhonov functional $\mathcal{M}_{\alpha,\eta}$ and the existing stability results for ℓ^1-Tikhonov regularization yields an elegant way to obtain the continuity of $\mathbf{M}_\kappa^+ \circ \mathbb{S}_{\alpha \boldsymbol{d}/\kappa}$. Verifying the continuity directly would also be possible but seems to be a harder task. A similar comment applies to the proof of Theorem 2 where we use the convergence of $\mathcal{M}_{\alpha,\eta}$ to show convergence the DFD soft-thresholding estimator $\mathbf{B}_\alpha^{\mathrm{DFD}}$.

4.2 Convergence Analysis

In this section, we show that $(\mathbf{B}_\alpha^{\mathrm{DFD}})_{\alpha>0}$ is well-posed and convergent.

Theorem 2 (Well-Posedness and Convergence) *Let (B1)–(B3) be satisfied, suppose* $\boldsymbol{y} \in \mathbb{Y}$ *and let* $(\boldsymbol{y}^k)_{k\in\mathbb{N}} \in \mathbb{Y}^{\mathbb{N}}$ *satisfy* $\boldsymbol{y}^k \to \boldsymbol{y}$.

(a) Existence: $\mathbf{B}_\alpha^{\mathrm{DFD}} \colon \mathbb{Y} \to \mathbb{X}$ *is well-defined for all* $\alpha > 0$.
(b) Stability: $\mathbf{B}_\alpha^{\mathrm{DFD}} \colon \mathbb{Y} \to \mathbb{X}$ *is continuous for all* $\alpha > 0$.
(c) Convergence: Assume $\boldsymbol{y} = \mathbf{A}\boldsymbol{x}$ *for some* $\boldsymbol{x} \in \mathbb{X}$ *with* $\|\mathbf{U}^* \boldsymbol{x}\|_{\boldsymbol{d},1} < \infty$, *suppose* $\|\boldsymbol{y}^k - \boldsymbol{y}\| \le \delta_k$ *with* $(\delta_k)_{k\in\mathbb{N}} \to 0$ *and consider a parameter choice* $(\alpha_k)_{k\in\mathbb{N}} \in (0,\infty)^{\mathbb{N}}$ *with* $\lim_{k\to\infty} \alpha_k = \lim_{k\to\infty} \delta_k^2/\alpha_k = 0$. *Then* $\|\mathbf{B}_{\alpha_k}^{\mathrm{DFD}}(\boldsymbol{x}^k) - \boldsymbol{x}^+\| \to 0$.

Proof (a), (b): According to Lemma 1, the mapping $\mathbf{M}_\kappa^+ \circ \mathbb{S}_{\alpha \boldsymbol{d}/\kappa} \colon \ell^2(\Lambda) \to \ell^2(\Lambda)$ is well-defined and continuous. Moreover, by definition we have $\mathbf{B}_\alpha^{\mathrm{DFD}} = \bar{\mathbf{U}} \circ (\mathbf{M}_\kappa^+ \circ \mathbb{S}_{\alpha \boldsymbol{d}/\kappa}) \circ \mathbf{V}^*$ which implies existence and stability of DFD thresholding.

(c): We have

$$\begin{aligned}\|\mathbf{B}_{\alpha_k}^{\mathrm{DFD}}(\boldsymbol{y}^k) - \boldsymbol{x}^+\| &= \|\bar{\mathbf{U}} \circ (\mathbf{M}_\kappa^+ \circ \mathbb{S}_{\alpha d/\kappa}) \circ \mathbf{V}^*(\boldsymbol{y}^k) - \bar{\mathbf{U}}\mathbf{U}^*\boldsymbol{x}^+\| \\ &\leq \|\mathbf{U}\| \; \|(\mathbf{M}_\kappa^+ \circ \mathbb{S}_{\alpha d/\kappa})(\mathbf{V}^*\boldsymbol{y}^k) - \mathbf{U}^*\boldsymbol{x}^+\| \,. \qquad (21)\end{aligned}$$

Moreover, $\|\mathbf{V}^*\boldsymbol{y}^k - \mathbf{M}_\kappa \mathbf{U}^*\boldsymbol{x}^+\| = \|\mathbf{V}^*\boldsymbol{y}^k - \mathbf{V}^*\mathbf{A}\boldsymbol{x}^+\| \leq \|\mathbf{V}\|\|\boldsymbol{y}^k - \mathbf{A}\boldsymbol{x}^+\| \leq \|\mathbf{V}\|\delta_k$. Therefore, Proposition 3 and the equality $\arg\min \mathcal{M}_{\alpha,\mathbf{V}^*\boldsymbol{y}^k} = (\mathbf{M}_\kappa^+ \circ \mathbb{S}_{\alpha d/\kappa})(\mathbf{V}^*\boldsymbol{y}^k)$ shown in Lemma 1 imply $\|(\mathbf{M}_\kappa^+ \circ \mathbb{S}_{\alpha d/\kappa})(\mathbf{V}^*\boldsymbol{y}^k) - \mathbf{U}^*\boldsymbol{x}^+\| \to 0$ for $k \to \infty$. Together with (21) this yields (c) and completes the proof. □

4.3 Convergence Rates

Next we derive linear convergence rates for sparse solutions. Let use denote by $\mathrm{supp}(\xi) := \{\lambda \in \Lambda : \xi_\lambda \neq 0\}$ the support of $\xi \in \ell^2(\Lambda)$. To derive the convergence rates, we assume the following for the exact solution $\boldsymbol{x}^+$ to be recovered.

(B4) $\mathbf{U}^*\boldsymbol{x}^+$ is sparse.
(B5) $\forall \lambda \in \mathrm{supp}(\mathbf{U}^*\boldsymbol{x}^+) \colon \kappa_\lambda \neq 0$.

Note that assumptions (B4) and (B5) imply the source condition

$$z \in \partial\| \cdot \|_{d,1} \cap \mathrm{ran}(\mathbf{M}_\kappa^*) \neq \emptyset \,.$$

is satisfied for some element $z \in \ell^2(\Lambda)$ that can be chosen such that $|z_\lambda| < 1$ for $\lambda \notin \mathrm{supp}(\mathbf{U}^*\boldsymbol{x}^+)$. Moreover, it follows that $\mathbf{M}_\kappa$ is injective on $\mathrm{span}\,\{e_\lambda \mid |z_\lambda| > t\}$ with $t := \max\left\{|\kappa_\lambda| \mid \lambda \notin \mathrm{supp}(\mathbf{U}^*\boldsymbol{x}^+)\right\}$. Because $\mathbf{U}^*\boldsymbol{x}^+ \in \ell^2(\Lambda)$, we have $t < 1$.

Theorem 3 (Convergence Rates) *Suppose that (B1)–(B5) hold. Then, for the parameter choice $\alpha = C\delta$ with $C \in (0,\infty)$, there is a constant $c_+ \in (0,\infty)$ such that for all $\boldsymbol{y}_\delta \in \mathbb{Y}$ with $\|\mathbf{A}\boldsymbol{x} - \boldsymbol{y}_\delta\| \leq \delta$ we have $\|\mathbf{B}_\alpha^{\mathrm{DFD}}(\boldsymbol{y}_\delta) - \boldsymbol{x}^+\| \leq c_+\delta$.*

Proof As in the proof of Theorem 2 one obtains

$$\|\mathbf{B}_\alpha^{\mathrm{DFD}}(\boldsymbol{y}^k) - \boldsymbol{x}^+\| \leq \|\mathbf{U}\| \; \|(\mathbf{M}_\kappa^+ \circ \mathbb{S}_{\alpha d/\kappa})(\mathbf{V}^*\boldsymbol{y}^k) - \mathbf{U}^*\boldsymbol{x}^+\| \qquad (22)$$

$$\|\mathbf{V}^*\boldsymbol{y}^k - \mathbf{M}_\lambda \mathbf{U}^*\boldsymbol{x}^+\| \leq \|\mathbf{V}\|\delta \,. \qquad (23)$$

According to the considerations below (B4) and (B5) the conditions (S4)–(S6) are satisfied for the operator $\mathbf{M}_\kappa$ in place of $\mathbf{A}$ and with $\mathbf{W} = \mathrm{Id}$. The convergence rates result in Proposition 2, estimate (23), and the identity $\arg\min \mathcal{M}_{\alpha,\mathbf{V}^*\boldsymbol{y}} = (\mathbf{M}_\kappa^+ \circ \mathbb{S}_{\alpha d/\kappa})(\mathbf{V}^*\boldsymbol{y})$ shown in Lemma 1 imply $\|(\mathbf{M}_\kappa^+ \circ \mathbb{S}_{\alpha d/\kappa})(\mathbf{V}^*\boldsymbol{y}) - \mathbf{U}^*\boldsymbol{x}^+\| \leq c\delta$. Together with (22) this implies $\|\mathbf{B}_\alpha^{\mathrm{DFD}}(\boldsymbol{y}_\delta) - \boldsymbol{x}^+\| \leq c\|\mathbf{U}\|\delta$ and concludes the proof. □

5 Conclusion

To overcome the inherent ill-posedness of inverse problems, regularization methods incorporate available prior information about the unknowns to be reconstructed. In this context, a useful prior is sparsity with respect to a certain frame. There are at least two different regularization strategies implementing sparsity with respect to a frame, namely ℓ^1-analysis regularization and ℓ^1-synthesis regularization. In this paper, we analyzed DFD-thresholding as a third variant of sparse regularization. One advantage of DFD-thresholding compared to other sparse regularization methods is its non-iterative nature leading to fast algorithms. Besides having a DFD, actually computing the DFD soft-thresholding estimator (19) requires a dual frame $\bar{\mathbf{U}}$. While in the general situation, the DFD and the dual frame have to be computed numerically, we have shown that for many practical examples (see Sect. 2) they are known explicitly and efficient algorithms are available for its numerical evaluation.

The DFD-approach presented in this paper is well studied in the context of statistical estimating using certain multi-scale systems. However, its analysis in the context of regularization theory has not been given so far. In this paper we closed this gap and presented a complete convergence analysis of DFD-thresholding as regularization method.

Acknowledgement The work of M.H has been supported by the Austrian Science Fund (FWF), project P 30747-N32.

References

1. Buhgeim, A.L., Kardakov, V.B.: Solution of an inverse problem for an elastic wave equation by the method of spherical means. Sibirsk. Mat. Z. **19**(4), 749–758 (1978)
2. Bürger, S., Flemming, J., Hofmann, B.: On complex-valued deautoconvolution of compactly supported functions with sparse fourier representation. Inverse Probl. **32**(10), 104006 (2016)
3. Candès, E.J., Donoho, D.: Recovering edges in ill-posed inverse problems: optimality of curvelet frames. Ann. Stat. **30**(3), 784–842 (2002)
4. Christensen, O.: An Introduction to Frames and Riesz Bases. Applied and Numerical Harmonic Analysis. Birkhäuser, Boston (2003)
5. Colonna, F., Easley, G., Guo, K., Labate, D.: Radon transform inversion using the shearlet representation. Appl. Comput. Harmon. Anal. **29**(2), 232–250 (2010). https://doi.org/10.1016/j.acha.2009.10.005
6. Daubechies, I., Defrise, M., De Mol, C.: An iterative thresholding algorithm for linear inverse problems with a sparsity constraint. Comm. Pure Appl. Math. **57**(11), 1413–1457 (2004)
7. Dicken, V., Maass, P.: Wavelet–Galerkin methods for ill-posed problems. J. Inverse Ill-Posed Probl. **4**, 203–221 (1996)
8. Donoho, D.L.: Nonlinear solution of linear inverse problems by Wavelet–Vaguelette decomposition. Appl. Comput. Harmon. Anal. **2**(2), 101–126 (1995)
9. Elad, M., Milanfar, P., Rubinstein, R.: Analysis versus synthesis in signal priors. Inveerse Prob. **23**(3), 947 (2007)
10. Engl, H.W., Hanke, M., Neubauer, A.: Regularization of Inverse Problems. Kluwer Academic Publishers, Dordrecht (1996)

11. Frikel, J.: Sparse regularization in limited angle tomography. Appl. Comput. Harmon. Anal. **34**(1), 117–141 (2013)
12. Frikel, J., Haltmeier, M.: Efficient regularization with wavelet sparsity constraints in photoacoustic tomography. Inverse Probl. **34**(2), 024006 (2018)
13. Grasmair, M., Haltmeier, M., Scherzer, O.: Sparse regularization with l^q penalty term. Inverse Probl. **24**(5), 055020 (2008)
14. Grasmair, M., Haltmeier, M., Scherzer, O.: Necessary and sufficient conditions for linear convergence of ℓ^1-regularization. Commun. Pure Appl. Math. **64**(2), 161–182 (2011)
15. Haltmeier, M.: Stable signal reconstruction via ℓ^1-Minimization in redundant, non-tight frames. IEEE Trans. Signal Process. **61**(2), 420–426 (2013)
16. Lorenz, D.A.: Convergence rates and source conditions for Tikhonov regularization with sparsity constraints. J. Inverse Ill-Posed Prob. **16**(5), 463–478 (2008)
17. Louis, A., Maass, P., Rieder, A.: Wavelets. Theorie und Anwendungen. Teubner, Stuttgart (1998)
18. Narayanan, E.K., Rakesh: Spherical means with centers on a hyperplane in even dimensions. Inverse Probl. **26**(3), 035014 (2010)
19. Natterer, F.: The Mathematics of Computerized Tomography. Classics in Applied Mathematics, vol. 32. SIAM, Philadelphia (2001)
20. Ramlau, R., Teschke, G.: A Tikhonov-based projection iteration for nonlinear ill-posed problems with sparsity constraints. Numer. Math. **104**(2), 177–203 (2006)
21. Rieder, A.: A wavelet multilevel method for ill-posed problems stabilized by Tikhonov regularization. Numer. Math. **75**, 501–522 (1997)
22. Scherzer, O., Grasmair, M., Grossauer, H., Haltmeier, M., Lenzen, F.: Variational methods in imaging. Appl. Math. Sci. **167** (2009). https://doi.org/10.1007/978-0-387-69277-7

Soliton Solutions for the Lugiato–Lefever Equation by Analytical and Numerical Continuation Methods

Janina Gärtner and Wolfgang Reichel

Abstract We investigate analytically and numerically soliton solutions of the stationary Lugiato–Lefever equation

$$-da'' - (\mathrm{i} - \zeta)a - |a|^2 a + \mathrm{i}f = 0$$

in the case of anomalous dispersion $d > 0$. By an analytical result based on the Implicit Function Theorem, we prove the existence of two families of soliton solutions on the real line when the parameters lie in certain regions of the ζ-f plane. We then use these ideas to compute numerically 2π-periodic approximate solutions and obtain a map of the ζ-f plane which shows numerical existence regions for soliton solutions. We also indicate the stability/instability of these solutions. The borders of the numerically computed existence regions in the ζ-f plane match very well heuristically obtained borders.

1 Introduction

In 1987, Lugiato and Lefever [14] considered a model for the description of the electromagnetic field inside a Kerr nonlinear microresonator which was coupled to a strong continuous wave (CW) laser of strength f. If one assumes the microresonator to be a ring of length 2π, then the optical field inside the resonator may be described by $a(x,t) = \sum_{k\in\mathbb{Z}} a_k(t)\mathrm{e}^{\mathrm{i}kx}$ where $a_k(t)$ denotes the complex amplitude of the k-th resonant mode in the microresonator. The frequency of the cw-laser has a detuning offset ζ with respect to the frequency of the 0-mode. The linear dispersion coefficient d of the resonator material is either normal ($d < 0$) or anomalous ($d > 0$). According to [14] the resulting model for the field a, which takes into

J. Gärtner · W. Reichel (✉)
Institute for Analysis (IANA), Karlsruhe Institute of Technology, Karlsruhe, Germany
e-mail: janina@jfgaertner.de; wolfgang.reichel@kit.edu

W. Dörfler et al. (eds.), *Mathematics of Wave Phenomena*, Trends in Mathematics,
https://doi.org/10.1007/978-3-030-47174-3_11

account Kerr-nonlinear effects inside the resonator, is given by

$$\mathrm{i}a_t = -(\mathrm{i}-\zeta)a - da_{xx} - |a|^2 a + \mathrm{i}f, \qquad a(x,t) = a(x+2\pi, t) \tag{1}$$

and derivations from first principles may be found in [2, 3, 12]. Stationary solutions of (1) satisfy

$$-da'' - (\mathrm{i}-\zeta)a - |a|^2 a + \mathrm{i}f = 0, \qquad a(x) = a(x+2\pi) \tag{2}$$

and the discrete Fourier transform $(a_k)_{k\in\mathbb{Z}}$ of a stable stationary solution of (2) is called a frequency comb. Spectrally broad octave spanning frequency combs have tremendous applicability in, e.g., time and frequency metrology [5, 27], high-speed optical data communications [17, 22, 23], and ultrafast optical ranging [25, 26]. The requirement of having a broad discrete Fourier transform is met by stationary solutions of (2) which are highly localized in space and therefore will be called solitons.

The opposite of soliton frequency combs are trivial frequency combs given by constant solutions of (2), since their discrete Fourier transform consists only of the 0-mode with no other modes being excited. A number of papers have studied how nontrivial frequency combs may bifurcate from trivial ones by varying either the detuning parameter ζ or the forcing parameter f, e.g., [4, 9, 10, 19–21] under a spatial dynamics point of view, and [15, 16, 18] where local and global bifurcations of 2π-periodic solutions were considered via the approaches of Crandall–Krasnoselski–Rabinowitz. Numerical continuation methods for bifurcations were considered in [16, 19–21], and recently, using the `Matlab` toolbox `pde2path`, in [8] with an emphasis on solitons and the evaluation of their quality measures such as band-width and power-conversion-efficiency.

The study of [16] on stationary solutions of (2) also included the global shape of bifurcation curves and explicit a-priori bounds for all stationary solutions of (1). This study was recently extended in [7] to a variant of the Lugiato–Lefever model (1) including the effect of two-photon absorption (TPA) which occurs, e.g., in silicon-based optical devices having the advantage of a much higher nonlinear Kerr effect compared to other materials. The extended model takes TPA into account by adding a further damping term into (1) which is proportional to $|a|^2 a$.

In [13] global well-posedness results as well as convergence results for the numerical Strang-splitting scheme for the time-dependent problem (1) can be found. Orbital asymptotic stability of 2π-periodic solutions was investigated in [24], and using the center manifold approach, spectral stability and instability results as well as nonlinear stability with respect to co-periodic or subharmonic perturbations were obtained in [4].

Our main interest in (2) is the detection of those parameters d, f, ζ where stable soliton solutions exist. Here the word "soliton" is understood as a highly localized 2π-periodic solution. There are two main results in this paper: the first result proves analytically the existence of soliton solutions of (2) where the 2π-periodicity of

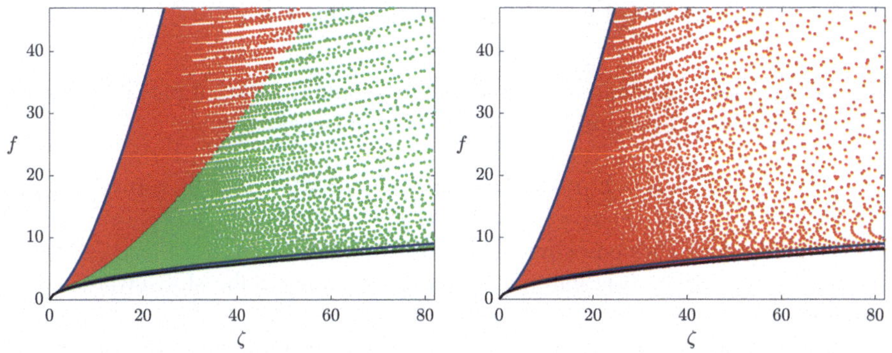

Fig. 1 ζ-f chart for $d = 0.01$ marking existence and stability of solitons found by continuation of an approximate solution of the NLS. The unstable solutions are marked in red, the stable solutions in green. Left: continuation starting from the symmetrically decreasing homoclinic orbit w^+. Right: continuation starting from the symmetrically increasing homoclinic orbit w^-. The blue and black curves are explained in Sect. 3.5

a is replaced by the requirement of a being a homoclinic solution on $\mathbb{R}$. The second result is shown in Fig. 1, where the ζ-f plane is drawn and green indicates the existence of spectrally stable solitons, red the existence of spectrally unstable solitons. This figure is the outcome of a numerical procedure, and its qualitative shape may be explained by our main analytical result. Also the spectral stability information is based on numerical approximations of the spectra of linearizations. In Sect. 3.4, we explain how spectral stability/instability transfers into nonlinear stability/instability due to the results of [24].

In Sect. 2, we explain by an analytical result given in Theorem 1, which is based on the Implicit Function Theorem, why one can expect ζ-f charts for the existence of solitons which qualitatively look like those shown in Fig. 1. The idea is to consider (2) on $\mathbb{R}$ and rescale the equation, the solutions and the parameters ζ, f in such a way that the damping term $-ia$ becomes $-\varepsilon ia$. For $\varepsilon = 0$, a phase-plane analysis shows the existence of two purely imaginary homoclinics iw^+ and iw^-. Based on the proof of the nondegeneracy of the linearized operator in Theorem 2, an application of the Implicit Function Theorem then allows to continue the purely imaginary homoclinics $iw^\pm$ into the range $\varepsilon > 0$ as two families of proper solution of (2) with $\zeta = \tilde{\zeta}\varepsilon^{-1}$, $f = \tilde{f}\varepsilon^{-3/2}$ and fixed $\tilde{\zeta}$, $\tilde{f}$. Similar results for a variant of the Lugiato–Lefever equation with two-photon absorption have been obtained in [7]. In Sect. 3, we give some details on how Fig. 1 was computed and we compare the regions of existence of stable solutions with previous results from the literature.

2 Existence of Bright Solitons on the Real Line

Let us consider the stationary Lugiato–Lefever equation on the entire real line in the case of anomalous dispersion $d > 0$. In this case the problem reads

$$-da'' - (\mathrm{i} - \zeta)a - |a|^2 a + \mathrm{i}f = 0 \quad \text{on } \mathbb{R}, \qquad a'(0) = 0 \tag{3}$$

and we are looking for homoclinic solutions a with $\lim_{|x|\to\infty} a(x) = a_\infty$. In (3), the condition $a'(0) = 0$ is introduced in order to remove the shift-invariance of solutions.

The following two theorems have been shown in the context of two-photon absorption in [7]. Here we provide additional details for the proof of Theorem 1 as well as a simplified proof of Theorem 2. Since the proof of Theorem 1 is based on the Implicit Function Theorem, the invertibility of a suitably defined linearized operator is necessary. This is provided by Theorem 2 together with the homogeneous Neumann condition at 0 in (3) which removes degenericity.

Theorem 1 *Let $d, \tilde{\zeta} > 0$ and $0 < |\tilde{f}| < \frac{2\sqrt{3}}{9}\tilde{\zeta}^{3/2}$. Then there exists $\varepsilon_0 = \varepsilon_0(d, \tilde{\zeta}, \tilde{f}) > 0$ such that for all $\varepsilon \in (0, \varepsilon_0)$ there are two even homoclinic solutions $a_\varepsilon^\pm$ of (3) with $\zeta = \tilde{\zeta}\varepsilon^{-1}$, $f = \tilde{f}\varepsilon^{-3/2}$ satisfying $\|a_\varepsilon^\pm - a_{\varepsilon,\infty}^\pm\|_{L^2} = O(\varepsilon^{-1/4})$ as $\varepsilon \to 0$.*

Remark 1 The two homoclinic solutions $a_\varepsilon^+, a_\varepsilon^-$ in the above theorem are distinguished as follows: for $\varepsilon > 0$ sufficiently small we have for all $x \in \mathbb{R}$

$$\operatorname{Im} a_\varepsilon^+(x) \geq \lim_{|y|\to\infty} \operatorname{Im} a_\varepsilon^+(y), \qquad \operatorname{Im} a_\varepsilon^-(x) \leq \lim_{|y|\to\infty} \operatorname{Im} a_\varepsilon^-(y).$$

Let us assume that $\varepsilon_0^\pm = \varepsilon_0^\pm(d, \tilde{\zeta}, \tilde{f})$ in the above theorem are chosen to be maximal with respect to the existence of the homoclinics $a_\varepsilon^\pm$. Then we can consider the curves $(0, \varepsilon_0^\pm) \ni \varepsilon \mapsto (\tilde{\zeta}\varepsilon^{-1}, \tilde{f}\varepsilon^{-3/2})$ in the (ζ, f)-plane. Each point on the curve(s) represents a soliton solution of (3) on the real line. By varying the parameters $\tilde{\zeta}$ and $\tilde{f}$, these curves cover regions in the (ζ, f)-plane, where existence of solitons for (3) is guaranteed. The lower envelope $(\tilde{\zeta}\varepsilon_0^{-1}, \tilde{f}\varepsilon_0^{-3/2})$ marks the boundary of the guaranteed existence region. A schematic view of such a curve (red) and the lower envelope (blue) is shown in Fig. 2. It resembles qualitatively the numerically obtained pictures in Fig. 1.

Before we give the proof of Theorem 1, we investigate (3) without damping which leads to the equation

$$-da'' + \zeta a - |a|^2 a + \mathrm{i}f = 0 \quad \text{on } \mathbb{R}, \qquad a'(0) = 0 \tag{4}$$

We obtain the following result on purely imaginary solutions of (4). The result is illustrated in Fig. 3.

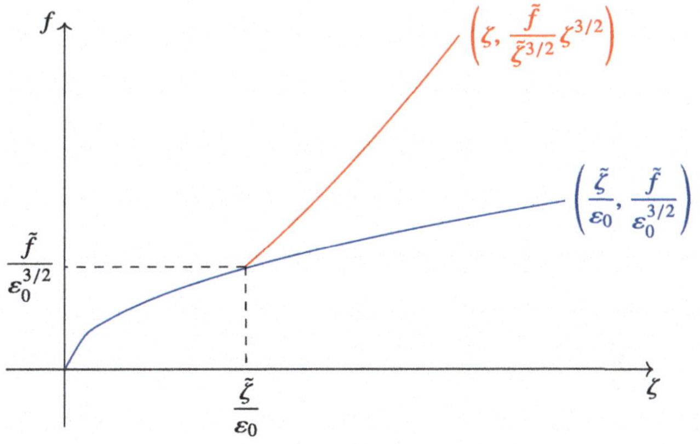

Fig. 2 Illustration of a curve $(\tilde{\zeta}\varepsilon^{-1}, \tilde{f}\varepsilon^{-\frac{3}{2}}) = (\zeta, \tilde{f}\tilde{\zeta}^{-\frac{3}{2}}\zeta^{\frac{3}{2}})$ on which solitons exists (red) and the lower envelope of all curves (blue) obtained by varying $\tilde{\zeta}$ and $\tilde{f}$

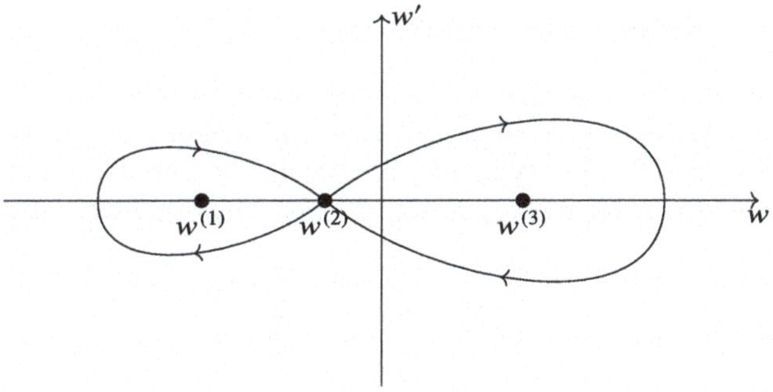

Fig. 3 Two homoclinic orbits w^+, w^- and three equilibria $w^{(1)} < w^{(2)} < w^{(3)}$ of (5) with $w^+(x)$ symmetrically decreasing (right loop) and $w^-(x)$ symmetrically increasing (left loop) towards $w^{(2)}$ as $x \to \pm\infty$

Lemma 1 *Let $d, \zeta > 0$ and $|f| < \frac{2\sqrt{3}}{9}\zeta^{3/2}$. There exist two even homoclinic solutions $\mathrm{i}w^+, \mathrm{i}w^-$ of (4) with $w^+(x)$ symmetrically decreasing and $w^-(x)$ symmetrically increasing as $x \to \pm\infty$.*

Proof Purely imaginary solutions $a = \mathrm{i}w$ of (4) satisfy

$$-dw'' + \zeta w - w^3 + f = 0. \tag{5}$$

The solutions can be studied in the phase plane (w, w') and trajectories of the phase plane correspond to the level sets of the first integral

$$I = -d(w')^2 + \zeta w^2 - \frac{1}{2}w^4 + 2fw. \tag{6}$$

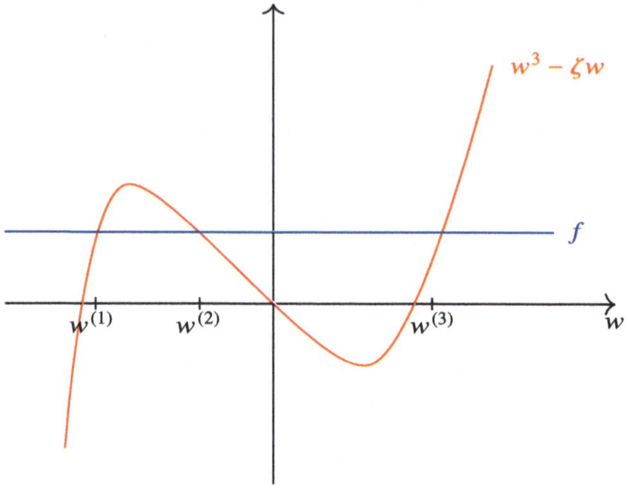

Fig. 4 $-\Delta$ is the slope of the graph at the three equilibria $w^{(1)} < w^{(2)} < w^{(3)}$

Since $d > 0$ one finds that all trajectories are bounded in the (w, w')-plane. Additionally, by the reflection symmetry around points with $w' = 0$ all trajectories of (5) are symmetric with respect to the w-axis. The equilibria of (5) solve the equation

$$\zeta w - w^3 + f = 0. \tag{7}$$

This equation has three distinct real-valued solutions $w^{(1)} < w^{(2)} < w^{(3)}$ for $|f| < \frac{2\sqrt{3}}{9}\zeta^{3/2}$, two real-valued solutions for $|f| = \frac{2\sqrt{3}}{9}\zeta^{3/2}$ and only one real valued solution for $|f| > \frac{2\sqrt{3}}{9}\zeta^{3/2}$. Due to the Ponicaré–Bendixson theorem, homoclinic orbits are only possible in the case $|f| < \frac{2\sqrt{3}}{9}\zeta^{3/2}$. The linearized operator $-d\frac{d^2}{dx^2} + \zeta - 3(w^{(j)})^2$ has a fundamental pair of solutions $\mathrm{e}^{\pm\sqrt{\Delta/d}x}$ with $\Delta = \zeta - 3(w^{(j)})^2$. Since $-\Delta$ is the slope of $w \mapsto w^3 - \zeta w$, we can read from Fig. 4 the sign of Δ for each equilibrium. Thus

$$\Delta < 0 \quad \text{for } j = 1, 3, \qquad \Delta > 0 \quad \text{for } j = 2$$

and linear stability analysis allows us to characterize the equilibria of the nonlinear system: for $j = 2$ we have an unstable saddle, and for $j = 1, 3$ we have stable centers surrounded by periodic orbits. Since the stable and the unstable manifold of the saddle $w^{(2)}$ are symmetric around the w-axis, they are identical and thus, provide the two homoclinic orbits. □

The homoclinic orbits from Lemma 1 represent soliton solutions $\mathrm{i}w^{\pm} = \mathrm{i}\tilde{w}^{\pm} + \mathrm{i}w^{\pm}_{\infty}$ of (4) with $\tilde{w}^{\pm} \in L^2(\mathbb{R})$ and $w^{\pm}_{\infty} = \lim_{|x|\to\infty} w^{\pm} = w^{(2)}$, where $w^{(2)}$ is the middle of the three zeroes of (7). Next we prove the nondegeneracy of the two homoclinics.

Theorem 2 *Let $d, \zeta > 0$, $|f| < \frac{2\sqrt{3}}{9}\zeta^{3/2}$ and define $g(a) = |a|^2 a - \mathrm{i}f$. If $w^\pm = \tilde{w}^\pm + w_\infty^\pm$ is one of the two homoclinic solutions of Lemma 1 then the linearized operator*

$$L : \begin{cases} H^2(\mathbb{R};\mathbb{C}) \mapsto L^2(\mathbb{R};\mathbb{C}), \\ \qquad \phi \mapsto -d\phi'' + \zeta\phi - Dg(\mathrm{i}w^\pm)\phi \end{cases}$$

is nondegenerate in the sense that $\ker_{H^2}(L) = \operatorname{span}\{\mathrm{i}\frac{d}{dx}w^\pm\}$.

Remark 2 Note that $Dg(a)z := \frac{d}{dt}g(a+tz)\big|_{t=0} = 2|a|^2 z + a^2\bar{z}$. If instead of $H^2(\mathbb{R};\mathbb{C})$ we set the domain of the differential operator as $\mathcal{H} := \{\phi \in H^2(\mathbb{R};\mathbb{C}) : \phi'(0) = 0\}$ then we get $\ker_{\mathcal{H}}(L) = \{0\}$. This is true since $\mathrm{i}\frac{d}{dx}w^\pm \notin \mathcal{H}$ because by (5) we have

$$d\frac{d^2}{dx^2}w^\pm(0) = \zeta w^\pm(0) - w^\pm(0)^3 + f$$

which is non-zero since the zeroes of the right-hand side are the equilibria $w^{(1)}, w^{(2)}, w^{(3)}$ and the homoclinic orbits approach $w^{(2)}$ only at infinity.

For the proof of Theorem 2 we split the linearized operator L into it real and imaginary part, i.e., if $\phi \in H^2(\mathbb{R};\mathbb{C})$ is split into $\phi = \phi_1 + \mathrm{i}\phi_2$ with real-valued functions $\phi_1, \phi_2 \in H^2(\mathbb{R};\mathbb{R})$ then we have $L\phi = L_1\phi_1 + \mathrm{i}L_2\phi_2$ with

$$L_1 = -d\frac{d^2}{dx^2} + \zeta - (w^\pm)^2, \quad L_2 = -d\frac{d^2}{dx^2} + \zeta - 3(w^\pm)^2.$$

The proof therefore amounts to showing the following two statements: $\ker_{H^2}(L_2) = \operatorname{span}\{\frac{d}{dx}w^\pm\}$ (Lemma 2) and $\ker_{H^2}(L_1) = \{0\}$ (Lemma 3).

Lemma 2 *Let $d, \zeta > 0$ and $|f| < \frac{2\sqrt{3}}{9}\zeta^{3/2}$. Then*

$$\ker_{H^2}(L_2) = \operatorname{span}\left\{\frac{d}{dx}w^\pm\right\}.$$

Proof Differentiating (5) and using the exponential decay of $w^\pm(x) - w^{(2)}$ and all of its derivatives to 0 as $|x| \to \infty$, we see that $\frac{d}{dx}w^\pm \in \ker_{H^2}(L_2)$. Since $u := \frac{d}{dx}w^\pm$ satisfies

$$-u'' + r(x)u = \text{ on } \mathbb{R} \text{ with } r(x) = d^{-1}(\zeta - 3w^\pm(x)^2)$$

and $\lim_{x\to\infty} r(x) = d^{-1}(\zeta - 3(w^{(2)})^2) > 0$ (cf. Fig. 4), we can apply Theorem 3.3 from [1] and find that up to multiplicative factors $u = \frac{d}{dx}w^\pm$ is the only solution of $L_2 u = 0$ which decays to 0 at ∞. □

Lemma 3 *Let $d, \zeta > 0$ and $0 < |f| < \frac{2\sqrt{3}}{9}\zeta^{3/2}$. Then* $\ker_{H^2}(L_1) = \{0\}$.

Proof Let $w = w^+$ or $w = w^-$. Suppose $\phi \in \ker_{H^2}(L_1)$ and split $\phi := \phi_{\text{even}} + \phi_{\text{odd}}$ into even and odd part. Then $\phi_{\text{even}}, \phi_{\text{odd}} \in \ker_{H^2}(L_1)$ with $\phi'_{\text{even}}(0) = 0$ and $\phi_{\text{odd}}(0) = 0$.

Let us first consider ϕ_{even}. We claim first that either $\phi_{\text{even}} \equiv 0$ or ϕ_{even} has no zero on $\mathbb{R}$. Assume for contraction that ϕ_{even} has a first positive zero at $x_0 > 0$ so that we can assume $\phi_{\text{even}} > 0$ on $(0, x_0)$. Since $\phi_{\text{even}}(x) \to 0$ as $x \to \infty$ there is $x_1 \in (x_0, \infty]$ such that $\phi_{\text{even}} < 0$ on (x_0, x_1) and $\lim_{x \to x_1} \phi_{\text{even}}(x) = 0$. We compute the identity

$$
\begin{aligned}
0 &= \int_{x_0}^{x_1} \left((L_2 w')\phi_{\text{even}} - (L_1\phi_{\text{even}})w'\right) dx \\
&= \int_{x_0}^{x_1} -d(w'''\phi_{\text{even}} - \phi''_{\text{even}} w'_1) - 2w^2 w'\phi_{\text{even}}\, dx \qquad (8) \\
&= -d\left[w''\phi_{\text{even}} - \phi'_{\text{even}} w'\right]_{x_0}^{x_1} - \int_{x_0}^{x_1} 2w^2 w'\phi_{\text{even}}\, dx \\
&= -d\Big(\underbrace{\phi'_{\text{even}}(x_0)}_{<0} w'(x_0) - \underbrace{\phi'_{\text{even}}(x_1)}_{\geq 0} w'(x_1)\Big) - \int_{x_0}^{x_1} 2w^2 w' \underbrace{\phi_{\text{even}}}_{<0}\, dx.
\end{aligned}
$$

In case $w = w^-$, we have $w' > 0$ on $(0, \infty)$ so that the latter expression is positive. In case $w = w^+$ we have $w' < 0$ on $(0, \infty)$ and hence the latter expression is negative. In both cases, we have reached a contradiction and thus ϕ_{even} has no zero on $\mathbb{R}$ or vanishes identically on $\mathbb{R}$. Testing (5) with ϕ_{even} and integrating twice by parts, we obtain

$$
\begin{aligned}
-\int_{\mathbb{R}} f\phi_{\text{even}}\, dx &= \int_{\mathbb{R}} -dw''\phi_{\text{even}} + (\zeta - w^2)w\phi_{\text{even}}\, dx \\
&= \int_{\mathbb{R}} -dw\phi''_{\text{even}} + (\zeta - w^2)w\phi_{\text{even}}\, dx = 0.
\end{aligned}
$$

Since ϕ_{even} does not change sign this implies $\phi_{\text{even}} \equiv 0$.

Now let us consider ϕ_{odd}. Suppose $\phi_{\text{odd}} \not\equiv 0$ so that with $x_0 = 0$ being the first zero of ϕ_{odd} on $[0, \infty)$ there is $x_1 \in (0, \infty]$ such that we may assume $\phi_{\text{odd}} < 0$ on $(0, x_1)$ and $\lim_{x \to x_1} \phi_{\text{odd}}(x) = 0$. Therefore ϕ_{odd} also satisfies identity (8) with $x_0 = 0$. The same case distinction $w = w^-$ and $w = w^+$ as for ϕ_{even} leads to a contradiction and directly shows that $\phi_{\text{odd}} \equiv 0$ on $\mathbb{R}$. This finishes the proof. □

We can now prove the main result of Theorem 1.

Proof of Theorem 1 For $\varepsilon > 0$ let us consider the rescaled version of (3) given by

$$
-du'' + (\tilde{\zeta} - \varepsilon \mathrm{i})u - |u|^2 u + \mathrm{i}\tilde{f} = 0 \quad \text{on } \mathbb{R}. \qquad (9)
$$

Here, u solves (9) with $\tilde{\zeta}$, $\tilde{f}$ if and only if $a(x) := \varepsilon^{-1/2}u(\varepsilon^{-1/2}x)$ solves (3) on $\mathbb{R}$ with $\zeta = \tilde{\zeta}\varepsilon^{-1}$ and $f = \tilde{f}\varepsilon^{-3/2}$. Recall the definition of the Hilbert-space $\mathcal{H} := \{\phi \in H^2(\mathbb{R}\ \mathbb{C}) : \phi'(0) = 0\}$ over $\mathbb{R}$ with scalar product $\langle \phi, \psi \rangle_{H^2} = \operatorname{Re} \int_{\mathbb{R}} \phi'' \bar{\psi}'' + \phi\bar{\psi}\, dx$ and define the real Hilbert-space $\mathcal{L} = \{\phi \in L^2(\mathbb{R}; \mathbb{C}) : \phi(x) = \phi(-x)\}$ with scalar product $\langle \phi, \psi \rangle_{L^2} = \operatorname{Re} \int_{\mathbb{R}} \phi\bar{\psi}\, dx$. For $z \in \mathbb{C}$ let $g(z) := |z|^2 z$. We use the notation $u = \tilde{u} + u_{\varepsilon,\infty}$ where $\tilde{u} \in \mathcal{H}$ and where $u_{\varepsilon,\infty} \in \mathbb{C}$ is a constant solution of (9) and a differentiable function of ε such that $u_{0,\infty} = \mathrm{i}w^{(2)}$. We can then rewrite (9) as

$$0 = -d\tilde{u}'' + \tilde{\zeta}\tilde{u} - \varepsilon \mathrm{i}\tilde{u} - \big(g(\tilde{u} + u_{\varepsilon,\infty}) - g(u_{\varepsilon,\infty})\big) \tag{10}$$

and we seek solutions $\tilde{u} \in \mathcal{H}$ by the Implicit Function Theorem. For this we set

$$F : \begin{cases} \mathcal{H} \times \mathbb{R} \to \mathcal{L} \\ (\tilde{u}, \varepsilon) \mapsto \left(-d\frac{d^2}{dx^2} + \tilde{\zeta}\right)\tilde{u} - \varepsilon \mathrm{i}\tilde{u} - g(\tilde{u} + u_{\varepsilon,\infty}) + g(u_{\varepsilon,\infty}). \end{cases}$$

In the following, we take $w = w^+$ or $w = w^-$ as one of the two homoclinics from Lemma 1. By definition of $w = \tilde{w} + w_\infty$ with $w_\infty = w^{(2)}$ and $\tilde{w} \in L^2(\mathbb{R}; \mathbb{R})$, we have

$$F(\mathrm{i}\tilde{w}, 0) = 0.$$

Consider the Fréchet-derivative $D_{\tilde{u}}F(\mathrm{i}\tilde{w}, 0) = -d\frac{d^2}{dx^2} + \tilde{\zeta} - Dg(\mathrm{i}w) : \mathcal{H} \to \mathcal{L}$. Clearly $D_{\tilde{u}}F(\mathrm{i}\tilde{w}, 0) = L$ with L being the operator from Theorem 2 and by Remark 2 we have

$$\ker_{\mathcal{H}} L = \{0\}$$

so that in particular L is injective. Let $L_0 := -d\frac{d^2}{dx^2} + (\zeta - \mathrm{i} - Dg(\mathrm{i}w_\infty))$ so that $L = L_0 + K$ where

$$K : \begin{cases} \mathcal{L} \to \mathcal{L} \\ \phi \mapsto (\tilde{w}^2 + 2\tilde{w}w_\infty)(2\phi - \bar{\phi}) \end{cases}$$

is a bounded and symmetric operator. If we consider $K : \mathcal{H} \to \mathcal{L}$ then it is compact due to the decay properties of $\tilde{w}$ and furthermore $L = L_0 + K : \mathcal{H} \to \mathcal{L}$ is self-adjoint. Finally, $L_0 : \mathcal{H} \to \mathcal{L}$ is an isomorphism since $L_0\phi = -d\phi'' + \alpha\phi + \beta\bar{\phi}$

with $\alpha = \tilde{\zeta} - 2w_\infty^2$, $\beta = w_\infty^2$ being two positive real numbers leads to

$$\|L_0\phi\|_{H^2}^2 = \int_{\mathbb{R}} d|\phi''|^2 + (\alpha^2+\beta^2)|\phi|^2 + 2\alpha\beta\,\mathrm{Re}(\phi^2) - 2d\,\mathrm{Re}(\phi''(\alpha\bar{\phi}+\beta\phi))\,dx$$
$$\geq \int_{\mathbb{R}} d|\phi''|^2 + (\alpha-\beta)^2|\phi|^2 + 2d(\alpha-\beta)|\phi'|^2\,dx.$$

Since $\alpha - \beta = \tilde{\zeta} - 3w_\infty^2 > 0$ the latter is equivalent to the H^2–norm of ϕ. Thus, L is an injective, compact perturbation of an isomorphism and thus has a bounded inverse $L^{-1} : \mathcal{L} \to \mathcal{H}$. The Implicit Function Theorem now yields a curve $\varepsilon \mapsto \tilde{u}_\varepsilon$ such that $F(\tilde{u}_\varepsilon, \varepsilon) = 0$, $\tilde{u}_0 = \mathrm{i}\tilde{w}$ and the triple $(\tilde{u}_\varepsilon + u_{\varepsilon,\infty}, \tilde{\zeta}, \tilde{f})$ solves (9) and hence the triple $(a_\varepsilon, \zeta, f)$ solves (3) with $a_\varepsilon(x) = \varepsilon^{-1/2}u(\varepsilon^{-1/2}x)$, $\zeta = \tilde{\zeta}\varepsilon^{-1}$ and $f = \tilde{f}\varepsilon^{-3/2}$. Finally we observe that

$$\|a_\varepsilon - a_{\varepsilon,\infty}\|_{L^2} = \varepsilon^{-\frac{1}{4}}\|u_\varepsilon - u_{\varepsilon,\infty}\|_{L^2} = \varepsilon^{-\frac{1}{4}}(1+o(1))\|\tilde{w}\|_{L^2}.$$

The two different choices $\tilde{u}_0 = \mathrm{i}\tilde{w}^+$ or $u_0 = \mathrm{i}\tilde{w}^-$ lead to $a_\varepsilon^\pm$. This finishes the proof of Theorem 1. □

3 Numerical Experiments

In this section, we show how to use the numerical path continuation software `pde2path` in order to compute numerical approximations to solutions of (3). Here the value $d > 0$ is fixed. The procedure is performed by numerically calculating 2π-periodic approximate solutions of (2) using the results of Sect. 2. For the `Matlab` toolbox `pde2path` we refer to [6, 28].

3.1 Continuation in $\tilde{f}$ and ε

We do this by considering the rescaled version (9) on $[\pi, 2\pi]$ with Neumann boundary conditions at the endpoints. We first fix $\tilde{\zeta} > 0$ and set $\tilde{f} = \varepsilon = 0$. In this case the functions

$$+\mathrm{i}\sqrt{2\tilde{\zeta}}\left(\cosh\left((x-\pi)\sqrt{\tilde{\zeta}/d}\right)\right)^{-1} \text{ and } -\mathrm{i}\sqrt{2\tilde{\zeta}}\left(\cosh\left((x-\pi)\sqrt{\tilde{\zeta}/d}\right)\right)^{-1} \quad (11)$$

solve the equation in (9) exactly. Moreover, they satisfy the Neumann boundary condition at $x = \pi$ exactly and (for sufficiently small $d > 0$) also approximately the Neumann boundary condition at $x = 2\pi$. We then continue both of these particular solutions in $\tilde{f} \in [0, \tilde{f}_0]$ and thus obtain solutions which correspond to the homoclinic solution iw^+ in case of the "+"-choice and iw^- in case of the "−"-choice in (11). Subsequently, each of these solutions for $\tilde{f} \in [0, \tilde{f}_0]$ is then continued in $\varepsilon \in [0, \varepsilon_0]$, so that we obtain a triply-indexed set $u^{\pm}_{\varepsilon,\tilde{f},\tilde{\zeta}}$ of approximate solutions of (9) with Neumann boundary conditions at $x = \pi$ and $x = 2\pi$.

3.2 Rescaling Back to (2) and Extending by a Constant

Next $u^{\pm}_{\varepsilon,\tilde{f},\tilde{\zeta}}$ is rescaled by

$$a^{\pm}_{\varepsilon,\tilde{f},\tilde{\zeta}}(x) := \varepsilon^{-1/2} u^{\pm}_{\varepsilon,\tilde{f},\tilde{\zeta}}(\varepsilon^{-1/2}(x-\pi)+\pi)$$

and thus $a^{\pm}_{\varepsilon,\tilde{f},\tilde{\zeta}}$ approximately solves

$$-da'' + (\zeta - \mathrm{i})a - |a|^2 a + \mathrm{i}f = 0, \quad a'(\pi) = a'((1+\sqrt{\varepsilon})\pi) = 0$$

with $\zeta = \frac{\tilde{\zeta}}{\varepsilon}$ and $f = \frac{\tilde{f}}{\varepsilon^{3/2}}$. These solutions are extended as constants to $[(1+\sqrt{\varepsilon})\pi, 2\pi]$ and then reflected vertically at $x = \pi$. Refining each of these functions with a Newton step yields approximate 2π-periodic solitons solving (2) for the parameters (ζ, f).

In Fig. 5, the continuation routine for $d = 0.01$ and $\tilde{\zeta} = 5$ is shown. First the approximate solution on the NLS is continued in $\tilde{f}$. Subsequently fixing $\tilde{f} = 2.9$, we continue in ε. The $\tilde{f}$-continuation and the ε-continuation are done both for the "+" (top row in Fig. 5) and for the "−" choice (bottom row in Fig. 5) of the starting soliton approximation in (11).

Let us briefly comment the shape of the curves in Fig. 5. Concerning the continuation in $\tilde{f}$, obviously the top left panel in Fig. 5 shows that for each $\tilde{f}$ (with $\varepsilon = 0$) we find two 2π-periodic solutions with different L^2-norm whereas the bottom left panel shows that we obtain only one 2π-periodic solution. This can be explained in the phase plane for (4) sketched in Fig. 3. For $0 \le \tilde{f} < \frac{2\sqrt{3}}{9}\tilde{\zeta}^{3/2}$, the two homoclinics separate periodic solutions inside the homoclinics from those outside the homoclinics. Seeking 2π-periodic solutions the numerical algorithm starts in both cases for $\tilde{f} = 0$ with a 2π-periodic approximate solution close to but lying inside one of the two homoclinics w^+ (top left) or w^- (bottom left). As $\tilde{f}$ approaches $\frac{2\sqrt{3}}{9}\tilde{\zeta}^{3/2}$ the two equilibria $w^{(1)}, w^{(2)}$ collide to one equilibrium $w^{(1)} = w^{(2)}$ while $w^{(3)}$ survives. Likewise the homoclinic w^- degenerates to a point while the homoclinic w^+ survives. Therefore, at the turning point of maximal $\tilde{f}$ as $\tilde{f}$ starts to decrease again one sees in the top left panel a switching from periodic solutions inside of w^+ to periodic solutions outside of w^+. The periodic solutions outside of w^+ are less localized, have a larger L^2-norm and can therefore be seen on the upper one of the two curves. In the lower left panel of Fig. 5 one finds only

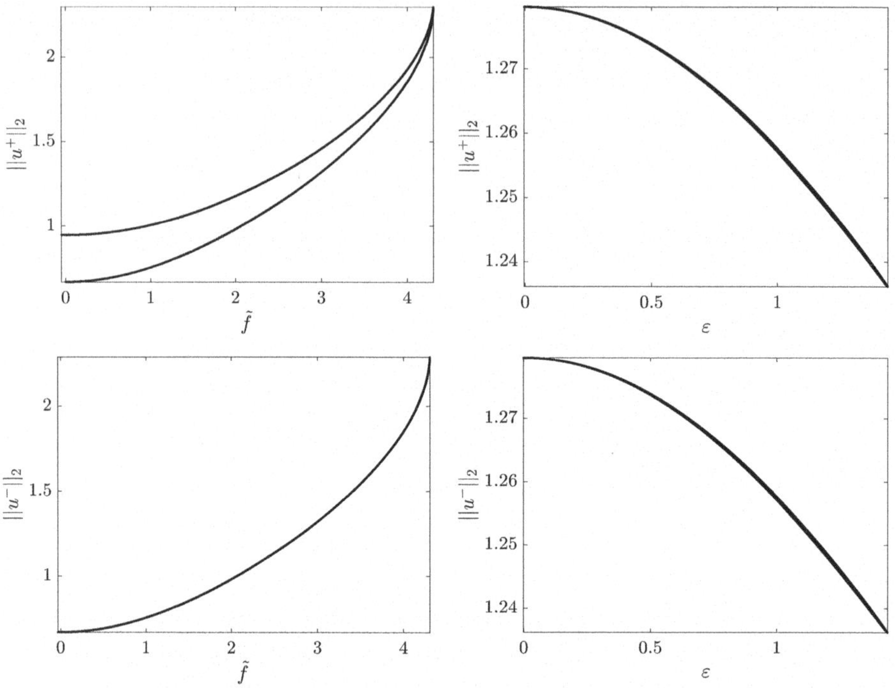

Fig. 5 Continuation of the approximate solution in $\tilde{f}$ and then in ε. Top row starts from "+" case of (11), bottom row from "−" case. Top left shows a switch at the turning point between periodic solutions inside the homoclinic w^+ (smaller L^2-norm) to periodic solutions outside the homoclinic w^+ (larger L^2-norm). Bottom left: no switch of periodic solutions since the homoclinic w^- degenerates to a point when f reaches the maximal value

one curve of periodic solutions. They lie inside the homoclinic w^-. And since the homoclinic w^- degenerates to a point as $\tilde{f}$ approaches the maximal possible value, no switching between periodic solutions inside and outside of w^- can happen.

For the subsequent continuation in ε (right panels in Fig. 5), only the more localized 2π-periodic solution lying on the lower curve was considered. It is an open problem to understand why the solutions obtained by continuing (both in $\tilde{f}$ and in ε) the starting approximation $1/\cosh$ or the $-1/\cosh$ have the same L^2-norm although they look qualitatively different, cf. Fig. 6 for a comparison of $|a|$ and Fig. 7 for a comparison of $\operatorname{Re} a$, $\operatorname{Im} a$.

3.3 An Example

As an example, for fixed $d = 0.01$, $\tilde{\zeta} = 5$ we initially set $\tilde{f} = \varepsilon = 0$, and first continued the $\pm 1/\cosh$-type soliton from (11) with respect to $\tilde{f} \in [0, 2.9]$. For fixed $\tilde{f} = 2.9$ the continuation is then done with respect to $\varepsilon \in [0, 1]$. With $\varepsilon = 0.5$ the corresponding detuning and forcing values are $\zeta = \tilde{\zeta}\varepsilon^{-1} = 10$ and $f = \tilde{f}\varepsilon^{-3/2} =$

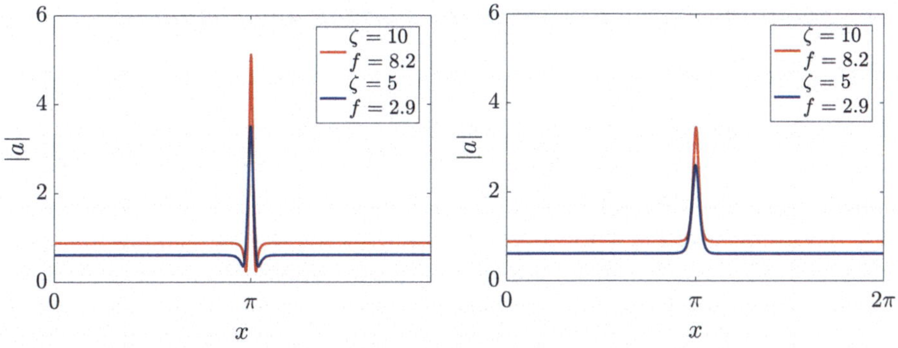

Fig. 6 Approximate solutions of (2). Continuation started from the $1/\cosh$-type soliton (left), $-1/\cosh$-type soliton (right)

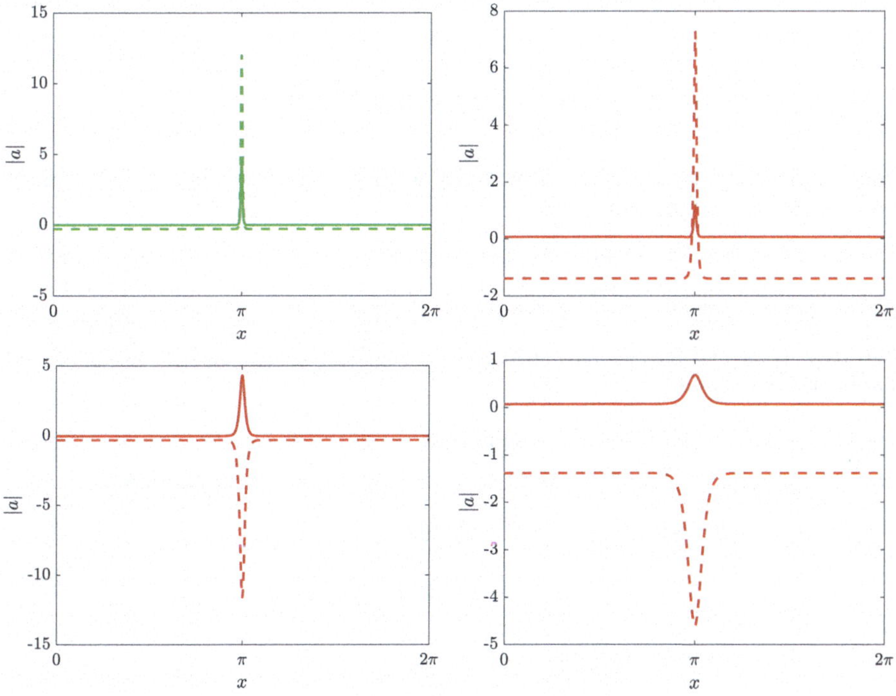

Fig. 7 Real part (solid line) and imaginary part (dashed line) of selected solitons. Top: two solutions of type a^+. Left (stable) for $\zeta = 80$, $f = 23$. Right (unstable) for $\zeta = 20$ and $f = 25$. Bottom: two unstable solutions of type a^-. Left for $\zeta = 80$, $f = 23$. Right for $\zeta = 20$, $f = 25$

8.20. Similarly with $\varepsilon = 1$ the corresponding detuning and forcing values are $\zeta = \tilde{\zeta} = 5$ and $f = \tilde{f} = 2.9$. The resulting solutions are shown in Fig. 6, where on the left the continuation was started with the $1/\cosh$-type soliton and on the right with the $-1/\cosh$-type soliton from (11).

3.4 Scanning Large Ranges of the Forcing and Detuning Variable

The procedure described in Sects. 3.1 and 3.2 is now repeated for all $\tilde{\zeta} \in (0, \tilde{\zeta}_0]$. Each value of $\tilde{f}$ and $\tilde{\zeta}$ generates two curves of solutions $a^{\pm}_{\varepsilon,\tilde{f},\tilde{\zeta}}$ with ε running through a maximal interval $[0, \varepsilon_0^{\pm}]$. The curves $\varepsilon \mapsto (\tilde{\zeta}\varepsilon^{-1}, \tilde{f}\varepsilon^{-3/2})$ with $\varepsilon \in [0, \varepsilon_0^{\pm}]$ are shown in Fig. 1 for $d = 0.01$ with "+" on the left and the "−" case on the right. We included the information on spectral stability by coloring points in green/red where spectrally stable/unstable solitons were found. Due to the result of [24], spectral stability/instability translates into nonlinear stability/instability. Note that the weaker orbital asymptotic stability result given in [24] is due to the shift-invariance which we have eliminated by the requirement $a'(0) = 0$. Observe that we can find both stable and unstable soliton solutions, cf. Fig. 7.

3.5 Border Curves in the ζ-f Stability Chart

Here we comment on the quantitative form of the border curves in Fig. 1. We begin with the analysis of the equilibria.

Existence Regions for Equilibria Our analysis in Sect. 2 consisted of an Implicit Function Theorem argument that continues a non-degenerate homoclinic solution of Eq. (4) into a solution of (3). For the existence of homoclinic solutions three equilibria (two stable, one unstable) are typically required. Let us compare the equilibria for (4) and (3):

(i) $|f| \le \frac{2\sqrt{3}}{9}\tilde{\zeta}^{\frac{3}{2}} \Leftrightarrow$ (4) has three equilibria

(ii) $\zeta > \sqrt{3}$, $|\frac{27}{2}f^2 - \zeta(\zeta^2+9)| < (\zeta^2-3)^{\frac{3}{2}} \Leftrightarrow$ (3) has three equilibria

Statement (i) follows from the observation that

$$x^3 - \zeta x - f = 0 \Leftrightarrow \mathrm{i}x \text{ solves (4)}$$

and (ii) follows from the fact that

$$x^3 f^2 - 2\zeta f x^2 + (1+\zeta^2)x - f = 0 \Leftrightarrow x + \mathrm{i}(-\zeta x + f x^2) \text{ solves (3).}$$

Analyzing the two cubic equations for x leads to the above restrictions in (i) and (ii), respectively.

A result from [11] states that highly localized solutions of (3) serve as good approximations for 2π-periodic solutions of (2). Therefore, one could conjecture from (ii) that f^2 should not exceed $(\frac{2}{27}\zeta(\zeta^2+9) - \frac{2}{27}(\zeta^2-3)^{\frac{3}{2}}, \frac{2}{27}\zeta(\zeta^2+9) + \frac{2}{27}(\zeta^2-3)^{\frac{3}{2}})$ and we included these bounds as blue curves in Fig. 1. Indeed, we observe that the upper bound $\frac{2}{27}\zeta(\zeta^2+9)+\frac{2}{27}(\zeta^2-3)^{\frac{3}{2}}$ coincides very well with our numerical results. However, the lower bound slightly exceeds the region of existence of solitons.

Comparison with Other Heuristic and Rigorous Bounds In [11] another lower bound

$$f^2 > \frac{8\zeta}{\pi^2} \tag{12}$$

is suggested, and its validity for large values of $\frac{f}{\sqrt{d}}$ is discussed. This bound is included in black in Fig. 1 and coincides well with our numerical results.

Next, let us mention that in [16], a bound for non-existence of nontrivial solutions is rigorously proven. It states that if

$$\zeta > 6f^2(1+12f^2\pi^2d^{-1})^2 \tag{13}$$

then nontrivial solutions of (2) cease to exist. As $\frac{f^2\pi^2}{8} < 6f^2(1+12f^2\pi^2d^{-1})^2$, these two bounds are compatible.

Finally, we compare our results to the ones obtained in [8] by bifurcation theory with the detuning ζ as a bifurcation parameter while keeping dispersion d and forcing f fixed. When running through the trivial branch of constant solutions beginning with small detuning, one finds branches of frequency combs bifurcating at designated points. The bifurcating branches can be classified by their pattern near the bifurcation point. For example, in the case of anomalous dispersion $d > 0$, it has been observed in [8] that one-solitons occur on the last of the bifurcating branches. This branch reaches its maximal value of detuning at a turning point where it returns and eventually meets other branches via secondary bifurcations. We observe (with a rigorous proof yet missing) that it is exactly this turning point of the bifurcating branch of one-solitions that coincides with the lower envelope of Fig. 1. In other words, if we draw a horizontal line in Fig. 1 at the height of the fixed forcing f and find the rightmost point of the existence region (lying on the black curve) then this point coincides with a turning point of the bifurcation diagram and represents the most localized one-soliton.

Acknowledgement Funded by the Deutsche Forschungsgemeinschaft (DFG, German Research Foundation)—Project-ID 258734477—SFB 1173.

References

1. Berezin, F.A., Shubin, M.A.: The Schrödinger equation. Math. Appl. (Soviet Ser.), vol. 66. Kluwer Academic Publishers Group, Dordrecht (1991). Translated from the 1983 Russian edition by Rajabov, Yu., Leĭtes, D.A., Sakharova, N.A., and revised by Shubin, with contributions by Litvinov, G.L., Leĭtes

2. Chembo, Y.K., Menyuk, C.R.: Spatiotemporal Lugiato–Lefever formalism for Kerr-comb generation in whispering-gallery-mode resonators. Phys. Rev. A **87**, 053852 (2013)
3. Chembo, Y.K., Yu, N.: Modal expansion approach to optical-frequency-comb generation with monolithic whispering-gallery-mode resonators. Phys. Rev. A **82**, 033801 (2010)
4. Delcey, L., Haragus, M.: Periodic waves of the Lugiato–Lefever equation at the onset of Turing instability. Philos. Trans. R. Soc. A **376**(2117), 20170188 (2018)
5. Diddams, S.A., Udem, T., Bergquist, J.C., Curtis, E.A., Drullinger, R.E., Hollberg, L., Itano, W.M., Lee, W.D., Oates, C.W., Vogel, K.R., Wineland, D.J., Reichert, J., Holzwarth, R.: An optical clock based on a single trapped 199 Hg+ Ion. Science (New York) **24**(13), 881–883 (1999)
6. Dohnal, T., Rademacher, J., Uecker, H., Wetzel, D.: pde2path 2.0: multi-parameter continuation and periodic domains (2014)
7. Gärtner, J., Mandel, R., Reichel, W.: The Lugiato–Lefever equation with nonlinear damping caused by two photon absorption. CRC 1173-Preprint 2018/44, Karlsruhe Institute of Technology (2018). http://www.waves.kit.edu/downloads/CRC1173_Preprint_2018-44.pdf
8. Gärtner, J., Trocha, P., Mandel, R., Koos, C., Jahnke, T., Reichel, W.: Bandwidth and conversion efficiency analysis of dissipative Kerr soliton frequency combs based on bifurcation theory. Phys. Rev. A **100**, 033819 (2019). https://link.aps.org/doi/10.1103/PhysRevA.100.033819
9. Godey, C.: A bifurcation analysis for the Lugiato–Lefever equation. Eur. Phys. J. D **71**(5), 131 (2017)
10. Godey, C., Balakireva, I.V., Coillet, A., Chembo, Y.K.: Stability analysis of the spatiotemporal Lugiato–Lefever model for Kerr optical frequency combs in the anomalous and normal dispersion regimes. Phys. Rev. A **89**, 063814 (2014)
11. Herr, T., Brasch, V., Jost, J., Wang, C., Kondratiev, N., Gorodetsky, M., Kippenberg, T.: Temporal solitons in optical microresonators. Nat. Photon. **8**, 145–152 (2014)
12. Herr, T., Hartinger, K., Riemensberger, J., Wang, C., Gavartin, E., Holzwarth, R., Gorodetsky, M., Kippenberg, T.: Universal formation dynamics and noise of Kerr-frequency combs in microresonators. Nat. Photon. **6**, 480–487 (2012)
13. Jahnke, T., Mikl, M., Schnaubelt, R.: Strang splitting for a semilinear Schrödinger equation with damping and forcing. J. Math. Anal. Appl. **455**(2), 1051–1071 (2017)
14. Lugiato, L.A., Lefever, R.: Spatial dissipative structures in passive optical systems. Phys. Rev. Lett. **58**, 2209–2211 (1987)
15. Mandel, R.: Global secondary bifurcation, symmetry breaking and period-doubling. Topol. Methods Nonlinear Anal. **53**(2), 779–800 (2019). https://doi.org/10.12775/tmna.2019.025
16. Mandel, R., Reichel, W.: A priori bounds and global bifurcation results for frequency combs modeled by the Lugiato–Lefever equation. SIAM J. Appl. Math. **77**(1), 315–345 (2017)
17. Marin-Palomo, P., Kemal, J.N., Karpov, M., Kordts, A., Pfeifle, J., Pfeiffer, M.H., Trocha, P., Wolf, S., Brasch, V., Anderson, M.H., Rosenberger, R., Vijayan, K., Freude, W., Kippenberg, T.J., Koos, C.: Microresonator-based solitons for massively parallel coherent optical communications. Nature **546**(7657), 274–279 (2017)
18. Miyaji, T., Ohnishi, I., Tsutsumi, Y.: Bifurcation analysis to the Lugiato–Lefever equation in one space dimension. Phys. D **239**(23-24), 2066–2083 (2010)
19. Parra-Rivas, P., Gomila, D., Gelens, L., Knobloch, E.: Bifurcation structure of localized states in the Lugiato–Lefever equation with anomalous dispersion. Phys. Rev. E **97**(4), 042204, 20 (2018)
20. Parra-Rivas, P., Gomila, D., Matias, M.A., Coen, S., Gelens, L.: Dynamics of localized and patterned structures in the Lugiato–Lefever equation determine the stability and shape of optical frequency combs. Phys. Rev. A **89**, 043813 (2014)
21. Parra-Rivas, P., Knobloch, E., Gomila, D., Gelens, L.: Dark solitons in the Lugiato–Lefever equation with normal dispersion. Phys. Rev. A **93**(6), 1–17 (2016)

22. Pfeifle, J., Brasch, V., Lauermann, M., Yu, Y., Wegner, D., Herr, T., Hartinger, K., Schindler, P., Li, J., Hillerkuss, D., Schmogrow, R., Weimann, C., Holzwarth, R., Freude, W., Leuthold, J., Kippenberg, T.J., Koos, C.: Coherent terabit communications with microresonator Kerr frequency combs. Nat. Photon. **8**, 375–380 (2014)
23. Pfeifle, J., Coillet, A., Henriet, R., Saleh, K., Schindler, P., Weimann, C., Freude, W., Balakireva, I.V., Larger, L., Koos, C., Chembo, Y.K.: Optimally coherent Kerr combs generated with crystalline whispering gallery mode resonators for ultrahigh capacity fiber communications. Phys. Rev. Lett. **114**(9), 1–5 (2015)
24. Stanislavova, M., Stefanov, A.G.: Asymptotic stability for spectrally stable Lugiato–Lefever solitons in periodic waveguides. J. Math. Phys. **59**(10), 101502, 12 (2018)
25. Suh, M.G., Vahala, K.: Soliton microcomb range measurement. Science **887**, 884–887 (2017)
26. Trocha, P., Karpov, M., Ganin, D., Pfeiffer, M.H., Kordts, A., Wolf, S., Krockenberger, J., Marin-Palomo, P., Weimann, C., Randel, S., Freude, W., Kippenberg, T.J., Koos, C.: Ultrafast optical ranging using microresonator soliton frequency combs. Science **359**(6378), 887–891 (2018)
27. Udem, T., Holzwarth, R., Hänsch, T.W.: Optical frequency metrology. Nature **416**(6877), 233–237 (2002)
28. Uecker, H., Wetzel, D., Rademacher, J.D.M.: pde2path—a Matlab package for continuation and bifurcation in 2D elliptic systems. Numer. Math. Theory Methods Appl. **7**(1), 58–106 (2014)

Error Analysis of Discontinuous Galerkin Discretizations of a Class of Linear Wave-type Problems

Marlis Hochbruck and Jonas Köhler

Abstract In this paper we consider central fluxes discontinuous Galerkin space discretizations of a general class of wave-type equations of Friedrichs' type. This class includes important examples such as Maxwell's equations and wave equations. We prove an optimal error bound which holds under suitable regularity assumptions on the solution. Our analysis is performed in a framework of evolution equations on a Hilbert space and thus allows for the combination with various time integration schemes.

1 Introduction

The aim of this paper is to provide a rigorous error analysis of central fluxes discontinuous Galerkin (dG) space discretizations of a large class of linear wave-type equations of the following form. For a given initial value u^0 we seek a solution u such that

$$\begin{cases} M\partial_t u = \widetilde{\mathcal{L}}u + g, & \mathbb{R}_+ \times \Omega, \quad (1a) \\ u(0) = u^0, & \Omega, \quad (1b) \end{cases}$$

supplied with suitable boundary conditions, which will be specified later. Here, Ω is an open, bounded and connected Lipschitz domain in $\mathbb{R}^d$ with boundary $\Gamma = \partial\Omega$, M is a symmetric positive definite material tensor, and g is a source term. Further, $\widetilde{\mathcal{L}}$ is a *Friedrichs' operator* [9] given by

$$\widetilde{\mathcal{L}}u = \sum_{i=1}^{d} L_i \partial_i u + L_0 u, \qquad L_i \in \mathbb{R}^{m\times m}, \quad i = 0, \ldots, d, \tag{2}$$

M. Hochbruck · J. Köhler (✉)
Karlsruhe Institute of Technology, Karlsruhe, Germany
e-mail: marlis.hochbruck@kit.edu; jonas.koehler@kit.edu

W. Dörfler et al. (eds.), *Mathematics of Wave Phenomena*, Trends in Mathematics,
https://doi.org/10.1007/978-3-030-47174-3_12

where $L_1, \dots, L_d$ are symmetric and the symmetric part of L_0 is negative semi-definite, i.e., $x^T(L_0 + L_0^T)x \le 0$ for all $x \in \mathbb{R}^m$. For the sake of presentation, we restrict ourselves to constant matrix coefficients L_i although our results also hold for space dependent coefficients under certain additional assumptions. We refer to [2, 15] for the more general case.

Important examples of this class of problems are the wave equation in first order formulation, Maxwell's equations, and the advection equation, see, e.g., [2–4, 15].

Partial differential equations governed by Friedrichs' operators have been studied intensively in the series of papers [6–8] and Chapter 7 of the book [4]. However, the results therein are only applicable to stationary problems or by treating the problem as a space-time problem, where the temporal variable is incorporated into the Friedrichs' operator.

In contrast to this work, our analysis is performed in a framework of evolution equations and can thus be combined with various time integration schemes. This then ultimately leads to full discretization error bounds, as has been shown in the thesis [15] for the particular choice of a Peaceman–Rachford ADI scheme.

A proof of the wellposedness of (1) supplied with suitable initial and boundary conditions was recently provided in [2]. This analysis covers the special case of $M = I$, which means that the material parameters are incorporated into the coefficients of the differential operator $\widetilde{\mathcal{L}}$. Unfortunately, this excludes materials with sharp interfaces as the coefficients of $\widetilde{\mathcal{L}}$ need to fulfill certain regularity restrictions, see the second remark after [2, Thm. 5].

Hence, we follow a slightly different approach by incorporating the material tensor into the inner product of the state space. This allows us to weaken the restrictions on the regularity of the material parameters. Moreover, we treat boundary conditions as in [6] since this fits better to the dG discretization than the approach in [2].

Semi-discretizations of more general hyperbolic problems were considered in a unified error analysis in [12]. In this analysis, error bounds are given in terms of various discretization defects, interpolation errors, errors in the approximations of the spatial domain, the bilinear forms, and starting values. To apply this analysis to a particular application and discretization, one has to check that the continuous and the discretized problem both fit into the very general framework and to provide bounds for all these approximation errors. This constitutes the main work in our paper. Although we could then apply the general result of [12], we present proofs of the final error bound in Sect. 5, since they are relatively short for our application and this keeps the paper self-contained.

To the best of our knowledge, such bounds for hyperbolic evolution equations in first order formulation are only available for Maxwell's equations, cf. [16]. For the wave equation in second order formulation similar results were derived in [10], where the Laplace operator was discretized by a symmetric interior penalty dG method.

Our main result shows that the solution of the spatially discrete evolution equation has an error of order h^k in the L^2-norm induced by the material tensor M.

The paper is organized as follows. In Sect. 3 we provide the analytical framework for our paper. In particular, we collect properties of Friedrichs' operators and show wellposedness of the linear wave-type problem (1) under suitable assumptions. Section 4 is devoted to the dG discretization of linear wave-type equations written as a Friedrichs' system. We show various properties of the discretized Friedrichs' operator which are crucial for the following error analysis given in Sect. 5.

2 Notation

Throughout this paper we use the following notation:

We use $d \in \mathbb{N}$ as the spatial dimension and $m \in \mathbb{N}$ as a generic positive integer, usually being the number of components of vector-valued functions. The indicator function of a set $K \subset \mathbb{R}^d$ is denoted as $\mathbb{1}_K$.

Let $(X, (\cdot \mid \cdot)_X)$ and $(Y, (\cdot \mid \cdot)_Y)$ be real Hilbert spaces. The identity operator on a Hilbert space X is denoted by $\mathcal{I}$ and by $\mathcal{B}(X, Y)$ we denote the set of all bounded operators from X to Y. The dual space of a Hilbert space X is denoted as X' and we use the notation $\langle \cdot \mid \cdot \rangle \colon X' \times X \to \mathbb{R}$ for the canonical dual pairing between a Hilbert space and its dual space.

Let $K \subset \mathbb{R}^d$ open. Then we denote the space of infinitely differentiable functions, which have compact support on K as $C_c^\infty(K)$ and the restrictions of functions in $C_c^\infty(\mathbb{R}^d)$ to K as $C^\infty_{c,\mathbb{R}^d}(K)$. For vector-valued functions $u, v \in L^2(K)^m$, the $L^2(K)$-inner product is denoted by

$$(u \mid v)_K = \int_K u \cdot v \, \mathrm{d}x,$$

and for $F \subset \partial K$ and $u|_F, v|_F \in L^2(F)^m$ we write

$$(u \mid v)_F = \int_F u|_F \cdot v|_F \, \mathrm{d}\sigma.$$

The norms induced by these inner products are denoted by $\|\cdot\|_K$ and $\|\cdot\|_F$.

For $i = 1, \ldots, d$, we denote the distributional partial derivative in ith coordinate direction of $\mathbb{R}^d$ by ∂_i. For $q \in \mathbb{N}$ we write $H^q(K)$ for the standard L^2-Sobolev spaces, which are Hilbert spaces if equipped with the $H^q(K)$-norm given by

$$\|v\|_{q,K}^2 = \sum_{j=0}^{q} |v|_{j,K}^2, \qquad |v|_{j,K}^2 = \sum_{|\alpha|=j} \|\partial^\alpha v\|_K^2, \quad j = 0, \ldots, q.$$

Here, $\alpha = (\alpha_1, \ldots, \alpha_d) \in \mathbb{N}_0^d$ is a multi-index with $|\alpha|$ being the ℓ^1-norm of α, and $\partial^\alpha v = \partial_1^{\alpha_1} \ldots \partial_d^{\alpha_d} v$ with the convention $\partial^{(0,\ldots,0)} v = v$.

Let $M \in L^\infty(K)^{m\times m}$ be a square matrix-valued field on K. We denote the essential supremum of the spectral norm of M by

$$\|M\|_{\infty,K} = \operatorname*{ess\,sup}_{x\in K} \|M(x)\|,$$

where $\|\cdot\|$ is the spectral norm.

3 Analytical Properties of Friedrichs' Systems

Throughout, we consider distributional derivatives in the Friedrichs' operator $\widetilde{\mathcal{L}}$ defined in (2). Hence, because of the boundedness of L_0, a function $v \in L^2(\Omega)^m$ satisfies $\widetilde{\mathcal{L}}v \in L^2(\Omega)^m$ if the linear form

$$C_c^\infty(\Omega)^m \to \mathbb{R}, \quad \varphi \mapsto \big(v \,\big|\, -\sum_{i=1}^d \partial_i(L_i\varphi)\big)_\Omega \tag{3}$$

is bounded in the L^2-norm (see [6, Sec. 2.1] for more insight). The *graph space of* $\widetilde{\mathcal{L}}$ is then given by

$$H(\mathcal{L}) = \{v \in L^2(\Omega)^m \mid \widetilde{\mathcal{L}}v \in L^2(\Omega)^m\},$$

and, endowed with the *graph norm* $\|\cdot\|_{\widetilde{\mathcal{L}}} = \|\cdot\|_\Omega + \|\widetilde{\mathcal{L}}\cdot\|_\Omega$, is a Hilbert space [4, Lem. 7.2]. Note that, by definition, we have $\widetilde{\mathcal{L}} \in \mathcal{B}(H(\mathcal{L}), L^2(\Omega)^m)$. The notation $H(\mathcal{L})$ is chosen based on the spaces $H(\mathrm{div})$ and $H(\mathrm{curl})$, which are the corresponding concepts for the divergence and curl operator, respectively.

Definition 3.1 We call $\widetilde{\mathcal{L}}^{\circledast} \in \mathcal{B}(H(\mathcal{L}), L^2(\Omega)^m)$ defined by

$$\widetilde{\mathcal{L}}^{\circledast} u = -\sum_{i=1}^d \partial_i(L_i u) + L_0^T u \tag{4}$$

the *formal adjoint of* $\widetilde{\mathcal{L}}$.

Note that, since the coefficients of $\widetilde{\mathcal{L}}$ are constant, we have

$$\widetilde{\mathcal{L}}^{\circledast} u = -\sum_{i=1}^d L_i \partial_i u + L_0^T u.$$

Functions in $H(\mathcal{L})$ are not necessarily smooth enough to admit L^2-traces on the boundary. To still obtain access to boundary values in this weak setting, we follow [6] and introduce the following abstract boundary operator.

Definition 3.2 We call $\mathcal{L}_\partial \colon H(\mathcal{L}) \to H(\mathcal{L})'$ defined by

$$\langle \mathcal{L}_\partial u \mid v \rangle = (\widetilde{\mathcal{L}} u \mid v)_\Omega - (u \mid \widetilde{\mathcal{L}}^{\circledast} v)_\Omega \qquad \text{for all } u, v \in H(\mathcal{L}) \tag{5}$$

the *boundary operator associated with* $\widetilde{\mathcal{L}}$.

Note that (5) can be seen as a generalization of the integration by parts formula. Further, by Ern and Guermond [6, Sec. 2.1] we have $\mathcal{L}_\partial \in \mathcal{B}(H(\mathcal{L}), H(\mathcal{L})')$ and that $\mathcal{L}_\partial$ is self-adjoint.

Next, we implement boundary conditions into the abstract setting. In particular, we consider a class of homogeneous conditions that can be treated by incorporating them into the space on which the wave-type problem (1) is considered. Again, we follow [6] and pose the following assumption.

Assumption 3.3 We assume there exists a bounded operator $\mathcal{L}_\Gamma \in \mathcal{B}(H(\mathcal{L}), H(\mathcal{L})')$ fulfilling

$$\langle \mathcal{L}_\Gamma v \mid v \rangle \leq 0 \quad \text{for all } v \in H(\mathcal{L}), \tag{6a}$$

$$H(\mathcal{L}) = \ker(\mathcal{L}_\partial - \mathcal{L}_\Gamma) + \ker(\mathcal{L}_\partial + \mathcal{L}_\Gamma). \tag{6b}$$

Note that both $\ker(\mathcal{L}_\partial - \mathcal{L}_\Gamma)$ and $\ker(\mathcal{L}_\partial + \mathcal{L}_\Gamma)$ are Hilbert spaces if endowed with the graph norm of $\widetilde{\mathcal{L}}$, as they are the kernels of bounded operators on $H(\mathcal{L})$.

Theorem 3.4 *The restriction of* $\widetilde{\mathcal{L}}$ *to* $\ker(\mathcal{L}_\partial - \mathcal{L}_\Gamma)$ *is maximal dissipative.*

Proof Let $v \in \ker(\mathcal{L}_\partial - \mathcal{L}_\Gamma)$. By Definitions 3.1 and 3.2 of the formal adjoint and the boundary operator, respectively, we have

$$\begin{aligned} 2(\widetilde{\mathcal{L}} v \mid v)_\Omega &= (\widetilde{\mathcal{L}} v \mid v)_\Omega + (\widetilde{\mathcal{L}}^{\circledast} v \mid v)_\Omega + (\widetilde{\mathcal{L}} v \mid v)_\Omega - (\widetilde{\mathcal{L}}^{\circledast} v \mid v)_\Omega \\ &= ((L_0 + L_0^T) v \mid v)_\Omega + \langle \mathcal{L}_\partial v \mid v \rangle \\ &\leq \langle (\mathcal{L}_\partial - \mathcal{L}_\Gamma) v \mid v \rangle + \langle \mathcal{L}_\Gamma v \mid v \rangle \\ &\leq 0, \end{aligned}$$

where the first inequality follows since the symmetric part of L_0 is negative semi-definite and the second because of $v \in \ker(\mathcal{L}_\partial - \mathcal{L}_\Gamma)$ and (6a). Hence, $\widetilde{\mathcal{L}}$ is dissipative on $\ker(\mathcal{L}_\partial - \mathcal{L}_\Gamma)$. The maximality is a direct consequence of [6, Thm. 2.5], which shows that $(\mathcal{I} - \lambda \widetilde{\mathcal{L}}) \colon \ker(\mathcal{L}_\partial - \mathcal{L}_\Gamma) \to L^2(\Omega)^m$ is an isomorphism for $\lambda > 0$. □

Next, we show wellposedness of (1) by using semigroup theory. To this end, we define the *domain of* $\widetilde{\mathcal{L}}$ as $D(\mathcal{L}) := \ker(\mathcal{L}_\partial - \mathcal{L}_\Gamma)$. We assume that $M \in$

$L^\infty(\Omega)^{m\times m}$ is symmetric positive definite a.e. on Ω, and that the source term satisfies $g \in C(\mathbb{R}_+; L^2(\Omega)^m)$. Then, Theorem 3.4 already yields wellposedness of (1) on $D(\mathcal{L})$ for suitable initial conditions if $M = I$. This is due to the fact that $\widetilde{\mathcal{L}}|_{D(\mathcal{L})}$ is the generator of a contraction semigroup w.r.t. $\|\cdot\|_\Omega$ by the Lumer–Phillips Theorem [5, Thm. II.3.15, Cor. II.3.20].

If we have $M \neq I$, we define the weighted inner product $(\cdot \mid \cdot)_M$ by

$$\left(u \mid v\right)_M = \left(Mu \mid v\right)_\Omega, \qquad u, v \colon \Omega \to \mathbb{R}^d,$$

and denote the induced norm by $\|\cdot\|_M$. Note that, since this inner product is equivalent to the standard L^2 inner product, $L^2(\Omega)^m$ is again a Hilbert space if endowed with $(\cdot \mid \cdot)_M$.

By multiplying (1a) with M^{-1}, we obtain the (equivalent) abstract evolution problem

$$\begin{cases} \partial_t u = \mathcal{L}u + f, & \mathbb{R}_+ \times \Omega, \qquad (7a) \\ u(0) = u^0, & \qquad (7b) \end{cases}$$

with $\mathcal{L} := M^{-1}\widetilde{\mathcal{L}}$ and $f := M^{-1}g$. We can now use Theorem 3.4 to show that the restriction of $\mathcal{L}$ to $D(\mathcal{L})$ is maximal dissipative w.r.t. the weighted inner product $(\cdot \mid \cdot)_M$.

Theorem 3.5 *The restriction of $\mathcal{L}$ to $D(\mathcal{L})$ is maximal dissipative.*

Proof The dissipativity of $\mathcal{L}$ directly follows from the dissipativity of $\widetilde{\mathcal{L}}$ as we have

$$\left(\mathcal{L}u \mid u\right)_M = \left(M\mathcal{L}u \mid u\right)_\Omega = \left(\widetilde{\mathcal{L}}u \mid u\right)_\Omega \leq 0, \qquad \text{for all } u \in D(\mathcal{L}). \tag{8}$$

Maximality again follows as a consequence of [6, Thm. 2.5], which yields that $(M - \lambda\widetilde{\mathcal{L}}) \colon D(\mathcal{L}) \to L^2(\Omega)^m$ is an isomorphism for all $\lambda > 0$. Since M is positive definite and bounded, this is equivalent to $(I - \lambda\mathcal{L}) \colon D(\mathcal{L}) \to L^2(\Omega)^m$ being an isomorphism, yielding the desired range condition. □

Hence, by the Lumer–Phillips Theorem, the restriction of $\mathcal{L}$ to $D(\mathcal{L})$ generates a contraction semigroup w.r.t. $\|\cdot\|_M$, which we denote by $\left(e^{t\mathcal{L}}\right)_{t\geq 0}$.

Corollary 3.6 *Let $f \in C^1(\mathbb{R}_+; L^2(\Omega)^m) \cup C(\mathbb{R}_+; D(\mathcal{L}))$. Then, for given initial value $u^0 \in D(\mathcal{L})$, there exists a unique solution $u \in C^1(\mathbb{R}_+; L^2(\Omega)^m) \cap C(\mathbb{R}_+; D(\mathcal{L}))$ of (7) given by the variation-of-constants formula*

$$u(t) = e^{t\mathcal{L}}u^0 + \int_0^t e^{(t-s)\mathcal{L}} f(s)\,ds.$$

Remark 3.7 For the sake of presentation, we only consider Friedrichs' operators with constant coefficients. However, all of the above can be extended to more general coefficients, e.g., Lipschitz coefficients. See [15, Chap. 2] for the more general case.

Further, the assumption of negative semi-definiteness of L_0 can be dropped. This leads to $\widetilde{\mathcal{L}}$ and $\mathcal{L}$ being shift-dissipative rather than dissipative on $D(\mathcal{L})$, see also the remark on the positivity condition (F2) in [2, Sec. 2.1].

Let us also point out that the restriction of the formal adjoint operator $\widetilde{\mathcal{L}}^{\circledast}$ to $\ker(\mathcal{L}_\partial + \mathcal{L}_\Gamma^*)$ is maximal dissipative. This can be shown with the exact same strategy of proof. In fact, this is the Hilbert space adjoint of $\widetilde{\mathcal{L}}$ restricted to $D(\mathcal{L})$.

4 Spatial Discretization

To obtain a spatially discretized version of (7) we discretize the differential operator $\mathcal{L}$ using a central fluxes dG approximation [4, 11]. As $\mathcal{L}$ is defined via the Friedrichs' operator $\widetilde{\mathcal{L}}$ we start by discretizing the latter and then define the discrete version of $\mathcal{L}$ analogously to the continuous case.

To avoid technicalities, we assume that the domain Ω is a polyhedron, meaning we can decompose Ω into a polyhedral mesh. However, we refer to [12] for a way to take errors made by approximating non-polyhedral domains into account.

For the sake of readability, we postpone some of the longer proofs in this section. They can be found in the appendix.

4.1 Discrete Setting

Before we define discrete Friedrichs' operators, we introduce some notation and the discrete setting. Let $\mathcal{T}$ be a general mesh of Ω in the sense of [4, Def. 1.12]. For each (open) mesh element $K \in \mathcal{T}$, we denote the diameter of K by h_K. To write down mesh-dependent norms more concisely, we define the piecewise constant function $h \in L^\infty(\Omega)$ by $h_{|K} \equiv h_K$ for all $K \in \mathcal{T}$. The maximal diameter $\hbar = \max_{K \in \mathcal{T}} h_K$ of all elements in $\mathcal{T}$ is called the meshsize of $\mathcal{T}$ and we use the notation $\mathcal{T}_\hbar$ for a mesh with meshsize $\hbar$. In order to investigate the convergence of the method, we consider a mesh sequence $\mathcal{T}_\mathcal{H} = \left(\mathcal{T}_\hbar\right)_{\hbar \in \mathcal{H}}$, where $\mathcal{H}$ is a countable collection of positive numbers with 0 as only accumulation point. We assume that we have $\hbar < 1$ for all $\hbar \in \mathcal{H}$ and that $\mathcal{T}_\mathcal{H}$ is admissible in the sense of [4, Def. 1.57], meaning that it is shape and contact regular and has optimal polynomial approximation properties, cf., [4, Def. 1.38 and 1.55]. We denote the mesh regularity parameter by ρ.

The faces of a mesh $\mathcal{T}_\hbar$ are gathered in the set $\mathcal{F}_\hbar = \mathcal{F}_\hbar^{\mathrm{int}} \cup \mathcal{F}_\hbar^{\mathrm{bnd}}$, where $\mathcal{F}_\hbar^{\mathrm{int}}$ contains the interior faces and $\mathcal{F}_\hbar^{\mathrm{bnd}}$ contains the boundary faces. For each $K \in \mathcal{T}_\hbar$, we denote the faces composing the boundary of an element K by $\mathcal{F}_\hbar^K = \mathcal{F}_\hbar^{K,\mathrm{int}} \cup \mathcal{F}_\hbar^{K,\mathrm{bnd}}$, again decomposed into interior and boundary faces. The maximum number of faces per element in $\mathcal{T}_{\hbar}$ is denoted by $N_\partial = \max_{K\in\mathcal{T}_{\hbar}} |\mathcal{F}_{\hbar}^K|$. Note that by Di Pietro and Ern [4, Lem. 1.41], N_∂ is bounded independently of $\hbar \in \mathcal{H}$.

The outward unit normal vector to an element $K \in \mathcal{T}_{\hbar}$ is denoted by $\mathbf{n}^K$. Further, for each interface $F \in \mathcal{F}_{\hbar}^{\text{int}}$, we arbitrarily denote the two neighboring elements, whose boundaries contain F, as K_1^F and K_2^F. We fix this choice and define the face normal vector $\mathbf{n}^F$ as the outward unit normal vector to K_1^F. For all boundary faces $F \in \mathcal{F}_{\hbar}^{\text{bnd}}$, we define $\mathbf{n}^F$ as the outward unit normal vector to Γ.

To approximate functions in space, we consider the discrete approximation space

$$V_{\hbar} = \{\, \boldsymbol{v} \in L^2(\Omega) \mid \boldsymbol{v}|_K \in \mathbb{Q}_d^k(K) \text{ for all } K \in \mathcal{T}_{\hbar} \,\}^m,$$

where $\mathbb{Q}_d^k(K)$ is the set of polynomials on K of degree at most k in each variable.

Remark 4.1 For the sake of presentation, we use the same polynomial degree on all elements $K \in \mathcal{T}_{\hbar}$. However, the method is flexible enough to easily allow varying polynomial degrees on each element. Note also that other choices for the discrete approximation space are possible. We refer to [4, Sec. 1.2.4.3] for further details.

We will frequently need the L^2-orthogonal projection $\pi_h \colon L^2(\Omega)^m \to V_{\hbar}$ onto $V_{\hbar}$, defined such that for $v \in L^2(\Omega)^m$ we have

$$\left(v - \pi_h v \,\middle|\, \boldsymbol{\varphi}\right)_\Omega = 0 \qquad \text{for all } \boldsymbol{\varphi} \in V_{\hbar}. \tag{9}$$

Using the L^2-orthogonal projection, by

$$e_\pi^v = v - \pi_h v$$

we denote the projection error of a function $v \in L^2(\Omega)^m$.

Assumption 4.2 We assume that the material tensor M is piecewise constant and that for all $\hbar \in \mathcal{H}$, the mesh $\mathcal{T}_{\hbar}$ is matched to the material, i.e., for all $K \in \mathcal{T}_{\hbar}$ we have $M|_K \equiv M_K$ with constant $M_K \in \mathbb{R}^{m\times m}$.

It is easy to see that for $v \in L^2(\Omega)^m$ we have

$$\left(e_\pi^v \,\middle|\, \boldsymbol{\varphi}\right)_M = 0 \qquad \text{for all } \boldsymbol{\varphi} \in V_{\hbar}$$

because of Assumption 4.2.

In the following, let $\boldsymbol{v} \in V_{\hbar}$. Since we assumed the mesh sequence to be admissible, we can infer some important properties of the discrete spaces. Namely, the inverse inequality [4, Lem. 1.44]

$$\|\nabla \boldsymbol{v}\|_K \leq C_{\text{inv}}' \|h^{-1} \boldsymbol{v}\|_K, \tag{10}$$

and the discrete trace inequality [4, Lem. 1.46]

$$\|\boldsymbol{v}\|_F \leq C_{\text{tr}} \|h^{-1/2}\boldsymbol{v}\|_K \tag{11}$$

hold as a consequence of the shape and contact regularity. From the inverse inequality (10), we can easily deduct a similar inequality for the Friedrichs' operator $\widetilde{\mathcal{L}}$ instead of the gradient, namely

$$\|\widetilde{\mathcal{L}}\boldsymbol{v}\|_K \leq C_{\mathcal{L}} C_{\text{inv}} \|h^{-1}\boldsymbol{v}\|_K, \tag{12}$$

where $C_{\mathcal{L}} = \max_{i=0,\ldots,d} \|L_i\|$ and $C_{\text{inv}} = \sqrt{d}C'_{\text{inv}} + 1$.

Further, the mesh sequence $\mathcal{T}_{\mathcal{H}}$ has optimal polynomial approximation properties in the sense of [4, Def. 1.55]. This means that for all $\hbar \in \mathcal{H}$, $K \in \mathcal{T}_\hbar$, $F \in \mathcal{F}_\hbar^K$ and $v \in H^{q+1}(K)$ the projection error of v satisfies

$$\|e_\pi^v\|_K \leq C_\pi |h^{q+1}v|_{q+1,K}, \qquad \|e_\pi^v\|_F \leq C_{\pi,\partial} |h^{q+1/2}v|_{q+1,K}, \tag{13}$$

where C_π and $C_{\pi,\partial}$ are independent of both K and $\hbar$.

The space $V_\hbar$ consists of functions that are polynomials on the elements of $\mathcal{T}_\hbar$. Hence, they can be used to approximate functions that are sufficiently smooth on these elements. Such functions are gathered in the broken Sobolev spaces

$$H^q(\mathcal{T}_\hbar) = \{\, v \in L^2(\Omega) \mid v|_K \in H^q(K) \text{ for all } K \in \mathcal{T}_\hbar \,\}, \qquad q \in \mathbb{N},$$

which are Hilbert spaces if endowed with the norm

$$\|v\|_{q,\mathcal{T}_\hbar}^2 = \sum_{j=0}^{q} |v|_{j,\mathcal{T}_\hbar}^2, \qquad |v|_{j,\mathcal{T}_\hbar}^2 = \sum_{K\in\mathcal{T}_\hbar} |v|_{j,K}^2, \quad j = 0,\ldots,q.$$

Functions in both $V_\hbar$ and $H^q(\mathcal{T}_\hbar)$ are only piecewise smooth, i.e., smooth on every mesh element $K \in \mathcal{T}_\hbar$, but not necessarily on the whole domain Ω. Hence, they may have discontinuities across the faces of the mesh, which is why we define the average and the jump of a function v across an interior face $F \in \mathcal{F}_\hbar^{\text{int}}$ as

$$\{\!\{v\}\!\}_F = \frac{v|_{K_1^F} + v|_{K_2^F}}{2} \qquad \text{and} \qquad [\![v]\!]_F = v|_{K_1^F} - v|_{K_2^F},$$

respectively. Here and in the following, the restriction of v to an element $K \in \mathcal{T}_\hbar$ evaluated on a face $F \in \mathcal{F}_\hbar^K$ is understood as the limit of v approaching F from K. For vector and matrix fields these operations act componentwise.

4.2 Friedrichs' Operators in the Discrete Setting

Up until now, we only got hold of the boundary operators in an abstract way, since functions in the graph space of a Friedrichs' operator $\widetilde{\mathcal{L}}$ are not necessarily smooth enough to admit square-integrable traces. However, to define the discrete operators and to implement them in the full discretization scheme, it is convenient to access boundary values in a more explicit way. This can be achieved by assuming a bit more regularity, which enables us to use the integration by parts formula.

Lemma 4.3 *Let K be an open and Lipschitz subset of Ω with outward unit normal vector $\mathfrak{n}^K$ and let $\widetilde{\mathcal{L}}^{\circledast}$ be the formal adjoint of $\widetilde{\mathcal{L}}$ defined in* (4). *Then, for $v, w \in L^2(\Omega)^m$ satisfying $v|_K, w|_K \in H^1(K)^m$, we have*

$$\big(\widetilde{\mathcal{L}}v \,\big|\, w\big)_K - \big(v \,\big|\, \widetilde{\mathcal{L}}^{\circledast} w\big)_K = \Big(\sum_{i=1}^{d} \mathfrak{n}_i^K L_i v \,\Big|\, w\Big)_{\partial K}.$$

Proof The statement follows directly using the integration by parts formula. □

Note that, in particular, the assumptions of Lemma 4.3 are fulfilled for elements $K \in \mathcal{T}_h$ if $v, w \in H^1(\mathcal{T}_h)^m$. Further, Lemma 4.3 yields that, for $v, w \in H^1(\Omega)^m$, the boundary operator $\mathcal{L}_\partial$ can be represented in a more explicit way, namely

$$\big\langle \mathcal{L}_\partial v \,\big|\, w\big\rangle = \Big(\sum_{i=1}^{d} \mathfrak{n}_i^{\mathrm{dom}} L_i v \,\Big|\, w\Big)_\Gamma.$$

Definition 4.4 For $K \in \mathcal{T}_h$ and $F \in \mathcal{F}_h$ we define the *boundary fields* $L_\partial^K \in L^\infty(\partial K)^{m\times m}$ *associated with an element of the mesh* and $L_\partial^F \in L^\infty(F)^{m\times m}$ *associated with a face of the mesh* by

$$L_\partial^K = \sum_{i=1}^{d} \mathfrak{n}_i^K L_i \qquad \text{and} \qquad L_\partial^F = \sum_{i=1}^{d} \mathfrak{n}_i^F L_i,$$

respectively.

The definition of L_∂^K is motivated by Lemma 4.3, which relates it to the boundary term of the integration by parts formula on each element of the mesh. Further, the operator L_∂^F will allow for a more concise notation in the definition and handling of the discrete operator. The latter is well-defined as we consider constant coefficients L_i, $i = 1, \ldots, d$, and hence, their traces are single-valued on each face.

To get hold of the abstract boundary operator $\mathcal{L}_\Gamma$ defined in Assumption 3.3 in a similar way, we make the following assumption.

Assumption 4.5 We assume that the boundary operator $\mathcal{L}_\Gamma$ is associated with a matrix-valued boundary field $L_\Gamma \in L^\infty(\Gamma)^{m\times m}$ such that for v, w sufficiently smooth we have

$$\langle \mathcal{L}_\Gamma v \,|\, w \rangle = (L_\Gamma v \,|\, w)_\Gamma .$$

In particular, Assumption 4.5 is satisfied for $v, w \in H^1(\mathcal{T}_h)^m$, as these functions are smooth enough to admit L^2-traces on all boundary faces $F \in \mathcal{F}_h^{\mathrm{bnd}}$ and thus also on Γ. Further, we point out that this assumption is not very restricting as it is fulfilled in many applications, see [6, Sec. 5].

Before we define the discrete Friedrichs' operator, we prove two auxiliary results, which are needed to show crucial properties of the discrete operators. The first one relates the boundary fields L_∂^K and L_∂^F.

Lemma 4.6 *Let $v, w \in H^1(\mathcal{T}_h)^m$. Then we have*

$$\sum_{K \in \mathcal{T}_h} (L_\partial^K v \,|\, w)_{\partial K} = \sum_{F \in \mathcal{F}_h^{\mathrm{int}}} \Big(\big(L_\partial^F \{\!\{v\}\!\}_F \,|\, [\![w]\!]_F\big)_F + \big(L_\partial^F [\![v]\!]_F \,|\, \{\!\{w\}\!\}_F\big)_F \Big) + \sum_{\mathcal{F}_h \in \mathcal{F}_h^{\mathrm{bnd}}} (L_\partial^F v \,|\, w)_F .$$

Proof Using the definition of the boundary fields L_∂^K and L_∂^F and the directions of the element and face normals $\mathfrak{n}^K$ and $\mathfrak{n}^F$, respectively, we calculate

$$\begin{aligned}\sum_{K \in \mathcal{T}_h} (L_\partial^K v \,|\, w)_{\partial K} &= \sum_{F \in \mathcal{F}_h^{\mathrm{int}}} \Big((L_\partial^F v_{|K_1^F} \,|\, w_{|K_1^F})_F - (L_\partial^F v_{|K_2^F} \,|\, w_{|K_2^F})_F \Big) \\ &\quad + \sum_{F \in \mathcal{F}_h^{\mathrm{bnd}}} (L_\partial^F v \,|\, w)_F \\ &= \sum_{F \in \mathcal{F}_h^{\mathrm{int}}} \big([\![(L_\partial^F v) \cdot w]\!]_F \,|\, 1\big)_F + \sum_{F \in \mathcal{F}_h^{\mathrm{bnd}}} (L_\partial^F v \,|\, w)_F .\end{aligned}$$

Using the identity $[\![f \cdot g]\!]_F = \{\!\{f\}\!\}_F \cdot [\![g]\!]_F + [\![f]\!]_F \cdot \{\!\{g\}\!\}_F$ for all $f, g \colon \Omega \to \mathbb{R}^m$ concludes the proof. □

The next result characterizes functions in the graph space of $\widetilde{\mathcal{L}}$, or rather its intersection with the broken Sobolev space $H^1(\mathcal{T}_h)^m$. It states that the jumps of such functions corresponding to L_∂^F vanish across interfaces of the mesh. Further, if these functions are additionally contained in the domain $D(\mathcal{L})$, they fulfill the corresponding boundary condition. The proof can be found in the appendix.

Lemma 4.7 *Let $v \in H^1(\mathcal{T}_h)^m$. Then we have $v \in H(\mathcal{L})$ if and only if*

$$L_\partial^F [\![v]\!]_F = 0 \qquad \text{a.e. on } F \text{ for all } F \in \mathcal{F}_h^{\mathrm{int}}. \tag{14}$$

Additionally, for $v \in D(\mathcal{L}) \cap H^1(\mathcal{T}_{\hbar})^m$, *we have*

$$(L_{\partial}^{F} - L_{\Gamma})v = 0 \qquad \textit{a.e. on } F \textit{ for all } F \in \mathcal{F}_{\hbar}^{\mathrm{bnd}}. \tag{15}$$

4.3 Discrete Friedrichs' Operators

In this section we define and investigate the central fluxes dG discretization of a Friedrichs' operator $\widetilde{\mathcal{L}}$. Naturally, one would define this discrete operator on the discrete approximation space $V_{\hbar}$. However, in view of the error analysis, it is convenient to extend the definition to the space $D(\mathcal{L}) \cap H^1(\mathcal{T}_{\hbar})^m$. We combine both spaces in the *discrete operator domain* associated with $\widetilde{\mathcal{L}}$ given by

$$V_{\hbar}^{\mathcal{L}} = V_{\hbar} + (D(\mathcal{L}) \cap H^1(\mathcal{T}_{\hbar})^m).$$

In order to define the discrete operator, for $v \in V_{\hbar}^{\mathcal{L}}$, we introduce the linear form $\ell_v \colon V_{\hbar} \to \mathbb{R}$ defined as

$$\ell_v(\varphi) = \sum_{K \in \mathcal{T}_{\hbar}} \big(\widetilde{\mathcal{L}}v \,|\, \varphi\big)_K - \sum_{F \in \mathcal{F}_{\hbar}^{\mathrm{int}}} \big(L_{\partial}^{F} [\![v]\!]_F \,|\, \{\!\{\varphi\}\!\}_F\big)_F - \tfrac{1}{2} \sum_{F y \in \mathcal{F}_{\hbar}^{\mathrm{bnd}}} \big((L_{\partial}^{F} - L_{\Gamma})v \,|\, \varphi\big)_F.$$

Using the Cauchy–Schwarz inequality, the boundedness of the terms involving v and the discrete trace inequality (11) on the boundary terms involving φ, it is easy to see that ℓ_v is bounded on $V_{\hbar}$ for all $v \in V_{\hbar}^{\mathcal{L}}$. Hence, by the Riesz representation theorem, the following definition provides a well-defined discrete operator.

Definition 4.8 The *central fluxes dG discretization of* $\widetilde{\mathcal{L}}$ is the operator $\widetilde{\mathcal{L}} \colon V_{\hbar}^{\mathcal{L}} \to V_{\hbar}$ defined as

$$\begin{aligned}\big(\widetilde{\mathcal{L}}v \,|\, \varphi\big)_{\Omega} = &\sum_{K \in \mathcal{T}_{\hbar}} \big(\widetilde{\mathcal{L}}v \,|\, \varphi\big)_K - \sum_{F \in \mathcal{F}_{\hbar}^{\mathrm{int}}} \big(L_{\partial}^{F} [\![v]\!]_F \,|\, \{\!\{\varphi\}\!\}_F\big)_F \\ &- \tfrac{1}{2} \sum_{F y \in \mathcal{F}_{\hbar}^{\mathrm{bnd}}} \big((L_{\partial}^{F} - L_{\Gamma})v \,|\, \varphi\big)_F \qquad \text{for all } \varphi \in V_{\hbar}.\end{aligned} \tag{16}$$

Remark 4.9 The average used in (16) can be replaced by a weighted average $\{\!\{v\}\!\}_F^{\Lambda} = \{\!\{\Lambda\}\!\}_F^{-1}\{\!\{\Lambda v\}\!\}_F$ with $\Lambda \in L^{\infty}(\Omega)^{m \times m}$ being symmetric and uniformly positive a.e. on Ω. The following results then still hold, albeit with different constants involving the weights. If the weight is chosen in a suitable way this can improve the constants, see e.g., [16] for isotropic Maxwell's equations.

We next gather some important properties of the discrete Friedrichs' operator. The first one is a consistency property that shows that the discrete operator in some sense indeed approximates its continuous counterpart.

Proposition 4.10 *The discrete Friedrichs' operator $\widetilde{\mathcal{L}}$ fulfills the consistency property*

$$\widetilde{\mathcal{L}}v = \pi_h \widetilde{\mathcal{L}}v \qquad \text{for all } v \in D(\mathcal{L}) \cap H^1(\mathcal{T}_h)^m.$$

Proof Let $v \in D(\mathcal{L}) \cap H^1(\mathcal{T}_h)^m$. By Lemma 4.7 the interface and boundary terms in (16) vanish. Hence, we have

$$(\widetilde{\mathcal{L}}v \mid \varphi)_\Omega = (\widetilde{\mathcal{L}}v \mid \varphi)_\Omega = (\pi_h \widetilde{\mathcal{L}}v \mid \varphi)_\Omega \qquad \text{for all } \varphi \in V_h$$

by the definition of π_h in (9). □

In addition, the discrete Friedrichs' operator inherits the dissipativity of $\widetilde{\mathcal{L}}$ if restricted to the discrete approximation space. To show this, we proceed as in the continuous case.

Lemma 4.11 *The adjoint operator $\widetilde{\mathcal{L}}^{\circledast} : V_h \to V_h$ of $\widetilde{\mathcal{L}}_{|V_h}$ is given by*

$$(\widetilde{\mathcal{L}}^{\circledast} \boldsymbol{v} \mid \boldsymbol{\varphi})_\Omega = \sum_{K \in \mathcal{T}_h} (\widetilde{\mathcal{L}}^{\circledast} \boldsymbol{v} \mid \boldsymbol{\varphi})_K + \sum_{F \in \mathcal{F}_h^{\text{int}}} (L_\partial^F [\![\boldsymbol{v}]\!]_F \mid \{\!\{\boldsymbol{\varphi}\}\!\}_F)_F$$
$$+ \tfrac{1}{2} \sum_{F y \in \mathcal{F}_h^{\text{bnd}}} ((L_\partial^F + L_\Gamma^T)\boldsymbol{v} \mid \boldsymbol{\varphi})_F \qquad \text{for all } \boldsymbol{\varphi} \in V_h,$$

and satisfies

$$(\widetilde{\mathcal{L}}\boldsymbol{v} \mid \boldsymbol{\varphi})_\Omega + (\widetilde{\mathcal{L}}^{\circledast} \boldsymbol{v} \mid \boldsymbol{\varphi})_\Omega = ((L_0 + L_0^T)\boldsymbol{v} \mid \boldsymbol{\varphi})_\Omega + \tfrac{1}{2}((L_\Gamma + L_\Gamma^T)\boldsymbol{v} \mid \boldsymbol{\varphi})_\Gamma \tag{17}$$

for all $\boldsymbol{v}, \boldsymbol{\varphi} \in V_h$.

Proof Using the integration by parts formula from Lemma 4.3 on each element K and Lemma 4.6 on the arising interface terms readily yields that $\widetilde{\mathcal{L}}^{\circledast}$ is in fact the adjoint of $\widetilde{\mathcal{L}}$. Identity (17) follows by a straightforward calculation. □

Proposition 4.12 *The restriction of the discrete Friedrichs' operator $\widetilde{\mathcal{L}}$ to V_h is dissipative, i.e., we have*

$$(\widetilde{\mathcal{L}}\boldsymbol{v} \mid \boldsymbol{v})_\Omega \le 0 \qquad \text{for all } \boldsymbol{v} \in V_h.$$

Proof By the adjointness of $\widetilde{\mathcal{L}}$ and $\widetilde{\mathcal{L}}^{\circledast}$ we have

$$(\widetilde{\mathcal{L}}\boldsymbol{v} \mid \boldsymbol{v})_\Omega = \tfrac{1}{2}\Big((\widetilde{\mathcal{L}}\boldsymbol{v} \mid \boldsymbol{v})_\Omega + (\widetilde{\mathcal{L}}^{\circledast} \boldsymbol{v} \mid \boldsymbol{v})_\Omega\Big) \le 0,$$

where we have used (17) together with the negative semi-definiteness of L_0, and Assumption 4.5 and (6a). □

Similar to the continuous operator fulfilling (12), the discrete Friedrichs' operator satisfies an inverse inequality.

Proposition 4.13 *Let $v \in V_{\hbar}$. Then, the discrete Friedrichs' operator $\widetilde{\mathcal{L}}$ fulfills the inverse inequality*

$$\|\widetilde{\mathcal{L}} v\|_{\Omega} \leq C_{\mathrm{inv},\mathcal{L}} \|h^{-1} v\|_{\Omega}.$$

The constant is given by $C_{\mathrm{inv},\mathcal{L}} = C_{\mathcal{L}} C_{\mathrm{inv}} + \frac{1}{2} C_{\mathrm{tr}}^2 \big(C_{\Gamma,\mathcal{L}} + N_{\partial} C_{\mathcal{L}} (1 + \rho^{1/2})\big)$ *with* $C_{\Gamma,\mathcal{L}} = \max_{F \in \mathcal{F}_{\hbar}^{\mathrm{bnd}}} \|L_{\partial}^F - L_{\Gamma}\|_{\infty,F}$.

Lastly, we have a result on the approximation properties of the discrete Friedrichs' operator. It gives a bound on the application of $\widetilde{\mathcal{L}}$ to the projection error of a function in $D(\mathcal{L}) \cap H^{q+1}(\mathcal{T}_{\hbar})^m$.

Proposition 4.14 *Let $v \in D(\mathcal{L}) \cap H^{q+1}(\mathcal{T}_{\hbar})^m$ for $0 \leq q \leq k$. Then we have*

$$\|\widetilde{\mathcal{L}} e_{\pi}^{v}\|_{\Omega} \leq C_{\pi,\mathcal{L}} |h^q v|_{q+1,\mathcal{T}_{\hbar}}. \tag{18}$$

The constant is given by $C_{\pi,\mathcal{L}} = \frac{1}{2} N_{\partial} C_{\mathrm{tr}} C_{\pi,\partial} \big(C_{\Gamma,\mathcal{L}} + C_{\mathcal{L}} (1 + \rho^{1/2})\big)$.

The proofs of both Propositions 4.13 and 4.14 are given in the appendix.

4.4 Spatial Discretization of the Wave-type Problem

We are now able to formulate the spatially semi-discrete version of the wave-type problem (7). To this end, we define the operator $\mathcal{L}\colon V_{\hbar}^{\mathcal{L}} \to V_{\hbar}$ analogously to the continuous case by $\mathcal{L} = M^{-1} \widetilde{\mathcal{L}}$. Note that, owing to Assumption 4.2, $\mathcal{L}$ exhibits the same consistency property as $\widetilde{\mathcal{L}}$, namely

$$\mathcal{L} v = \pi_h \mathcal{L} v \qquad \text{for all } v \in D(\mathcal{L}) \cap H^1(\mathcal{T}_{\hbar})^m. \tag{19}$$

Using $\mathcal{L}$ we can state the spatially discrete wave-type problem

$$\begin{cases} \partial_t \boldsymbol{u} = \mathcal{L} \boldsymbol{u} + \boldsymbol{f}_{\pi}, & \mathbb{R}_+ \times \Omega, \quad \text{(20a)} \\ \boldsymbol{u}(0) = \boldsymbol{u}_{\pi}^0, & \quad \text{(20b)} \end{cases}$$

with $\boldsymbol{f}_{\pi} := \pi_h f$ and initial value $\boldsymbol{u}_{\pi}^0 := \pi_h u^0$.

Since $\widetilde{\mathcal{L}}$ inherits the dissipativity of the continuous operator on the discrete approximation space $V_{\hbar}$ by Proposition 4.12, $\mathcal{L}$ is dissipative w.r.t. the weighted inner product $(\cdot \mid \cdot)_M$. This can easily be seen since (8) also holds for the discrete operators. Further, both $\widetilde{\mathcal{L}}$ and $\mathcal{L}$ are maximal as $V_{\hbar}$ is finite-dimensional. Hence, by the Lumer–Phillips Theorem, the restriction of $\mathcal{L}$ to $V_{\hbar}$ generates a contraction semigroup w.r.t. $\|\cdot\|_M$, which we denote by $\left(e^{t\mathcal{L}}\right)_{t\geq 0}$.

Corollary 4.15 *There exists a unique solution* $\boldsymbol{u} \in C^1(\mathbb{R}_+; V_{\hbar})$ *of* (20) *given by the variation-of-constants formula*

$$\boldsymbol{u}(t) = e^{t\mathcal{L}}\boldsymbol{u}_\pi^0 + \int_0^t e^{(t-s)\mathcal{L}} \boldsymbol{f}_\pi(s)\,ds. \tag{21}$$

5 Error Analysis of the Spatially Semi-discrete Problem

We are now able to analyze the *spatially semi-discrete error*

$$e = u - \boldsymbol{u},$$

where u is the exact solution of (7) and $\boldsymbol{u}$ denotes the semi-discrete approximation given by (20). We split this error into

$$e = e_\pi + \boldsymbol{e}_h = u - \pi_h u + \pi_h u - \boldsymbol{u}, \tag{22}$$

where e_π is the *projection error* and $\boldsymbol{e}_h$ is the *space discretization error*. By (13) and the boundedness of M we have the following bound on the projection error

$$\|e_\pi(t)\|_M \leq C_{\pi,M}|h^{k+1}u(t)|_{k+1,\mathcal{T}_{\hbar}} \tag{23}$$

with $C_{\pi,M} = \|M\|_{\infty,\Omega}^{1/2} C_\pi$.

Hence, it remains to bound the space discretization error $\boldsymbol{e}_h$. We do this by showing that $\boldsymbol{e}_h$ satisfies the semi-discrete problem (20) with zero initial value and the right hand side given by a defect stemming from the spatial discretization. Using the variation-of-constants formula (21) and the stability of the semi-discrete scheme (owing to the contractivity of the semigroup $\left(e^{t\mathcal{L}}\right)_{t\geq 0}$) we can then bound the discretization error by this defect. The approximation property (18) of the discrete operator $\widetilde{\mathcal{L}}$ then provides a bound on the defect and thus on the discretization error.

Lemma 5.1 *Assume that the exact solution of* (7) *fulfills* $u \in C^1(\mathbb{R}_+; L^2(\Omega)^m) \cap C(\mathbb{R}_+; D(\mathcal{L}) \cap H^1(\mathcal{T}_{\hbar})^m)$. *Then the space discretization error* $\boldsymbol{e}_h = \pi_h u - \boldsymbol{u}$ *satisfies*

$$\begin{cases} \partial_t \boldsymbol{e}_h(t) = \mathcal{L}\boldsymbol{e}_h(t) + \boldsymbol{d}_\pi(t), & t \in \mathbb{R}_+, \\ \boldsymbol{e}_h(0) = 0, \end{cases} \tag{24}$$

where the defect $\boldsymbol{d}_\pi : \mathbb{R}_+ \to \mathbb{R}^m$ *is given by*

$$\boldsymbol{d}_\pi = \mathcal{L}e_\pi. \tag{25}$$

Proof We begin by inserting the projected exact solution $\pi_h u$ into the semi-discrete equation (20a) and defining the error made by this as $\boldsymbol{d}_\pi$, yielding

$$\partial_t \pi_h u = \mathcal{L}\pi_h u + \boldsymbol{f}_\pi + \boldsymbol{d}_\pi . \tag{26}$$

Subtracting the semi-discrete scheme (20a) from (26) readily implies (24).

To show (25), we use that ∂_t and the L^2-projection commute and that u solves the continuous problem (7) to obtain

$$\partial_t \pi_h u = \pi_h \partial_t u = \pi_h(\mathcal{L}u + f) = \mathcal{L}u + \boldsymbol{f}_\pi .$$

Here, we have used the consistency property (19) in the last step. Equating this with (26) and solving for $\boldsymbol{d}_\pi$ yields

$$\boldsymbol{d}_\pi = \mathcal{L}u - \mathcal{L}\pi_h u = \mathcal{L}e_\pi ,$$

concluding the proof. □

Having derived an evolution equation for the error, we can now solve it to obtain a bound on the space discretization error. Together with the already mentioned bound on the projection error (23) we can thus bound the spatially semi-discrete error.

Theorem 5.2 *Assume that the exact solution u of the wave-type problem* (7) *satisfies $u \in C^1(\mathbb{R}_+; L^2(\Omega)^m) \cap C(\mathbb{R}_+; D(\mathcal{L}) \cap H^{k+1}(\mathcal{T}_h)^m)$. Then, for $t \in \mathbb{R}_+$, the spatially semi-discrete error satisfies*

$$\begin{aligned}\|u(t) - \boldsymbol{u}(t)\|_M &\le C_{\pi,M} |h^{k+1} u(t)|_{k+1,\mathcal{T}_h} + C_{\pi,\mathcal{L},M} \int_0^t |h^k u(s)|_{k+1,\mathcal{T}_h} \,\mathrm{d}s \\ &\le C h^k ,\end{aligned}$$

where $C_{\pi,\mathcal{L},M} = \|M^{-1}\|_{\infty,\Omega}^{1/2} C_{\pi,\mathcal{L}}$ and C only depends on $C_{\pi,M}$, $C_{\pi,\mathcal{L},M}$ and $|u(s)|_{k+1,\mathcal{T}_h}$, $s \in [0,t]$.

Proof We use Corollary 4.15 to solve the error equation (24), which yields

$$\boldsymbol{e}_h(t) = \int_0^t \mathrm{e}^{(t-s)\mathcal{L}} \boldsymbol{d}_\pi(s) \,\mathrm{d}s .$$

By the contractivity of the semigroup $\left(\mathrm{e}^{t\mathcal{L}}\right)_{t\ge 0}$ in the $\|\cdot\|_M$-norm we obtain

$$\|\boldsymbol{e}_h(t)\|_M \le \int_0^t \|\boldsymbol{d}_\pi(s)\|_M \,\mathrm{d}s = \int_0^t \|\mathcal{L}e_\pi(s)\|_M \,\mathrm{d}s .$$

It remains to bound $\|\mathcal{L}e_\pi(s)\|_M$. To do so, we use the boundedness of M and the approximation property from Proposition 4.14 applied to $\widetilde{\mathcal{L}}$, yielding

$$\begin{aligned}\|\mathcal{L}e_\pi(s)\|_M &= \|M^{-1/2}\widetilde{\mathcal{L}}e_\pi(s)\|_\Omega \\ &\le \|M^{-1}\|_{\infty,\Omega}^{1/2}\,\|\widetilde{\mathcal{L}}e_\pi(s)\|_\Omega \\ &\le \|M^{-1}\|_{\infty,\Omega}^{1/2}\,C_{\pi,\mathcal{I}}\,|h^k u(s)|_{k+1,\mathcal{T}_h}.\end{aligned}$$

Taking norms and using the triangle inequality in the error splitting (22) together with the already established bound (23) on e_π proves the claim. □

6 Concluding Remarks

In this paper we presented a rigorous error analysis of the spatial discretization of a large class of wave-type problems by discontinuous Galerkin methods. This class includes Maxwell's equations and the acoustic wave equation, for instance. It has been shown in [13] that such a space discretization on cuboids and tensorial grids can be combined with a Peaceman–Rachford (ADI) time integration scheme in such a way that it has optimal (linear) complexity for suitable problems. The full discretization error of the resulting scheme is studied in [15].

Acknowledgments We thank the referee for the comments which helped us to improve on a previous version of this paper. Moreover, we thank Benjamin Dörich for his careful reading of this manuscript and helpful discussions.

Funded by the Deutsche Forschungsgemeinschaft (DFG, German Research Foundation)—Project-ID 258734477—SFB 1173.

Appendix: Proofs from Sect. 4

In this appendix we collect the proofs we postponed in Sect. 4.

Proof *(Lemma 4.7)* Let $v \in H^1(\mathcal{T}_h)^m$.

(i) We first prove that $v \in H(\mathcal{L})$ follows from (14). By (3), the definition of $\widetilde{\mathcal{L}}^{\circledast}$ in (4) and the boundedness of L_0, this is the case if the mapping

$$C_c^\infty(\Omega)^m \to \mathbb{R}, \quad \varphi \mapsto \left(v \,\middle|\, \widetilde{\mathcal{L}}^{\circledast}\varphi\right)_\Omega \tag{27}$$

is bounded in the L^2-norm. Let $\varphi \in C_c^\infty(\Omega)^m$ so that $[\![\varphi]\!]_F = 0$ and $\{\{\varphi\}\}_F = \varphi|_F$ for all $F \in \mathcal{F}_h^{\mathrm{int}}$ and $\varphi|_F = 0$ for all $F \in \mathcal{F}_h^{\mathrm{bnd}}$. By applying the integration by parts formula from Lemma 4.3 on each element we have

$$\begin{aligned}
(v \,|\, \widetilde{\mathcal{L}}^{\circledast}\varphi)_\Omega &= \sum_{K\in\mathcal{T}_h} (v \,|\, \widetilde{\mathcal{L}}^{\circledast}\varphi)_K \\
&= \sum_{K\in\mathcal{T}_h} (\widetilde{\mathcal{L}}v \,|\, \varphi)_K + \sum_{K\in\mathcal{T}_h} (L_\partial^K v \,|\, \varphi)_{\partial K} \\
&= \sum_{K\in\mathcal{T}_h} (\widetilde{\mathcal{L}}v \,|\, \varphi)_K + \sum_{F\in\mathcal{F}_h^{\mathrm{int}}} (L_\partial^F [\![v]\!]_F \,|\, \varphi)_F \\
&= \sum_{K\in\mathcal{T}_h} (\widetilde{\mathcal{L}}v \,|\, \varphi)_K,
\end{aligned} \tag{28}$$

where we have used Lemma 4.6 and the aforementioned properties of φ in the third and (14) in the last step. Applying the Cauchy–Schwarz and Young's inequality we obtain the boundedness of (27) and hence $v \in H(\mathcal{L})$.

(ii) Next, let $v \in H(\mathcal{L}) \cap H^1(\mathcal{T}_h)^m$. Then, [14, Thm. 1.4] or [1, Thm. 4] imply that $H(\mathcal{L}) \cap C^\infty_{c,\mathbb{R}^d}(\Omega)^m$ is dense in $H(\mathcal{L})$. Hence, we can choose a sequence $(v_n)_{n\in\mathbb{N}}$ in $H(\mathcal{L}) \cap C^\infty_{c,\mathbb{R}^d}(\Omega)^m$ with

$$v_n \to v, \qquad \widetilde{\mathcal{L}}v_n \to \widetilde{\mathcal{L}}v \qquad \text{in } L^2(\Omega)^m.$$

For arbitrary $\varphi \in C_c^\infty(\Omega)^m$ and with $\mathfrak{n}^{\mathrm{dom}}$ denoting the outward unit normal vector to Γ, Lemma 4.3 yields

$$\begin{aligned}
(\widetilde{\mathcal{L}}v \,|\, \varphi)_\Omega &= \lim_{n\to\infty} (\widetilde{\mathcal{L}}v_n \,|\, \varphi)_\Omega \\
&= \lim_{n\to\infty} \Big((v_n \,|\, \widetilde{\mathcal{L}}^{\circledast}\varphi)_\Omega + \Big(\sum_{i=1}^d \mathfrak{n}_i^{\mathrm{dom}} L_i v_n \,|\, \varphi\Big)_\Gamma \Big) \\
&= (v \,|\, \widetilde{\mathcal{L}}^{\circledast}\varphi)_\Omega.
\end{aligned}$$

Comparing this with the third line in (28) yields

$$\sum_{F\in\mathcal{F}_h^{\mathrm{int}}} (L_\partial^F [\![v]\!]_F \,|\, \varphi)_F = 0.$$

Since φ was arbitrary, this in particular holds for $\operatorname{supp}\varphi$ only intersecting a single interface, which implies (14).

(iii) To show the last assertion, let $v \in D(\mathcal{L}) \cap H^1(\mathcal{T}_h)^m$ and $\varphi \in C^\infty_{c,\mathbb{R}^d}(\Omega)^m$. Proceeding similarly to (28) (additionally using $v \in H(\mathcal{L})$) it is easy to see that

$$\langle \mathcal{L}_\partial v \mid \varphi \rangle = \sum_{F \in \mathcal{F}_h^{\mathrm{bnd}}} \left(L_\partial^F v \mid \varphi \right)_F,$$

by the definition of $\mathcal{L}_\partial$ in (5). Together with Assumption 4.5 this yields

$$\sum_{F \in \mathcal{F}_h^{\mathrm{bnd}}} \left((L_\partial^F - L_\Gamma) v \mid \varphi \right)_F = \langle (\mathcal{L}_\partial - \mathcal{L}_\Gamma) v \mid \varphi \rangle = 0 \qquad \text{for all } \varphi \in C^\infty_{c,\mathbb{R}^d}(\Omega)^m,$$

where we have used $v \in D(\mathcal{L})$ for the last equality. Again, this particularly holds for all φ with $\operatorname{supp} \varphi$ only intersecting a single boundary face, showing (15). □

Proof *(Proposition 4.13)* We begin by deriving an elementwise representation of $\widetilde{\mathfrak{L}}$. Namely, since we have $\left(\widetilde{\mathfrak{L}} \boldsymbol{v} \mid \boldsymbol{\varphi} \right)_K = \left(\widetilde{\mathfrak{L}} \boldsymbol{v} \mid \mathbb{1}_K \boldsymbol{\varphi} \right)_\Omega$ for all $\boldsymbol{\varphi} \in V_h$, a straightforward calculation yields

$$\begin{aligned} \left(\widetilde{\mathfrak{L}} \boldsymbol{v} \mid \boldsymbol{\varphi} \right)_K &= \left(\widetilde{\mathcal{L}} \boldsymbol{v} \mid \boldsymbol{\varphi} \right)_K - \tfrac{1}{2} \sum_{F \in \mathcal{F}_h^{K,\mathrm{int}}} \left(L_\partial^F [\![\boldsymbol{v}]\!]_F \mid \boldsymbol{\varphi}|_K \right)_F \\ &\quad - \tfrac{1}{2} \sum_{F \in \mathcal{F}_h^{K,\mathrm{bnd}}} \left((L_\partial^F - L_\Gamma) \boldsymbol{v} \mid \boldsymbol{\varphi} \right)_F. \end{aligned} \tag{29}$$

We now bound the element, interface and boundary face terms individually, beginning with the former. Using the Cauchy–Schwarz inequality and the inverse inequality (12) yields

$$\left(\widetilde{\mathcal{L}} \boldsymbol{v} \mid \boldsymbol{\varphi} \right)_K \le \| \widetilde{\mathcal{L}} \boldsymbol{v} \|_K \| \boldsymbol{\varphi} \|_K \le C_{\mathcal{L}} C_{\mathrm{inv}} \| h^{-1} \boldsymbol{v} \|_K \| \boldsymbol{\varphi} \|_K.$$

The boundary terms are treated similarly by again using the Cauchy–Schwarz inequality and this time the boundedness of L_∂^F and L_Γ and the trace inequality (11) to obtain

$$\begin{aligned} \left((L_\partial^F - L_\Gamma) \boldsymbol{v} \mid \boldsymbol{\varphi} \right)_F &\le C_{\Gamma,\mathcal{L}} \| \boldsymbol{v} \|_F \| \boldsymbol{\varphi} \|_F \\ &\le C_{\Gamma,\mathcal{L}} C_{\mathrm{tr}} \| h^{-1/2} \boldsymbol{v} \|_K \, C_{\mathrm{tr}} \| h^{-1/2} \boldsymbol{\varphi} \|_K \\ &= C_{\Gamma,\mathcal{L}} C_{\mathrm{tr}}^2 \| h^{-1} \boldsymbol{v} \|_K \| \boldsymbol{\varphi} \|_K, \end{aligned}$$

where we used that h is piecewise constant. To bound the interface terms, we first rewrite the jump $[\![\boldsymbol{v}]\!]_F$ as

$$\big(L_\partial^F [\![\boldsymbol{v}]\!]_F \mid \boldsymbol{\varphi}_{|K}\big)_F = \varepsilon_{K,F}\Big(\big(L_\partial^F \boldsymbol{v}_{|K} \mid \boldsymbol{\varphi}_{|K}\big)_F - \big(L_\partial^F \boldsymbol{v}_{|K_F} \mid \boldsymbol{\varphi}_{|K}\big)_F\Big),$$

where here and in the following, K_F denotes the neighbor of K w.r.t. F and $\varepsilon_{K,F} = \mathfrak{n}^K \cdot \mathfrak{n}^F = \pm 1$. The first term can be bounded completely analogously to the boundary term by

$$\big(L_\partial^F \boldsymbol{v}_{|K} \mid \boldsymbol{\varphi}_{|K}\big)_F \leq C_{\mathcal{L}} C_{\mathrm{tr}}^2 \|h^{-1}\boldsymbol{v}\|_K \|\boldsymbol{\varphi}\|_K .$$

To bound the second term, we additionally use $h_K^{-1} \leq \rho h_{K_F}^{-1}$ (see [4, Lem. 1.43]) and thus

$$\big(L_\partial^F \boldsymbol{v}_{|K_F} \mid \boldsymbol{\varphi}_{|K}\big)_F \leq \rho^{1/2} C_{\mathcal{L}} C_{\mathrm{tr}}^2 \|h^{-1}\boldsymbol{v}\|_{K_F} \|\boldsymbol{\varphi}\|_K .$$

Assembling all these bounds and taking into account that each element has at most N_∂ neighboring elements and at most one boundary face yields

$$\big(\widetilde{\mathcal{L}}\boldsymbol{v} \mid \boldsymbol{\varphi}\big)_K \leq \Big(C_{\widetilde{\mathcal{L}},\mathrm{el},1} \|h^{-1}\boldsymbol{v}\|_K + C_{\widetilde{\mathcal{L}},\mathrm{el},2} \sum_{F \in \mathcal{F}_h^{K,\mathrm{int}}} \|h^{-1}\boldsymbol{v}\|_{K_F}\Big) \|\boldsymbol{\varphi}\|_K \tag{30}$$

for all $\boldsymbol{\varphi} \in V_h$. The constants are given as $C_{\widetilde{\mathcal{L}},\mathrm{el},1} = C_{\mathcal{L}} C_{\mathrm{inv}} + \frac{1}{2} C_{\mathrm{tr}}^2 (C_{\Gamma,\mathcal{L}} + N_\partial C_{\mathcal{L}})$ and $C_{\widetilde{\mathcal{L}},\mathrm{el},2} = \frac{1}{2} \rho^{1/2} C_{\mathrm{tr}}^2 C_{\mathcal{L}}$.

It remains to put these elementwise bounds together to obtain a bound w.r.t. the whole domain Ω. Summing (30) over all elements $K \in \mathcal{T}_h$ yields

$$\big(\widetilde{\mathcal{L}}\boldsymbol{v} \mid \boldsymbol{\varphi}\big)_\Omega \leq C_{\widetilde{\mathcal{L}},\mathrm{el},1} \sum_{K \in \mathcal{T}_h} \|h^{-1}\boldsymbol{v}\|_K \|\boldsymbol{\varphi}\|_K + C_{\widetilde{\mathcal{L}},\mathrm{el},2} \sum_{K \in \mathcal{T}_h} \sum_{F \in \mathcal{F}_h^{K,\mathrm{int}}} \|h^{-1}\boldsymbol{v}\|_{K_F} \|\boldsymbol{\varphi}\|_K .$$

From here, the assertion follows by straightforward applications of the Cauchy–Schwarz and Young's inequality, respectively. □

Proof *(Proposition 4.14)* We proceed similarly to the proof of Proposition 4.13, meaning that we first work on the element-based formulation (29). Using integration by parts yields

$$\begin{aligned}
\big(\widetilde{\mathcal{L}} e_\pi^v \mid \boldsymbol{\varphi}\big)_K &= \big(\widetilde{\mathcal{I}} e_\pi^v \mid \boldsymbol{\varphi}\big)_K - \tfrac{1}{2} \sum_{F \in \mathcal{F}_h^{K,\mathrm{int}}} \big(L_\partial^F [\![e_\pi^v]\!]_F \mid \boldsymbol{\varphi}_{|K}\big)_F \\
&\quad - \tfrac{1}{2} \sum_{F \in \mathcal{F}_h^{K,\mathrm{bnd}}} \big((L_\partial^F - L_\Gamma) e_\pi^v \mid \boldsymbol{\varphi}\big)_F \\
&= \tfrac{1}{2} \sum_{F \in \mathcal{F}_h^{K,\mathrm{int}}} \big(L_\partial^F \{\!\{e_\pi^v\}\!\}_F \mid \boldsymbol{\varphi}_{|K}\big)_F + \tfrac{1}{2} \sum_{F \in \mathcal{F}_h^{K,\mathrm{bnd}}} \big((L_\partial^F + L_\Gamma) e_\pi^v \mid \boldsymbol{\varphi}\big)_F
\end{aligned}$$

for all $\boldsymbol{\varphi} \in V_{\tilde{h}}$, where the element term vanishes because of the defining property of the L^2-projection (9) since $\widetilde{\mathcal{L}}^{\circledast} \boldsymbol{\varphi}|_K \in \mathbb{Q}_d^k(K)$.

The rest of the proof is completely analogous to the corresponding part of the proof of Proposition 4.13. The only difference is that we use the second bound in (13) instead of the discrete trace inequality. □

References

1. Antonić, N., Burazin, K.: Graph spaces of first-order linear partial differential operators. Math. Commun. **14**(1), 135–155 (2009). https://hrcak.srce.hr/37441
2. Burazin, K., Erceg, M.: Non-stationary abstract Friedrichs systems. Mediterr. J. Math. **13**(6), 3777–3796 (2016). https://doi.org/10.1007/s00009-016-0714-8
3. Burman, E., Ern, A., Fernández, M.A.: Explicit Runge-Kutta schemes and finite elements with symmetric stabilization for first-order linear PDE systems. SIAM J. Numer. Anal. **48**(6), 2019–2042 (2010). https://dx.doi.org/10.1137/090757940
4. Di Pietro, D.A., Ern, A.: Mathematical aspects of discontinuous Galerkin methods. In: Mathématiques & Applications (Berlin) [Mathematics & Applications], vol. 69. Springer, Heidelberg (2012). https://dx.doi.org/10.1007/978-3-642-22980-0
5. Engel, K.J., Nagel, R.: One-parameter semigroups for linear evolution equations. In: Graduate Texts in Mathematics, vol. 194. Springer, New York (2000). https://dx.doi.org/10.1007/b97696
6. Ern, A., Guermond, J.L.: Discontinuous Galerkin methods for Friedrichs' systems. I. General theory. SIAM J. Numer. Anal. **44**(2), 753–778 (2006). https://doi.org/10.1137/050624133
7. Ern, A., Guermond, J.L.: Discontinuous Galerkin methods for Friedrichs' systems. II. Second-order elliptic PDEs. SIAM J. Numer. Anal. **44**(6), 2363–2388 (2006). https://doi.org/10.1137/05063831X
8. Ern, A., Guermond, J.L.: Discontinuous Galerkin methods for Friedrichs' systems. III. Multifield theories with partial coercivity. SIAM J. Numer. Anal. **46**(2), 776–804 (2008). https://doi.org/10.1137/060664045
9. Friedrichs, K.O.: Symmetric positive linear differential equations. Commun. Pure Appl. Math. **11**, 333–418 (1958). https://doi.org/10.1002/cpa.3160110306
10. Grote, M.J., Schneebeli, A., Schötzau, D.: Interior penalty discontinuous Galerkin method for Maxwell's equations: energy norm error estimates. J. Comput. Appl. Math. **204**(2), 375–386 (2007). https://dx.doi.org/10.1016/j.cam.2006.01.044
11. Hesthaven, J.S., Warburton, T.: Nodal discontinuous Galerkin methods: algorithms, analysis, and applications. In: Texts in Applied Mathematics, vol. 54. Springer, New York (2008). https://dx.doi.org/10.1007/978-0-387-72067-8
12. Hipp, D., Hochbruck, M., Stohrer, C.: Unified error analysis for nonconforming space discretizations of wave-type equations. IMA J. Numer. Anal. (2018). https://dx.doi.org/10.1093/imanum/dry036
13. Hochbruck, M., Köhler, J.: On the efficiency of the Peaceman–Rachford ADI-dG method for wave-type problems. In: Radu, F., Kumar, K., Berre, I., Nordbotten, J., Pop, I. (eds.) Numerical Mathematics and Advanced Applications ENUMATH 2017. Lecture Notes in Computational Science and Engineering, vol. 126, pp. 135–144. Springer, Berlin (2019). https://dx.doi.org/10.1007/978-3-319-96415-7
14. Jensen, M.: Discontinuous Galerkin methods for Friedrichs systems with irregular solutions. Ph.D. thesis, University of Oxford, Oxford (2004). http://sro.sussex.ac.uk/45497/

15. Köhler, J.: The Peaceman–Rachford ADI-dG method for linear wave-type problems. Ph.D. thesis, Karlsruhe Institute of Technology, Karlsruhe (2018). https://dx.doi.org/10.5445/IR/1000089271
16. Sturm, A.: Locally implicit time integration for linear Maxwell's equations. Ph.D. thesis, Karlsruhe Institute of Technology, Karlsruhe (2017). https://dx.doi.org/10.5445/IR/1000069341

Ill-posedness of the Third Order NLS with Raman Scattering Term in Gevrey Spaces

Nobu Kishimoto and Yoshio Tsutsumi

Abstract The Cauchy problem for the third order nonlinear Schrödinger equation with Raman scattering term on the one dimensional torus is shown to be ill-posed in Gevrey class G^σ for any $\sigma > 1$.

1 Introduction

We consider the nonlinear Schrödinger equation with third order dispersion and intrapulse Raman scattering term:

$$\partial_t u = \alpha_1 \partial_x^3 u + i\alpha_2 \partial_x^2 u + i\gamma_1 |u|^2 u + \gamma_2 \partial_x\big(|u|^2 u\big) - i\Gamma u \partial_x\big(|u|^2\big), \tag{1}$$

$$t \in (-T, T), \quad x \in \mathbb{T} = \mathbb{R}/2\pi\mathbb{Z},$$

$$u(0, x) = u_0(x), \qquad x \in \mathbb{T}, \tag{2}$$

where α_j, γ_j $(j = 1, 2)$ are real constants with $\alpha_1 \neq 0$, Γ is a complex constant with non-zero real part, and T is a positive constant.

Equation (1) is the mathematical model for the signal propagation in a crystal optical fiber (see, e.g., [1]). Throughout this article, we assume that

$$\mathrm{Re}(\Gamma) > 0, \qquad \frac{2\alpha_2}{3\alpha_1} \notin \mathbb{Z}.$$

The sign of $\mathrm{Re}(\Gamma)$ is actually not important in our analysis, whereas in deriving the key equation (4) below we need to assume the latter condition $\frac{2\alpha_2}{3\alpha_1} \notin \mathbb{Z}$, which has

N. Kishimoto
Research Institute for Mathematical Sciences, Kyoto University, Kyoto, Japan
e-mail: nobu@kurims.kyoto-u.ac.jp

Y. Tsutsumi (✉)
Department of Mathematics, Graduate School of Science, Kyoto University, Kyoto, Japan
e-mail: tsutsumi@math.kyoto-u.ac.jp

W. Dörfler et al. (eds.), *Mathematics of Wave Phenomena*, Trends in Mathematics,
https://doi.org/10.1007/978-3-030-47174-3_13

never appeared in the physical literature. It is not clear whether the latter condition has relevance to physical phenomena.

In the previous paper [12], the authors prove the ill-posedness of the Cauchy problem (1)–(2) in the Sobolev space and the well-posedness in a certain space of analytic functions. It is natural to ask whether or not the Cauchy problem (1)–(2) is well-posed in the Gevrey class, which is the intermediate space between the Sobolev space and the analytic space. To make the setting of our problem clear, we begin with the definition of the Gevrey class.

Definition 1 For $\sigma \geq 1$, $s \geq 0$ and $a > 0$, we define Gevrey space $G^{\sigma}_{s,a}$ as the space of all functions $f \in C^{\infty}(\mathbb{T}; \mathbb{C})$ such that

$$\hat{f}_k := (2\pi)^{-1/2} \int_{\mathbb{T}} f(x) e^{-ikx}\, dx = O(|k|^{-s} e^{-a|k|^{1/\sigma}})$$

as $|k| \to \infty$. The space $G^{\sigma}_{s,a}$ is a Banach space with norm

$$\|f\|_{G^{\sigma}_{s,a}} := \sup_{k \in \mathbb{Z}} e^{a|k|^{1/\sigma}} \langle k \rangle^{s} |\hat{f}_k|,$$

where $\langle k \rangle := \max\{1, |k|\}$. Then, the Gevrey class of order σ, which is denoted by G^{σ}, is defined as the union $\bigcup_{a>0} G^{\sigma}_{0,a}$ equipped with the inductive limit topology.

Remark 1

(i) Although Definition 1 is slightly different from the standard one (see, e.g., [17, Definition 1.4.1 on page 19]), our definition gives the identical space. See Proposition 2 (ii) in Appendix for a proof.
(ii) We have continuous embeddings $G^{\sigma}_{s,a} \hookrightarrow G^{\sigma}_{r,a}$ $(s > r \geq 0)$, $G^{\sigma}_{0,a} \hookrightarrow G^{\sigma}_{s,b}$ $(a > b > 0, s \geq 0)$, and $G^{\sigma}_{0,a} \hookrightarrow G^{\tau}_{0,b}$ $(\tau > \sigma \geq 1, a, b > 0)$. It can be shown that the embeddings $G^{\sigma}_{0,a} \hookrightarrow G^{\sigma} \hookrightarrow G^{\tau}_{0,b}$ $(\tau > \sigma \geq 1, a, b > 0)$ are also continuous. Moreover, it turns out that a function u belongs to $C([-T,T]; G^{\sigma})$ if and only if $u \in C([-T,T]; G^{\sigma}_{0,a})$ for some $a > 0$. We will give proofs of these facts in Appendix, Propositions 2 and 3.

Our goal is to show the following smoothing effect:

Theorem 1 *Let $\sigma > 1$, and let $u(t) \in C([-T,T]; G^{\sigma})$ be a solution to* (1) *on $(-T,T)$ for some $T > 0$. Then, $u(t) \in \bigcap_{a'>0} G^{\sigma}_{0,a'}$ for all $t \in (-T,T)$.*

Theorem 1 implies the ill-posedness of the Cauchy problem in Gevrey class:

Corollary 1 *Let $\sigma > 1$. Then, the Cauchy problem* (1)–(2) *is ill-posed in G^{σ}. More precisely, for any $u_0 \in G^{\sigma} \setminus \bigcap_{a'>0} G^{\sigma}_{0,a'}$ there exists no $T > 0$ such that the Cauchy problem* (1)–(2) *has a solution in $C([-T,T]; G^{\sigma})$.*

In linear partial differential equations, it is well known that the Gevrey class provides a clear boundary between the existence and the non-existence of solution for the Cauchy problem. For example, we consider the following Cauchy problem of the heat operator:

$$\frac{\partial^2 u}{\partial x^2} - \frac{\partial u}{\partial t} = 0, \quad t \in \mathbb{R}, \quad x \in (-\delta, \delta),$$
$$u(0,t) = \phi_0(t), \quad \frac{\partial u}{\partial x}(0,t) = \phi_1(t), \quad t \in \mathbb{R},$$

where δ is a positive constant. This Cauchy problem of the heat operator admits a solution in G^σ for $\sigma \leq 2$, while it has no solution in G^σ for $\sigma > 2$ with ϕ_0 or $\phi_1 \notin G^2$ (see, e.g., [3, Section 1.2]). For this reason, it seems very interesting to study the ill-posedness of (1)–(2) in the Gevrey class. In [12, equation (4) in Section 1], it is shown that the Cauchy-Riemann type operator appears from the Raman scattering term, that is, from the last term on the right hand side of (1). This yields the ill-posedness of the Cauchy problem (1)–(2). From the theory of linear partial differential equations, the Cauchy-Riemann operator is analytic-hypoelliptic, which implies the non-existence of solution for the initial value problem of the Cauchy-Riemann operator in the Gevrey class as well as in the Sobolev space (see, e.g., [3] and [17]). Thus, it is presumed that the Cauchy problem (1)–(2) may not admit a solution in the Gevrey class. But it is not obvious at all, because the principal part of Eq. (1) is not the Cauchy-Riemann operator. In fact, Eq. (1) is of third order and it includes the nonlinear terms dependent on the first derivative of the solution itself. We use the Gevrey smoothing estimate for nonlinear interactions to overcome this difficulty (see Proposition 1, (10) and Lemma 3 below). Here, we emphasize that we do not consider the local solvability of the Ehrenpreis and Malgrange theorem type but consider the well-posedness of the Cauchy problem (1)–(2) (see, e.g., [10, Theorem 0.1, on page1779], [15, Example on page 2926] and [7, Section 4] for the local solvability).

There are several papers which study the ill-posedness nature of other nonlinear evolution equations within the framework of regular function spaces. In [4], Colin and Métivier show the ill-posedness of the degenerate Zakharov equations in the Sobolev space. In [9], Gérard-Varet and Dormy show the strong instability in the Sobolev space for the Cauchy problem of the Prandtl equation, while Dietert and Gérard-Varet [8] prove the well-posedness in the Gevrey class (see [16] and [2] for the well-posedness of the Prandtl equation with particular initial data in Sobolev spaces). These papers suggest that the space G^2 may be critical in the degenerate Zakharov equations and the Prandtl equation concerning the ill-posedness, while Eq. (1) admits no solution in the Gevrey class. The well-posedness issue has been also studied in the Gevrey class for many other nonlinear evolution equations (see, e.g., Leray and Ohya [14] and Kajitani [11] for nonlinear hyperbolic equations and Ukai [18] for the Boltzmann equation).

The plan of our paper is as follows. In Sect. 2, we reduce Eq. (1) to the system of ordinary differential equations, which makes the analysis in the Gevrey class more transparent. In Sect. 3, we show the multilinear estimates, which are useful for the analysis of nonlinear interactions. In Sect. 4, we give a proof of Theorem 1 by using the results obtained in Sects. 2 and 3.

2 Reduction to ODE

Let u be a non-trivial solution to (1) on $(-T, T)$ which belongs to $C([-T, T]; H^1(\mathbb{T}))$. Then, the $L^2(\mathbb{T})$-norm of $u(t)$ is conserved (see [12, Lemma 2.5]), and hence, $\|u(0)\|_{L^2(\mathbb{T})} > 0$. We define the new unknown function $w \in C([-T, T]; H^1(\mathbb{T}))$ by

$$\hat{w}_k(t) := \exp\Big[i\Big(\alpha_1 k^3 + \alpha_2 k^2 - \frac{2\gamma_2 + \mathrm{Im}(\Gamma)}{2\pi}\|u(0)\|_{L^2}^2 k\Big)t\Big]\hat{u}_k(t),$$
$$(t, k) \in [-T, T] \times \mathbb{Z}.$$

Then, $\hat{w} = \{\hat{w}_k(\cdot)\}_k \in C([-T, T]; \ell^2_1)$, where

$$\ell^p_s(\mathbb{Z}) := \{c \in \ell^p(\mathbb{Z}) \mid \|c\|_{\ell^p_s} := \|\langle\cdot\rangle^s c\|_{\ell^p} < \infty\},$$

and $\hat{w}_k(t)$ solves

$$\partial_t \hat{w}_k = \sum_{\substack{k_1,k_2,k_3\in\mathbb{Z}\\ k_{123}=k}} \mu(k_1, k_2, k_3) e^{it\Phi(k_1,k_2,k_3)} \hat{w}_{k_1} \hat{\bar{w}}_{k_2} \hat{w}_{k_3}, \quad t \in (-T, T), \tag{3}$$

where we have introduced the notation $k_{ij\cdots} := k_i + k_j + \cdots$ and

$$\mu(k_1, k_2, k_3) := \frac{i\gamma_1 + i\gamma_2 k_{123} + \Gamma k_{12}}{2\pi} - \chi_{\{k_{23}=0\}}(k_1, k_2, k_3)\frac{2i\gamma_2 + i\mathrm{Im}(\Gamma)}{2\pi}k_1,$$
$$\Phi(k_1, k_2, k_3) := 3\alpha_1 k_{12} k_{23}(k_{31} + \frac{2\alpha_2}{3\alpha_1}).$$

Following the argument in [12, the derivation process of (2.11) from (2.2)], we compute the right-hand side of the equation and, under the condition $\frac{2\alpha_2}{3\alpha_1} \notin \mathbb{Z}$, obtain that

$$\partial_t \hat{w}_k = \frac{\mathrm{Re}(\Gamma)}{2\pi}\|u(0)\|_{L^2(\mathbb{T})}^2 k\hat{w}_k - \Big[\frac{\Gamma}{2\pi}\Big(\sum_{k'\in\mathbb{Z}} k'|\hat{w}_{k'}|^2\Big)\hat{w}_k + \frac{i\gamma_2 k}{2\pi}|\hat{w}_k|^2\hat{w}_k\Big]$$
$$+ \frac{i\gamma_1}{2\pi}\sum_{k_{123}=k} e^{it\Phi}\hat{w}_{k_1}\hat{\bar{w}}_{k_2}\hat{w}_{k_3} + \sum_{D_1(k)} \frac{i\gamma_2 k + \Gamma k_{12}}{2\pi} e^{it\Phi}\hat{w}_{k_1}\hat{\bar{w}}_{k_2}\hat{w}_{k_3}$$

$$
\begin{aligned}
&+\partial_t\Big[\sum_{D_2(k)} \frac{i\gamma_2 k+\Gamma k_{12}}{2\pi i\Phi} e^{it\Phi}\hat{w}_{k_1}\hat{\bar{w}}_{k_2}\hat{w}_{k_3}\Big] \\
&-\sum_{D_2(k)} \frac{i\gamma_2 k+\Gamma k_{12}}{2\pi i\Phi} e^{it\Phi}\partial_t\big[\hat{w}_{k_1}\hat{\bar{w}}_{k_2}\hat{w}_{k_3}\big],
\end{aligned} \tag{4}
$$

where we define $D(k) := \{(k_1,k_2,k_3)\in \mathbf{Z}^3 \mid k_{123}=k,\ k_{12}k_{23}\neq 0\}$ and

$$
\begin{aligned}
D_1(k) &:= \{(k_1,k_2,k_3)\in D(k)\mid \tfrac{1}{4}|k_2|\le |k_1|,|k_3|\le 4|k_2|\},\\
D_2(k) &:= D(k)\setminus D_1(k).
\end{aligned}
$$

Inserting (3) in the last term on the right side of (4), we rewrite the above Eq. (4) as

$$
\partial_t \hat{w} = Ak\hat{w}_k + \hat{F}[w]_k + \partial_t\hat{G}[w]_k + \hat{H}[w]_k, \qquad t\in(-T,T), \tag{5}
$$

where $A := \frac{\mathrm{Re}(\Gamma)}{2\pi}\|u(0)\|_{L^2}^2$ is a positive constant and

$$
\begin{aligned}
\hat{F}[w]_k &:= \sum_{k_{123}=k} M_3\hat{w}_{k_1}\hat{\bar{w}}_{k_2}\hat{w}_{k_3} - \frac{\Gamma}{2\pi}\Big(\sum_{k'\in\mathbb{Z}} k'|\hat{w}_{k'}|^2\Big)\hat{w}_k,\\
\hat{G}[w]_k &:= \sum_{k_{123}=k} M_3^0\hat{w}_{k_1}\hat{\bar{w}}_{k_2}\hat{w}_{k_3},\\
\hat{H}[w]_k &:= \sum_{k_{12345}=k} M_5\hat{w}_{k_1}\hat{\bar{w}}_{k_2}\hat{w}_{k_3}\hat{\bar{w}}_{k_4}\hat{w}_{k_5},
\end{aligned}
$$

$$
\begin{aligned}
M_3 := \Big[&\frac{i\gamma_1}{2\pi} + \chi_{D_1(k)}(k_1,k_2,k_3)\frac{i\gamma_2 k+\Gamma k_{12}}{2\pi}\\
&- \chi_{\{k_{12}=k_{23}=0\}}(k_1,k_2,k_3)\frac{i\gamma_2 k}{2\pi}\Big]e^{it\Phi(k_1,k_2,k_3)},\\
M_3^0 := {}&\chi_{D_2(k)}(k_1,k_2,k_3)\frac{i\gamma_2 k+\Gamma k_{12}}{2\pi i\Phi(k_1,k_2,k_3)}e^{it\Phi(k_1,k_2,k_3)},\\
M_5 := {}&-\chi_{D_2(k)}(k_{123},k_4,k_5)\frac{i\gamma_2 k+\Gamma k_{1234}}{2\pi i\Phi(k_{123},k_4,k_5)}\\
&\qquad\times \mu(k_1,k_2,k_3)e^{it[\Phi(k_{123},k_4,k_5)+\Phi(k_1,k_2,k_3)]}\\
&-\chi_{D_2(k)}(k_1,k_{234},k_5)\frac{i\gamma_2 k+\Gamma k_{1234}}{2\pi i\Phi(k_1,k_{234},k_5)}\\
&\qquad\times \overline{\mu(-k_2,-k_3,-k_4)}e^{it[\Phi(k_1,k_{234},k_5)-\Phi(-k_2,-k_3,-k_4)]}\\
&-\chi_{D_2(k)}(k_1,k_2,k_{345})\frac{i\gamma_2 k+\Gamma k_{12}}{2\pi i\Phi(k_1,k_2,k_{345})}\\
&\qquad\times \mu(k_3,k_4,k_5)e^{it[\Phi(k_1,k_2,k_{345})+\Phi(k_3,k_4,k_5)]}.
\end{aligned}
$$

We see from (5) that

$$\partial_t\Big[e^{-Akt}\Big(\hat{w}_k - \hat{G}[w]_k\Big)\Big] = e^{-Akt}\Big[\hat{F}[w]_k + Ak\hat{G}[w]_k + \hat{H}[w]_k\Big], \qquad t \in (-T, T).$$

Integrating the both sides in t, we verify that

$$\begin{aligned}\hat{w}_k(t_2) &= e^{Ak(t_2-t_1)}\hat{w}_k(t_1) + \Big[\hat{G}[w(t_2)]_k - e^{Ak(t_2-t_1)}\hat{G}[w(t_1)]_k\Big] \\ &\quad + \int_{t_1}^{t_2} e^{Ak(t_2-t)}\Big[\hat{F}[w(t)]_k + Ak\hat{G}[w(t)]_k + \hat{H}[w(t)]_k\Big]\,dt\end{aligned} \tag{6}$$

for any $t_1, t_2 \in [-T, T]$ and $k \in \mathbb{Z}$.

3 Multilinear Estimates

In this section, we show the following multilinear estimates on $F[w]$, $G[w]$ and $H[w]$ in Gevrey spaces.

Proposition 1 *Let $\sigma > 1$, $a > 0$. There exist some constants $\theta = \theta(\sigma) \in (0, 1)$, $C_0 = C_0(\sigma, a) > 0$ such that for any $s \geq 0$ the following estimates hold:*

$$\|P_{\geq C_0 s^\sigma} F[w]\|_{G^\sigma_{s,a}} \lesssim \|w\|^2_{G^\sigma_{3,\theta a}} \|w\|_{G^\sigma_{s,a}},$$

$$\|P_{\geq C_0 s^\sigma} G[w]\|_{G^\sigma_{s+1,a}} \lesssim \|w\|^2_{G^\sigma_{2,\theta a}} \|w\|_{G^\sigma_{s,a}},$$

$$\|P_{\geq C_0 s^\sigma} H[w]\|_{G^\sigma_{s,a}} \lesssim \|w\|^4_{G^\sigma_{3,\theta a}} \|w\|_{G^\sigma_{s,a}},$$

where $[P_{\geq N} f]\hat{}_k := \chi_{\{|k|\geq N\}}(k)\hat{f}_k$ and the implicit constants are independent of σ, a and s.

Remark 2 Based on the inequality $|k_{123}|^s \leq 3^{s-1}(|k_1|^s + |k_2|^s + |k_3|^s)$ for $s \geq 1$, we can easily show multilinear estimates with Sobolev regularity such as

$$\|\hat{F}[w]\|_{\ell^\infty_s} \leq C3^s \|\hat{w}\|^2_{\ell^\infty_3} \|\hat{w}\|_{\ell^\infty_s}$$

for any s. Combined with the inequality $e^{a|k_{123}|^{1/\sigma}} \leq e^{a|k_1|^{1/\sigma}} e^{a|k_2|^{1/\sigma}} e^{a|k_3|^{1/\sigma}}$, this yields a crude estimate in Gevrey space

$$\|F[w]\|_{G^\sigma_{s,a}} \leq C3^s \|w\|^2_{G^\sigma_{3,a}} \|w\|_{G^\sigma_{s,a}}.$$

The estimates restricted onto high frequencies in Proposition 1 are better than the above estimate covering all frequencies in the following two respects: The constant can be made s-independent, and the exponent a can be replaced by the strictly smaller one θa except for one function. Both of these improvements will play important roles in the proof of Theorem 1. We also note that these improvements are possible only in the case $\sigma > 1$, since we rely heavily on the strict concavity of the map $k \mapsto |k|^{1/\sigma}$. Indeed, the situation is different in the case $\sigma = 1$, where we can show well-posedness of the Cauchy problem (see [12, Proposition 4.3]).

We begin with the following bounds on M_3, M_3^0 and M_5, which means that no derivative loss occurs in $F[w]$, $H[w]$, and that $G[w]$ gains one derivative.

Lemma 1 *The following holds, with the implicit constant being independent of $k \in \mathbb{Z}$.*

1. *If $k_1, k_2, k_3 \in \mathbb{Z}$ satisfy $k_{123} = k$, then*

$$|M_3| \lesssim \langle k_{(2)} \rangle, \qquad |k||M_3^0| \lesssim 1,$$

where $\langle k_{(2)} \rangle$ denotes the second largest quantity of all $\langle k_j \rangle$, $j = 1, 2, 3$.

2. *If $k_1, \cdots, k_5 \in \mathbb{Z}$ satisfy $k_{12345} = k$, then*

$$|M_5| \lesssim \langle k'_{(2)} \rangle,$$

where $\langle k'_{(2)} \rangle$ denotes the second largest quantity of all $\langle k_j \rangle$, $j = 1, \ldots, 5$.

Proof Recall that

$$(k_1, k_2, k_3) \in D_1(k) \quad \Rightarrow \quad |k| \lesssim |k_1| \sim |k_2| \sim |k_3|,$$

$$(k_1, k_2, k_3) \in D_2(k) \quad \Rightarrow \quad |\Phi(k_1, k_2, k_3)| \gtrsim \max_{1 \le j \le 3} \langle k_j \rangle^2.$$

The estimates on M_3 and M_3^0 follow easily from these facts. We focus on estimating the first term in M_5. It is easy to see that

$$\left| \text{the first term in } M_5 \right| \lesssim \frac{|\mu(k_1, k_2, k_3)|}{\langle k_{123} \rangle} \lesssim 1 + \frac{|k_{12}|}{\langle k_{123} \rangle} \lesssim 1 + \frac{|k_3|}{\langle k_{123} \rangle}.$$

If $|k_3| \gg |k_{123}|$, then one of $|k_1|$ and $|k_2|$ must be greater than or comparable to $|k_3|$, and hence $|k_3| \lesssim \langle k'_{(2)} \rangle$. The other two terms in M_5 can be estimated in a similar manner, and we have the desired bound. □

The following inequality is the key to establishing Proposition 1.

Lemma 2 *Let $\sigma > 1$, $a > 0$, and $p \in \mathbb{Z}$ with $p \ge 2$. There exists $C = C(\sigma, a, p) > 0$ such that*

$$e^{a|k|^{1/\sigma}} \langle k \rangle^s \le \max_{1 \le q \le p} e^{a|k_q|^{1/\sigma}} \langle k_q \rangle^s \prod_{\substack{1 \le r \le p \\ r \ne q}} e^{\theta(p) a |k_r|^{1/\sigma}}$$

for any $s \geq 0$ *and* $k_1, \dots, k_p \in \mathbb{Z}$ *satisfying* $|k| := |k_1 + \cdots + k_p| \geq Cs^\sigma$, *where* $\theta(p) := (1 - \frac{1}{p})^{1-\frac{1}{\sigma}} \in (1 - \frac{1}{p}, 1)$.

Proof By symmetry, we may assume that $|k_1| = \max_{1 \leq q \leq p} |k_q|$, which implies $|k'| := |k_2 + \cdots + k_p| \leq (p-1)|k_1|$. Since $|k'|^{1/\sigma} \leq |k_2|^{1/\sigma} + \cdots + |k_p|^{1/\sigma}$, it suffices to show that

$$e^{a(|k_1|+|k'|)^{1/\sigma}} \langle |k_1| + |k'| \rangle^s \leq e^{a|k_1|^{1/\sigma}} e^{\theta(p)a|k'|^{1/\sigma}} \langle k_1 \rangle^s \tag{7}$$

for $k_1, k' \in \mathbb{Z}$ with $|k_1| \geq \frac{C}{p} s^\sigma$ and $0 < |k'| \leq (p-1)|k_1|$ (the inequality is trivial when $k' = 0$).

Let $f(\alpha) := 1 + \theta(p)\alpha^{1/\sigma} - (1+\alpha)^{1/\sigma}$. Setting $|k'| = \alpha|k_1|$, we see that

$$\frac{e^{a(|k_1|+|k'|)^{1/\sigma}} \langle |k_1| + |k'| \rangle^s}{e^{a|k_1|^{1/\sigma}} e^{\theta(p)a|k'|^{1/\sigma}} \langle k_1 \rangle^s} = e^{-a|k_1|^{1/\sigma} f(\alpha)} (1+\alpha)^s = e^{s \log(1+\alpha) - a|k_1|^{1/\sigma} f(\alpha)}.$$

Hence, (7) holds if

$$\frac{s}{a|k_1|^{1/\sigma}} \leq \frac{f(\alpha)}{\log(1+\alpha)}.$$

Now, for $\alpha \in (0, p-1)$ we see that

$$f'(\alpha) = \frac{\theta(p)}{\sigma} \alpha^{\frac{1}{\sigma}-1} - \frac{1}{\sigma}(1+\alpha)^{\frac{1}{\sigma}-1} > \frac{\theta(p)}{\sigma}\alpha^{\frac{1}{\sigma}-1} - \frac{1}{\sigma}\Big(\frac{\alpha}{p-1} + \alpha\Big)^{\frac{1}{\sigma}-1} = 0,$$

and thus f is increasing and $f(\alpha) > f(0) = 0$ on $(0, p-1]$. Since

$$\frac{f(\alpha)}{\log(1+\alpha)} = \frac{\theta(p)\alpha^{1/\sigma} + O(\alpha)}{\alpha + O(\alpha^2)} \to \infty \qquad (\alpha \to +0),$$

it holds that $\eta = \eta(\sigma, p) := \inf_{0 < \alpha \leq p-1} f(\alpha)/\log(1+\alpha) > 0$. Therefore, (7) holds whenever $|k_1| \geq (\frac{s}{a\eta})^\sigma$, and the claim follows if we set $C := p(a\eta)^{-\sigma}$. □

We are ready to show the multilinear estimates.

Proof *(Proof of Proposition 1)* The estimate on the second term in $F[w]$ follows from

$$\| |\cdot|^{1/2} \hat{w} \|_{\ell^2} \lesssim \|\hat{w}\|_{\ell^\infty_2} \leq \|w\|_{G^\sigma_{2,a'}}, \qquad \sigma \geq 1, \quad a' > 0.$$

All the other estimates are easy consequences of Lemmas 1 and 2 (with $p = 3, 5$) together with the embedding $\ell^\infty_2(\mathbb{Z}) \hookrightarrow \ell^1(\mathbb{Z})$. □

4 Proof of Theorem 1

We first establish the following smoothing effect:

Lemma 3 *Let $\sigma > 1$, and let $w(t) \in L^\infty([-T,T]; G^\sigma_{0,a})$ for some $T > 0$ and $a > 0$ such that (6) holds for every $t_1, t_2 \in [-T,T]$. Then, there exists $\varepsilon > 0$ depending only on σ, a, A and $\sup_{t\in[-T,T]} \|w(t)\|_{G^\sigma_{3,\theta a}}$ such that $w(t) \in G^\sigma_{0,a+\varepsilon}$ for any $t \in (-T,T)$ and $\|w(t)\|_{G^\sigma_{0,a+\varepsilon}}$ is bounded on any compact subinterval of $(-T,T)$. Here, $\theta \in (0,1)$ is the constant given in Proposition 1.*

Remark 3 In the above lemma, the solution gets smoother in index a only by a finite amount. However, we note that this smoothing occurs instantaneously and ε does not shrink as $t \to \pm T$. This property will imply that the solution belongs to $\bigcap_{a>0} G^\sigma_{0,a}$ on $(-T,T)$; see the proof of Theorem 1 below.

Proof Let t_0 and $(1 \geq)\ \delta > 0$ satisfy $t_0 \pm \delta \in [-T,T]$. From (6), we have

$$\hat{w}_k(t_0) = e^{-\delta Ak}\hat{w}_k(t_0+\delta) + \Big[\hat{G}[w(t_0)]_k - e^{-\delta Ak}\hat{G}[w(t_0+\delta)]_k\Big] - \int_{t_0}^{t_0+\delta} e^{-Ak(t-t_0)}\Big[\hat{F}[w(t)]_k + Ak\hat{G}[w(t)]_k + \hat{H}[w(t)]_k\Big]\,dt \tag{8}$$

$$= e^{\delta Ak}\hat{w}_k(t_0-\delta) + \Big[\hat{G}[w(t_0)]_k - e^{\delta Ak}\hat{G}[w(t_0-\delta)]_k\Big] + \int_{t_0-\delta}^{t_0} e^{Ak(t_0-t)}\Big[\hat{F}[w(t)]_k + Ak\hat{G}[w(t)]_k + \hat{H}[w(t)]_k\Big]\,dt. \tag{9}$$

We use (8) and (9) for positive and negative k's, respectively. By Proposition 1, we have

$$\begin{aligned}
&\|P_{\geq C_0 s^\sigma} w(t_0)\|_{G^\sigma_{s+1,a}}\\
&\leq \|w\|_{L^\infty_{[t_0-\delta,t_0+\delta]}G^\sigma_{s,a}} \sup_{k\in\mathbb{Z}}\big[|k|e^{-\delta A|k|}\big] + C\|w\|^2_{L^\infty_{[t_0-\delta,t_0+\delta]}G^\sigma_{2,\theta a}}\|w\|_{L^\infty_{[t_0-\delta,t_0+\delta]}G^\sigma_{s,a}}\\
&\quad + C\Big(\|w\|^2_{L^\infty_{[t_0-\delta,t_0+\delta]}G^\sigma_{3,\theta a}} + A\|w\|^2_{L^\infty_{[t_0-\delta,t_0+\delta]}G^\sigma_{2,\theta a}} + \|w\|^4_{L^\infty_{[t_0-\delta,t_0+\delta]}G^\sigma_{3,\theta a}}\Big)\\
&\qquad\qquad \times \|w\|_{L^\infty_{[t_0-\delta,t_0+\delta]}G^\sigma_{s,a}} \sup_{k\in\mathbb{Z}}\int_0^\delta |k|e^{-A|k|t'}\,dt'\\
&\leq \frac{C_1}{\delta}\|w\|_{L^\infty_{[t_0-\delta,t_0+\delta]}G^\sigma_{s,a}},
\end{aligned} \tag{10}$$

where $C_1 > 0$ is a constant depending only on A and $\|w\|_{L^\infty_{[-T,T]}G^\sigma_{3,\theta a}}$, and we write $\|w\|_{L^\infty_I G^\sigma_{s,a}}$ to denote $\sup_{t\in I}\|w(t)\|_{G^\sigma_{s,a}}$. On the other hand,

$$\begin{aligned}\|P_{<C_0 s^\sigma} w(t_0)\|_{G^\sigma_{s+1,a}} &\le \|w(t_0)\|_{G^\sigma_{1,\theta a}} \sup_{|k|<C_0 s^\sigma} e^{(1-\theta)a|k|^{1/\sigma}}|k|^s \\ &\le C_1 e^{(1-\theta)aC_0^{1/\sigma}s}\big[C_0 s^\sigma\big]^s = C_1\big[C_0 e^{(1-\theta)aC_0^{1/\sigma}} s^\sigma\big]^s.\end{aligned}$$

Hence, we have

$$\begin{aligned}\|w(t_0)\|_{G^\sigma_{s+1,a}} &= \max\Big\{\|P_{<C_0 s^\sigma} w(t_0)\|_{G^\sigma_{s+1,a}},\ \|P_{\ge C_0 s^\sigma} w(t_0)\|_{G^\sigma_{s+1,a}}\Big\} \\ &\le \max\Big\{C_2\big[C_2 s^\sigma\big]^s,\ \frac{C_2}{\delta}\|w\|_{L^\infty_{[t_0-\delta,t_0+\delta]}G^\sigma_{s,a}}\Big\},\end{aligned}$$

where $C_2 := \max\{C_1,\ C_0 e^{(1-\theta)aC_0^{1/\sigma}}\}$ is a constant depending only on σ, a, A and $\|w\|_{L^\infty_{[-T,T]}G^\sigma_{3,\theta a}}$. (Note that we have only the second term when $s = 0$.)

For any sufficiently large positive integer s (satisfying $T(s+1)^{\sigma-1} \ge 1$), we apply the above estimate with $\delta = T/(s+1)$ repeatedly and obtain

$$\begin{aligned}&\|w(0)\|_{G^\sigma_{s+1,a}} \\ &\le \max\Big\{\max_{1\le r\le s} C_2\big[C_2 r^\sigma\big]^r\Big(\frac{C_2(s+1)}{T}\Big)^{s-r},\ \Big(\frac{C_2(s+1)}{T}\Big)^{s+1}\|w\|_{L^\infty_{[-T,T]}G^\sigma_{0,a}}\Big\} \\ &\le \max\Big\{\max_{1\le r\le s} C_2\big[C_2(s+1)^\sigma\big]^r\Big(\frac{C_2(s+1)^\sigma}{T(s+1)^{\sigma-1}}\Big)^{s-r}, \\ &\qquad\qquad \Big(\frac{C_2(s+1)^\sigma}{T(s+1)^{\sigma-1}}\Big)^{s+1}\|w\|_{L^\infty_{[-T,T]}G^\sigma_{0,a}}\Big\} \\ &\le C_2^{s+1}(s+1)^{\sigma s}\max\Big\{1,\ (s+1)^\sigma\|w\|_{L^\infty_{[-T,T]}G^\sigma_{0,a}}\Big\} \\ &\le C_3^{s+1}(s+1)^{\sigma s}\max\Big\{1,\ \|w\|_{L^\infty_{[-T,T]}G^\sigma_{0,a}}\Big\},\end{aligned}$$

where $C_3 := 2^\sigma C_2$. At the last inequality, we have used $s+1 \le 2^{s+1}$. For $s \notin \mathbb{Z}$, take $s_0 \in \mathbb{Z}$ with $s_0 < s < s_0+1$ and observe that

$$\begin{aligned}\|w(0)\|_{G^\sigma_{s_0,a}} &\le C_3^{s_0} s_0^{\sigma s_0}\max\Big\{1,\ \|w\|_{L^\infty_{[-T,T]}G^\sigma_{0,a}}\Big\}, \\ \|w(0)\|_{G^\sigma_{s_0+1,a}} &\le C_3^{s_0+1}(s_0+1)^{\sigma s_0}\max\Big\{1,\ \|w\|_{L^\infty_{[-T,T]}G^\sigma_{0,a}}\Big\}\end{aligned}$$

whenever s_0 is sufficiently large. By interpolation, we have

$$\|w(0)\|_{G^\sigma_{s,a}} \le C_3^s s^{\sigma s}\max\Big\{1,\ \|w\|_{L^\infty_{[-T,T]}G^\sigma_{0,a}}\Big\}$$

for any $s > 0$ sufficiently large. This Sobolev smoothing estimate leads to the Gevrey smoothing as follows. For $\varepsilon > 0$ and any sufficiently large positive integer n, it follows that

$$\begin{aligned}\left\| \frac{(\varepsilon|k|^{1/\sigma})^n}{n!} e^{a|k|^{1/\sigma}} \hat{w}(0) \right\|_{\ell^\infty} &\le \frac{\varepsilon^n}{n!} \|w(0)\|_{G^\sigma_{n/\sigma,a}} \\ &\le \frac{\varepsilon^n}{n!} C_3^{n/\sigma} \left(\frac{n}{\sigma}\right)^{\sigma\cdot\frac{n}{\sigma}} \max\left\{1,\ \|w\|_{L^\infty_{[-T,T]}G^\sigma_{0,a}}\right\} \\ &\le \left(\frac{\varepsilon e C_3^{1/\sigma}}{\sigma}\right)^n \max\left\{1,\ \|w\|_{L^\infty_{[-T,T]}G^\sigma_{0,a}}\right\},\end{aligned}$$

where we have used the fact $n^n \le e^n n!$. Therefore, the series

$$\|e^{\varepsilon|k|^{1/\sigma}} e^{a|k|^{1/\sigma}} \hat{w}(0)\|_{\ell^\infty} \le \sum_{n=0}^{\infty} \left\| \frac{(\varepsilon|k|^{1/\sigma})^n}{n!} e^{a|k|^{1/\sigma}} \hat{w}(0) \right\|_{\ell^\infty}$$

converges and $w(0) \in G^\sigma_{0,a+\varepsilon}$ if $0 < \varepsilon < \frac{\sigma}{e} C_3^{-1/\sigma}$. □

We are now in a position to prove the main theorem.

Proof *(Proof of Theorem 1)* Let $u(t) \in C([-T,T]; G^\sigma)$ be a solution to (1) on $(-T,T)$ for some $T > 0$. As we mentioned in Remark 1 (ii), u actually belongs to $C([-T,T]; G^\sigma_{0,a_0})$ for some $a_0 > 0$. We define the function $\tilde{a} : (0,T] \to [0,\infty]$ by

$$\tilde{a}(T') := \sup\left\{ a > 0 \,\middle|\, \begin{array}{c} u(t) \in G^\sigma_{0,a} \text{ for every } t \in [-T',T'], \\ \|u\|_{L^\infty_{[-T',T']}G^\sigma_{0,a}} < \infty \end{array} \right\}.$$

By the definition, we see that $\tilde{a}(T) \ge a_0$ and $\tilde{a}$ is non-increasing.

Assume that $\tilde{a}(T') < \infty$ for some $T' \in (0,T]$. From the definition, for any $\frac{1}{2}\tilde{a}(T') < a < \tilde{a}(T')$, u is a bounded function on $[-T',T']$ with values in $G^\sigma_{0,a}$, and moreover, $\|u\|_{L^\infty_{[-T',T']}G^\sigma_{3,\theta a}} \le \|u\|_{L^\infty_{[-T',T']}G^\sigma_{3,\theta\tilde{a}(T')}} =: M < \infty$. Noticing that $\|u(t)\|_{G^\sigma_{0,a}} = \|w(t)\|_{G^\sigma_{0,a}}$ and $w(t)$ satisfies (6) for all $t_1, t_2 \in [-T,T]$, we invoke Lemma 3 to conclude that $\tilde{a}(t) \ge a + \varepsilon$ for any $0 < t < T'$ and $\frac{1}{2}\tilde{a}(T') < a < \tilde{a}(T')$, with ε depending only on σ, $\tilde{a}(T')$, M and being uniform as a tends to $\tilde{a}(T')$. Letting $t \to T' - 0$ and $a \to \tilde{a}(T') - 0$ implies $\tilde{a}(T' - 0) \ge \tilde{a}(T') + \varepsilon$; i.e., $\tilde{a}$ is discontinuous at T'. Therefore, $\tilde{a}$ would have to be discontinuous at any point where it is finite. Since $\tilde{a}$ is non-increasing, the only possibility is that $\tilde{a} \equiv \infty$ on $(0,T)$.

This completes the proof. □

Acknowledgments The authors would like to thank Professor Tej-edinne Ghoul, New York University in Abu Dhabi and Professor Takayoshi Ogawa, Tohoku University for drawing their attention to the ill-posedness in the Gevrey class. The first author N.K is partially supported by

JSPS KAKENHI Grant-in-Aid for Young Researchers (B) (16K17626). The second author Y.T is partially supported by JSPS KAKENHI Grant-in-Aid for Scientific Research (B) (17H02853) and Grant-in-Aid for Exploratory Research (16K13770).

Appendix: Topology on G^σ

Let us recall that the inductive limit topology $\mathscr{T}$ on G^σ defined through the inductive system $\{G^\sigma_{0,a}\}_{a>0}$ is the finest locally convex topology that makes the embedding $G^\sigma_{0,a} \subset G^\sigma$ continuous for all $a > 0$.

The following properties are verified by the general theory of the inductive limits:

Lemma 4

(i) *For a locally convex space X, a linear map from G^σ into X is continuous if and only if its restriction on $G^\sigma_{0,a}$ is continuous as a map from $G^\sigma_{0,a}$ into X for any $a > 0$.*

(ii) *Let $\{\tilde{G}^\sigma_b\}_b$ be another inductive system with $G^\sigma = \bigcup_b \tilde{G}^\sigma_b$. If for every $a > 0$ there exists b such that $\tilde{G}^\sigma_b$ is continuously embedded into $G^\sigma_{0,a}$, and vice versa, then the inductive limit topology on G^σ defined by $\{\tilde{G}^\sigma_b\}$ coincides with $\mathscr{T}$.*

Proof See, for instance, [5, Section IV.5]. □

From these properties we are able to show:

Proposition 2

(i) *For $\tau > \sigma \geq 1$ and $a, b > 0$, the embeddings $G^\sigma_{0,a} \hookrightarrow G^\sigma \hookrightarrow G^\tau_{0,b}$ are continuous.*

(ii) *Define the Banach spaces $\tilde{G}^\sigma_b$ for $b > 0$ by*

$$\tilde{G}^\sigma_b := \left\{ f \in C^\infty(\mathbb{T}) : \|f\|_{\tilde{G}^\sigma_b} := \sup_n b^n (n!)^{-\sigma} \|\partial_x^n f\|_{L^\infty(\mathbb{T})} < \infty \right\}.$$

Then, $G^\sigma = \bigcup_{b>0} \tilde{G}^\sigma_b$, and the inductive limit topology on G^σ defined by $\{\tilde{G}^\sigma_b\}$ is identical with $\mathscr{T}$.

Proof

(i) The first embedding is continuous by the definition of the inductive limit topology. The continuity of the second embedding follows from that of $G^\sigma_{0,a} \hookrightarrow G^\tau_{0,b}$ ($a > 0$) and Lemma 4(i).

(ii) By Lemma 4(ii), it suffices to show that for every $a, b > 0$ there exist $\alpha, \beta > 0$ such that

$$\|f\|_{G^\sigma_{0,a}} \lesssim \|f\|_{\tilde{G}^\sigma_\beta}, \qquad \|f\|_{\tilde{G}^\sigma_b} \lesssim \|f\|_{G^\sigma_{0,\alpha}}.$$

First, for given $a > 0$ we take $\beta = (2a)^\sigma$. Then, we have

$$\|f\|_{G^\sigma_{0,a}} \le \sum_{n\ge 0} \frac{a^n}{n!} \left\| |k|^{n/\sigma} \hat{f}_k \right\|_{\ell^\infty} \le \sum_{n\ge 0} \frac{a^n}{n!} \left\| \hat{f}_k \right\|_{\ell^\infty}^{1-1/\sigma} \left\| |k|^n \hat{f}_k \right\|_{\ell^\infty}^{1/\sigma}$$

$$\lesssim \sum_{n\ge 0} \frac{a^n}{n!} \|f\|_{L^\infty}^{1-1/\sigma} \left\|\partial_x^n f\right\|_{L^\infty}^{1/\sigma} \le \|f\|_{\tilde{G}^\sigma_\beta} \sum_{n\ge 0} \frac{a^n}{n!} \left(\beta^{-n}(n!)^\sigma\right)^{1/\sigma}$$

$$= \|f\|_{\tilde{G}^\sigma_\beta} \sum_{n\ge 0} \left(\frac{a}{\beta^{1/\sigma}}\right)^n \lesssim \|f\|_{\tilde{G}^\sigma_\beta}.$$

Next, for $b > 0$ and $n \ge 0$ we see that

$$b^n (n!)^{-\sigma} \|\partial_x^n f\|_{L^\infty} \lesssim \left\| \left(\frac{b^{n/\sigma}|k|^{n/\sigma}}{n!}\right)^\sigma \hat{f}_k \right\|_{\ell^1}$$

$$\lesssim \left\| \left(e^{b^{1/\sigma}|k|^{1/\sigma}}\right)^\sigma \langle k\rangle^2 \hat{f}_k \right\|_{\ell^\infty} = \|f\|_{G^\sigma_{2,\alpha/2}},$$

where we have chosen $\alpha = 2\sigma b^{1/\sigma}$. The desired estimate follows from the embedding $G^\sigma_{0,\alpha} \hookrightarrow G^\sigma_{2,\alpha/2}$. □

From the continuous embedding $G^\sigma \hookrightarrow G^{2\sigma}_{0,1}$, the topology $\mathscr{T}$ on G^σ is Hausdorff. However, the inductive limit topology is in general not necessarily Hausdorff. To exclude such a pathological situation, various restrictive assumptions are imposed on the inductive system. A typical example is the *strict* inductive system, which requires the topology of each space to coincide with the relative topology induced as a subspace of any larger space in the system. For instance, this is the case for the inductive limit topology of the test function space $\mathscr{D}(\Omega)$ on a nonempty open set $\Omega \subset \mathbb{R}^n$. Unfortunately, the inductive limit of the system $\{G^\sigma_{0,a}\}_{a>0}$ is not strict, as easily seen by the existence of a sequence $\{f_n\} \subset G^\sigma_{0,a}$, such as

$$f_n(x) = ne^{-a|n|^{1/\sigma}} e^{inx},$$

which is unbounded but converges to zero in any larger space $G^\sigma_{0,a'}$, $a' \in (0, a)$.

Here, we notice that $\{G^\sigma_{0,a}\}_{a>0}$ is an inductive system with *compact* embeddings. In fact, the embedding $G^\sigma_{0,a} \hookrightarrow G^\sigma_{0,b}$ is compact for each $a > b > 0$ by Rellich's theorem. The following general result is known in this situation:

Lemma 5 *Let $\{X_j\}$ be an increasing sequence of locally convex spaces such that each embedding $X_j \subset X_{j+1}$ is compact. Then, the inductive limit X of $\{X_j\}$ is a complete locally convex space. Moreover, a bounded set in X is included and bounded in some X_j. Furthermore, a sequence in X converges to zero if and only if it is included in some X_j and converges to zero in X_j.*

Proof See [13, Theorem 6' on page 375] (also [6, Section 4.2.3]). □

We are now able to prove the required property of G^σ mentioned in Remark 1(ii).

Proposition 3 *A function u defined on* $[-T, T] \times \mathbb{T}$ *belongs to* $C([-T, T]; G^\sigma)$ *if and only if* $u \in C([-T, T]; G^\sigma_{0,a})$ *for some* $a > 0$.

Proof The *if* part follows from the continuous embedding $G^\sigma_{0,a} \hookrightarrow G^\sigma$. To show the *only if* part, we assume $u \in C([-T, T]; G^\sigma)$. Then, $\{u(t) : t \in [-T, T]\}$ is bounded in G^σ, and hence included and bounded in some $G^\sigma_{0,a}$ by Lemma 5. (In view of Lemma 4 (ii) we may regard G^σ as the inductive limit of the sequence of Banach spaces $\{G^\sigma_{0,1/j}\}$, and hence Lemma 5 can be applied.) On the other hand, the continuous embedding $G^\sigma \hookrightarrow G^{2\sigma}_{0,1}$ implies $u \in C([-T, T]; G^{2\sigma}_{0,1})$. Therefore, the interpolation inequality:

$$\|f\|_{G^\sigma_{0,a'}} \le \|f\|^{a'/a}_{G^\sigma_{0,a}} \|\hat{f}\|^{1-a'/a}_{\ell^\infty} \le \|f\|^{a'/a}_{G^\sigma_{0,a}} \|f\|^{1-a'/a}_{G^{2\sigma}_{0,1}} \qquad (0 < a' < a)$$

shows that $u \in C([-T, T]; G^\sigma_{0,a'})$ for any $a' \in (0, a)$. □

References

1. Agrawal, G.: Nonlinear Fiber Optics, 4th edn. Elsevier/Academic Press, Burlington (2007)
2. Alexandre, R., Wang, Y.G., Xu, C.J., Yang, T.: Well-posedness of the Prandtl equation in Sobolev spaces. J. Am. Math. Soc. **28**(3), 745–784 (2015). https://doi.org/10.1090/S0894-0347-2014-00813-4
3. Chen, H., Rodino, L.: General theory of PDE and Gevrey classes. In: General Theory of Partial Differential Equations and Microlocal Analysis (Trieste, 1995). Pitman Research Notes in Mathematics Series, vol. 349, pp. 6–81. Longman, Harlow (1996)
4. Colin, T., Métivier, G.: Instabilities in Zakharov equations for laser propagation in a plasma. In: Phase space analysis of partial differential equations. Progress in Nonlinear Differential Equations and their Applications, vol. 69, pp. 63–81. Birkhäuser Boston, Boston (2006). https://doi.org/10.1007/978-0-8176-4521-2_6
5. Conway, J.B.: A course in functional analysis. In: Graduate Texts in Mathematics, vol. 96. Springer, New York (1985). https://doi.org/10.1007/978-1-4757-3828-5
6. Cristescu, R.: Topological vector spaces. In: Editura Academiei, Bucharest. Noordhoff International Publishing, Leyden (1977). Translated from the Romanian by Mihaela Suliciu
7. De Donno, G., Oliaro, A., Rodino, L.: Analytic and Gevrey solutions of non-linear partial differential equations. Far East J. Appl. Math. **15**(3), 403–425 (2004)
8. Dietert, H., Gérard-Varet, D.: Well-posedness of the Prandtl equation without any structural assumption. Ann. PDE **5**(1), 51 (2019). https://doi.org/10.1007/s40818-019-0063-6
9. Gérard-Varet, D., Dormy, E.: On the ill-posedness of the Prandtl equation. J. Am. Math. Soc. **23**(2), 591–609 (2010). https://doi.org/10.1090/S0894-0347-09-00652-3
10. Hounie, J., Santiago, P.: On the local solvability of semilinear equations. Comm. Partial Diff. Equat. **20**(9–10), 1777–1789 (1995). https://doi.org/10.1080/03605309508821151
11. Kajitani, K.: Local solution of Cauchy problem for nonlinear hyperbolic systems in Gevrey classes. Hokkaido Math. J. **12**(3, part 2), 434–460 (1983). https://doi.org/10.14492/hokmj/1525852966

12. Kishimoto, N., Tsutsumi, Y.: Ill-posedness of the third order NLS equation with Raman scattering term. Math. Res. Lett. **25**(5), 1447–1484 (2018). https://doi.org/10.4310/MRL.2018.v25.n5.a5
13. Komatsu, H.: Projective and injective limits of weakly compact sequences of locally convex spaces. J. Math. Soc. Japan **19**, 366–383 (1967). https://doi.org/10.2969/jmsj/01930366
14. Leray, J., Ohya, Y.: Équations et systèmes non-linéaires, hyperboliques nonstricts. Math. Ann. **170**, 167–205 (1967). https://doi.org/10.1007/BF01350150
15. Messina, F., Rodino, L.: Local solvability for nonlinear partial differential equations. In: Proceedings of the Third World Congress of Nonlinear Analysts, Part 5 (Catania, 2000), vol. 47, pp. 2917–2927 (2001). https://doi.org/10.1016/S0362-546X(01)00413-8
16. Oleinik, O.A., Samokhin, V.N.: Mathematical models in boundary layer theory. In: Applied Mathematics and Mathematical Computation, vol. 15. Chapman & Hall/CRC, Boca Raton (1999)
17. Rodino, L.: Linear Partial Differential Operators in Gevrey Spaces. World Scientific, Singapore (1993). https://doi.org/10.1142/9789814360036
18. Ukai, S.: Local solutions in Gevrey classes to the nonlinear Boltzmann equation without cutoff. Jpn. J. Appl. Math. **1**(1), 141–156 (1984). https://doi.org/10.1007/BF03167864

Invariant Measures for the DNLS Equation

Renato Lucà

Abstract We describe invariant measures associated to the integrals of motion of the periodic derivative nonlinear Schrödinger equation (DNLS) constructed in Genovese et al. (Selecta Math. (N.S.) 22(3):1663–1702, 2016. https://doi.org/10.1007/s00029-016-0225-2; Math Ann 2018. https://doi.org/10.1007/s00208-018-1754-0). The construction works for small L^2 data. The measures are absolutely continuous with respect to suitable weighted Gaussian measures supported on Sobolev spaces of increasing regularity. These results have been obtained in collaboration with Giuseppe Genovese (University of Zürich) and Daniele Valeri (University of Glasgow).

1 Introduction

We consider the periodic DNLS equation

$$\begin{cases} i\partial_t \psi + \psi'' = i\beta \left(\psi|\psi|^2\right)' \\ \quad \psi(x,0) = \psi_0(x)\,, \qquad x \in \mathbb{T}\,, \end{cases} \tag{1}$$

where $\psi(x,t) : \mathbb{T} \times \mathbb{R} \to \mathbb{C}$, $\psi_0(x) : \mathbb{T} \to \mathbb{C}$, $\psi'(x,t)$ is the derivative of ψ with respect to x, and $\beta \in \mathbb{R}$. We write Φ_t for the associated flow-map. A (Gibbs) measure associated to the energy functional

$$E_1[\psi] = \frac{1}{2}\|\psi\|^2_{\dot{H}^1} + \frac{3i}{4}\beta \int |\psi|^2 \psi' \bar{\psi} + \frac{\beta^2}{4}\|\psi\|^6_{L^6}\,, \tag{2}$$

R. Lucà (✉)
BCAM—Basque Center for Applied Mathematics, Bilbao, Spain

Ikerbasque, Basque Foundation for Science, Bilbao, Spain
e-mail: rluca@bcamath.org

W. Dörfler et al. (eds.), *Mathematics of Wave Phenomena*, Trends in Mathematics,
https://doi.org/10.1007/978-3-030-47174-3_14

has been constructed in [8], while in [7] and [6] this measure has been proved to be invariant under Φ_t.

Let $k \geq 2$ be an integer and γ_k be the Gaussian measure induced by the random Fourier series

$$f^{\omega}(x) = \sum_{n \in \mathbb{Z}} \frac{e^{inx}}{\sqrt{1 + n^{2k}}} g_n^{\omega},$$

where g_n^{ω} are normalized independent complex Gaussian random variables. Notice that

$$f^{\omega} \in H^s \quad \omega\text{-almost surely for} \quad s < k - \frac{1}{2}$$

and

$$\|f^{\omega}\|_{H^{k-\frac{1}{2}}} = \infty \quad \omega\text{-almost surely},$$

thus we may think that γ_k “concentrates” on $\bigcap_{s<k-\frac{1}{2}} H^s$. We fix $R_m > 0$, for $m = 0, \ldots, k-1$, and define the k-th Gibbs measure associated to the DNLS equation by

$$\rho_k(A) := \int_A \left(\prod_{m=0}^{k-1} \chi_{R_m}(E_m[\psi]) \right) e^{-Q_k[\psi]} \gamma_k(d\psi), \tag{3}$$

where χ_{R_m} are smooth cut–off of the interval $[-R_m, R_m]$,

$$Q_k[\psi] := E_k[\psi] - \frac{1}{2}\|\psi\|^2_{\dot{H}^k} \tag{4}$$

and $E_1, \ldots, E_k$ are integrals of motion of the DNLS equation; see (5), (6)

Recall from [5] (see also [1, 2]) that there exists an infinite sequence of integrals of motion $\{E_k[\psi]\}_{k \in \frac{1}{2}\mathbb{N}_0}$ for the DNLS equation (1). We will use the form of E_k found in [2]. In particular

$$E_0[\psi] = \frac{1}{2}\|\psi\|^2_{L^2},$$

$$E_{\frac{1}{2}}[\psi] = \frac{i}{2}\int \psi'\bar{\psi} + \frac{\beta}{4}\|\psi\|^4_{L^4},$$

$$E_1[\psi] = \frac{1}{2}\|\psi\|^2_{\dot{H}^1} + \frac{3i}{4}\beta \int |\psi|^2\psi'\bar{\psi} + \frac{\beta^2}{4}\|\psi\|^6_{L^6},$$

$$E_{\frac{3}{2}}[\psi] = \frac{i}{2}\int \psi''\bar{\psi}' + \frac{\beta}{4}\int \left((\psi')^2\bar{\psi}^2 + 8|\psi|^2\psi'\bar{\psi}' + \psi^2(\bar{\psi}')^2\right)$$
$$+ \frac{5i}{4}\beta^2\int |\psi|^4\psi'\bar{\psi} + \frac{5}{16}\beta^3\|\psi\|_{L^8}^8,$$
$$E_2[\psi] = \frac{1}{2}\|\psi\|_{\dot{H}^2}^2 + \frac{5i}{4}\beta\int |\psi|^2\left(\psi''\bar{\psi}' - \psi'\bar{\psi}''\right) + \frac{5}{4}\beta^2\int |\psi|^2\left((\psi')^2\bar{\psi}^2 + \psi^2(\bar{\psi}')^2\right)$$
$$+ \frac{25}{4}\beta^2\int |\psi|^4\psi'\bar{\psi}' + \frac{35i}{16}\beta^3\int |\psi|^6\psi'\bar{\psi} + \frac{7}{16}\beta^4\|\psi\|_{L^{10}}^{10}. \tag{5}$$

Remark 1 These integrals of motion are slightly different from those appearing in the introduction of [2], where there is a typo in the coefficient of β of $E_1[\psi]$.

While we have in general (see agian [2]) that, for $k \geq 2$ and integer[1]

$$E_k[\psi] = \frac{1}{2}\|\psi\|_{\dot{H}^k}^2 + i\frac{2k+1}{4}\beta\int |\psi|^2\left(\psi^{(k)}\bar{\psi}^{(k-1)} - \psi^{(k-1)}\bar{\psi}^{(k)}\right) + \int \mathcal{R}_k, \tag{6}$$

where $\mathcal{R}_k$ are polynomials in $\psi, \psi^{(1)}, \dots, \psi^{(k-1)}$ and their conjugate. We have denoted

$$\psi^{(m)} := \partial_x^m \psi.$$

2 Main Results

The weighted gaussian measure (3) is well defined and absolutely continuous with respect to γ_k, as long as we restrict to functions with small L^2 norm. This is formalized in the next theorem.

Theorem 1 ([2]) *We denote with*

$$F_k := \left(\prod_{m=0}^{k-1}\chi_{R_m}(E_m[\psi])\right)e^{-Q_k[\psi]}$$

the density of ρ_k. Let $k \geq 2$ and integer, $p \in (1,\infty)$. If $R_0 = R_0(p, |\beta|, k)$ is sufficiently small, then

$$F_k \in L^p(\gamma_k).$$

[1] In this survey we never focus on half-integer conservation laws, on which almost no analytical results are available.

We can not infer $F_k \in L^\infty(\gamma_k)$. Indeed $R_0 = R_0(p, |\beta|) \to 0$ as $p \to \infty$. The small mass assumption $R_0 \ll 1$ is not surprising since the problem is L^2 critical. Regarding the well-posedness, the periodic DNLS equation has been shown to be locally well-posed for initial data in $H^{s \geq 1/2}$ in [4]. It is not clear if this result may extend to $s < 1/2$. Then, a standard procedure allows to globalise the local H^1 solutions with small mass $R_0 < \delta$ (notice (5) that the mass is conserved by Φ_t). taking advantage of the fact that in this regime E_1 dominates the H^1 norm.

Theorem 2 ([3]) *Let $k \geq 2$ and integer, $p \in (1, \infty)$. Let $R_0 = R_0(p, |\beta|, k)$ be sufficiently small. There exists a probability measure $\tilde{\rho}_k$, absolutely continuous with respect to γ_k, which is preserved by Φ_t.*

This result can be used to deduce informations on the long time behavior of solutions to DNLS. This follows by a direct application of the Poincaré recurrence theorem. Thus we have obtained

Corollary 1 *Let $k \geq 2$ and integer and let ψ a solution of the DNLS equation with initial datum $\psi(\cdot, 0) \in H^s$ with $s < k - \frac{1}{2}$. For $\tilde{\rho}_k$-a.e. (and so γ_k-a.e.) initial datum $\psi(\cdot, 0)$ there exists a divergent sequence $\{t_n\}_{n \in \mathbb{N}}$ such that*

$$\lim_{n \to \infty} \|\psi(\cdot, t_n) - \psi(\cdot, 0)\|_{H^s} = 0 .$$

An analog conclusion for $k = 1$ follows from combining the contributions in [6–8]. To the best of our knowledge, these are the sole known results on the long-time behavior of the DNLS equation.

2.1 Strategy of the Proof

An alternative formulation of our main Theorem 2 is that the DNLS equation has the structure of an infinite-dimensional (Hamiltonian) dynamical system. Since the earlier works of Bourgain and Zhidkov this concept has been made rigorous using a suitable finite dimensional approximation. In our case, this is done considering, for all $N \in \mathbb{N}$, the following equation

$$i\partial_t \psi_N + \psi_N'' = i\beta P_N(|\psi_N|^2 \psi_N)' , \tag{7}$$

with initial datum

$$\psi_N(\cdot, 0) := P_N \psi(\cdot, 0) , \tag{8}$$

where P_N is the projection on functions with less than N Fourier modes, namely

$$P_N\left(\sum_{n\in\mathbb{Z}} e^{inx} a_n\right) = \sum_{|n|\leq N} e^{inx} a_n$$

Notice that

$$\psi_N(\cdot, t) = P_N \psi_N(\cdot, t), \qquad \text{for all } t \geq 0\,.$$

Moreover, this system is actually Hamiltonian, but in does not preserve all the integrals of motion. This is often a major issue in this class of problems. However one expects the integrals of motion to be conserved in the limit $N \to \infty$. Following an approach developed by Tzvetkov and Visciglia for the Benjamin-Ono equation, we will show that the derivative of the integrals of motion along the flow of the truncated systems vanishes in the $L^2(\gamma_k)$ mean. Actually, as first observed in [11], one can reduce to consider only the derivative at the initial time, which is a crucial simplification.

It is helpful to recall that the integrals of motion of DNLS have the following form

$$E_k[\psi] = \frac{1}{2}\|\psi\|^2_{\dot{H}^k} - \frac{1}{2}\beta(2k+1)\Im \int \psi^{(k)}\bar{\psi}^{(k-1)}|\psi|^2 + \text{remainders}\,, \qquad k \geq 2\,, \tag{9}$$

where we consider as remainders all the terms that are in $L^\infty(\gamma_k)$. The difficulty to show the asymptotic (w.r.t. N) conservation of E_k comes form the second addendum in the r.h.s. of Eq. (9). Notably the integrals of motion of the Benjamin-Ono equation have an analog structure. However in that case a convenient cancellation coming from the symmetries of the problem simplifies substantially the computations [10, 11]. We cannot find a similar cancellation here. Nevertheless it is possible to eliminate the troubling term using the following gauge transformation.

$$(\Gamma_\alpha f)(x) := e^{i\alpha\mathcal{I}[f(x)]} f(x)\,. \tag{10}$$

where

$$\mathcal{I}[f(x)] := \frac{1}{2\pi}\int_0^{2\pi} d\theta \int_\theta^x \left(|f(y)|^2 - \frac{\|f\|^2_{L^2}}{2\pi}\right) dy\,. \tag{11}$$

One can easily check that the (real) function $\mathcal{I}[f(x)]$ is the unique zero average (2π-periodic) primitive of $|f(x)|^2 - (2\pi)^{-1}\|f\|^2_{L^2}$. Note that $|f| = |\Gamma_\alpha f|$ and $\Gamma_\alpha f$ is 2π-periodic. Hence, Γ_α maps L^2 into L^2 preserving the norm (namely $\|\Gamma_\alpha f\|_{L^2} = \|f\|_{L^2}$). Using that $\cup f] = \mathcal{I}[\Gamma_\alpha(f)]$ one can easily show that the map $\alpha \to \Gamma_\alpha$ is a one parameter group of transformations on $(\mathbb{R}, +)$, namely

$$\Gamma_0 = \mathbb{I} \qquad \text{and} \qquad \Gamma_{\alpha_1} \circ \Gamma_{\alpha_2} = \Gamma_{\alpha_1+\alpha_2}\,, \qquad \text{for any } \alpha_1, \alpha_2 \in \mathbb{R}\,. \tag{12}$$

For any $s \geq 0$ the gauge transformation Γ_α is also an homeomorphism of H^s into itself.

A generic gauge choice (without specifying the value of α) yields the following expression for the integrals of motion of the gauged equation

$$\begin{aligned}\mathcal{E}_k[\phi] &= \frac{1}{2}\|\phi\|^2_{\dot{H}^k} + ik\alpha\mu \int \bar{\phi}^{(k)}\phi^{(k-1)} \\ &- \frac{1}{2}((2k+2)\alpha + (2k+1)\beta)\,\Im \int \phi^{(k)}\bar{\phi}^{(k-1)}|\phi|^2 + \text{remainders},\end{aligned}$$

where $\phi = \Gamma_\alpha \psi$ is the solution of the gauged equation and we shortened

$$\mu := \frac{1}{2\pi}\|\phi\|^2_{L^2} = \frac{1}{2\pi}\|\psi\|^2_{L^2}.$$

In general we will use the notation $\mu[f] := \frac{1}{2\pi}\|f\|^2_{L^2}$ and we will simply write μ if there will be no ambiguity. We recover (9) as $\alpha = 0$. Setting

$$\alpha = -\frac{2k+1}{2k+2}\beta, \tag{13}$$

we reduce to

$$\mathcal{E}_k[\phi] = \frac{1}{2}\|\phi\|^2_{\dot{H}^k} - ik\frac{2k+1}{2k+2}\beta\mu \int \bar{\phi}^{(k)}\phi^{(k-1)} + \text{remainders}.$$

This form of the integrals of motion is much more suitable in order to prove the asymptotic conservation property and such a reduction is the main step in our proof. Of course also the flow of DNLS changes accordingly to the gauge transformation: indeed our gauge choice leads to a somewhat more involved form for the equation

$$i\partial_t\phi + \phi'' + 2i\alpha\mu\phi' = ic_1|\phi|^2\phi' + ic_2\phi^2\bar{\phi}' + c_3|\phi|^4\phi + c_4\mu|\phi|^2\phi + L[\phi]\phi, \tag{14}$$

where

$$c_1 = 2(\alpha+\beta), \qquad c_2 = 2\alpha + \beta, \qquad c_3 = -\alpha^2 - \frac{\alpha\beta}{2}, \qquad c_4 = -\alpha\beta \tag{15}$$

and

$$L[f] = \left(\frac{3\alpha\beta}{4\pi} + \frac{\alpha^2}{\pi}\right)\|f\|^4_{L^4} - \alpha^2\mu[f]^2 + \frac{i\alpha}{\pi}\int_{\mathbb{T}} f'\bar{f}. \tag{16}$$

However, this does not introduce significant difficulties, as the form of the nonlinearity is essentially the same and we are working with rather regular solutions (let say in $H^{s>5/4}$). It is worthy to point out the difference with what is usually done in the low regularity theory for DNLS, where the choice of the gauge parameter $\alpha = -\beta$ aims to simplify the equation.

The next step is to define for any $k \geq 2$ a gauged Gibbs measure starting from the gauge-transformed (or *gauged*) integrals of motion and to prove its invariance w.r.t. the gauged flow. This requires some groundwork, namely a careful analysis of the DNLS-flow and gauge-flow maps, in order to adapt the strategy of [11].

The invariance of the gauged Gibbs measure under the gauged flow easily implies the invariance of its push-forward through Γ_α, under the DNLS flow. This will be our invariant measure $\tilde{\rho}_k$. We stress that in principle one expects $\rho_k = \tilde{\rho}_k$. The missing step to show the invariance of the Gibbs measures is the proof of absolute continuity of the pull-back $\gamma_k \circ \Gamma_\alpha$ w.r.t. γ_k with the explicit density. So far what we can prove is the following theorem:

Theorem 3 *Let $R_0 > 0$ small enough and*

$$\tilde{\gamma}_k(A) = \gamma_k(A \cap \{f \in L^2 : \mu[f] \leq R_0\}) .$$

Then for any $k \geq 2$ and integer and $\alpha \in \mathbb{R}$ the measure $\tilde{\gamma}_k \circ \Gamma_\alpha$ is absolutely continuous w.r.t. γ_k.

The absolute continuity of $\tilde{\rho}_k$ w.r.t. γ_k is a direct consequence of Theorem 3. The change of variable formula for $k = 1$ was established in [6]. This is however a very special case, as the typical trajectories for γ_1 are complex Brownian bridges, whose properties are crucially employed in the argument of [6]. For more regular processes one cannot expect to reproduce the same proof and some new idea is needed.

Recently Tzvetkov [9] proposed a strategy for proving quasi-invariance of the Gaussian measure under a one-parameter group of transformations via a *soft* argument, which does not provide the explicit density. We use this approach to prove Theorem 3, from which we deduce the absolutely continuity of $\tilde{\rho}_k$ w.r.t. γ_k. When $k \geq 2$, to prove that the Gibbs measures are invariant, one should know the exact form of the densities after the change of variables given by the gauge. As it happens in the case $k = 1$, these densities should complete exactly the part of the integrals of motion missing in the gauged Gibbs measure. We do not give here further details, leaving the discussion of this problem to future works.

Acknowledgments R. Lucà is supported by BERC 2018–2021, by BCAM Severo Ochoa SEV-2017-0718 and IHAIP project PGC2018-094528-B-I00 (AEI/FEDER, UE).

References

1. De Sole, A., Kac, V.G.: Non-local Poisson structures and applications to the theory of integrable systems. Jpn. J. Math. **8**(2), 233–347 (2013). https://doi.org/10.1007/s11537-013-1306-z
2. Genovese, G., Lucà, R., Valeri, D.: Gibbs measures associated to the integrals of motion of the periodic derivative nonlinear Schrödinger equation. Selecta Math. (N.S.) **22**(3), 1663–1702 (2016). https://doi.org/10.1007/s00029-016-0225-2

3. Genovese, G., Lucà, R., Valeri, D.: Invariant measures for the periodic derivative nonlinear Schrödinger equation. Math. Ann. (2018). https://doi.org/10.1007/s00208-018-1754-0
4. Herr, S.: On the Cauchy problem for the derivative nonlinear Schrödinger equation with periodic boundary condition. Int. Math. Res. Not. **33**, Art. ID 96763 (2006). https://doi.org/10.1155/IMRN/2006/96763
5. Kaup, D.J., Newell, A.C.: An exact solution for a derivative nonlinear Schrödinger equation. J. Math. Phys. **19**(4), 798–801 (1978). https://doi.org/10.1063/1.523737
6. Nahmod, A.R., Rey-Bellet, L., Sheffield, S., Staffilani, G.: Absolute continuity of Brownian bridges under certain gauge transformations. Math. Res. Lett. **18**(5), 875–887 (2011). https://doi.org/10.4310/MRL.2011.v18.n5.a6
7. Nahmod, A.R., Oh, T., Rey-Bellet, L., Staffilani, G.: Invariant weighted Wiener measures and almost sure global well-posedness for the periodic derivative NLS. J. Eur. Math. Soc. (JEMS) **14**(4), 1275–1330 (2012). https://doi.org/10.4171/JEMS/333
8. Thomann, L., Tzvetkov, N.: Gibbs measure for the periodic derivative nonlinear Schrödinger equation. Nonlinearity **23**(11), 2771–2791 (2010). https://doi.org/10.1088/0951-7715/23/11/003
9. Tzvetkov, N.: Quasiinvariant gaussian measures for one-dimensional hamiltonian partial differential equations. Forum Math. Sigma **3**, e28 (2015). https://doi.org/10.1017/fms.2015.27
10. Tzvetkov, N., Visciglia, N.: Gaussian measures associated to the higher order conservation laws of the Benjamin-Ono equation. Ann. Sci. Éc. Norm. Supér. (4) **46**(2), 249–299 (2013). https://doi.org/10.24033/asens.2189
11. Tzvetkov, N., Visciglia, N.: Invariant measures and long-time behavior for the Benjamin?Ono equation. Int. Math. Res. Not. **2014**(17), 4679–4714 (2013). https://doi.org/10.1093/imrn/rnt094

A Global div-curl-Lemma for Mixed Boundary Conditions in Weak Lipschitz Domains

Dirk Pauly

Abstract We prove a global version of the so-called div-curl-lemma, a crucial result for compensated compactness and in homogenization theory, for mixed tangential and normal boundary conditions in bounded weak Lipschitz domains in 3D and weak Lipschitz interfaces. The crucial tools and the core of our arguments are the de Rham complex and Weck's selection theorem, the essential compact embedding result for Maxwell's equations.

1 Introduction and Main Results

We shall prove a global (and hence also a local) version of the so-called div-curl-lemma, with mixed tangential and normal boundary conditions for bounded weak Lipschitz domains Ω in 3D, more precisely for admissible pairs (Ω, Γ_t) of a bounded weak Lipschitz domain $\Omega \subset \mathbb{R}^3$ and a part Γ_t of its boundary Γ, see Definition 1 for details.

Theorem 1 (Global div-curl-**Lemma)** *Let* (Ω, Γ_t) *be admissible and let*

(i) $E_n, E \in D(\mathrm{curl}_{\Gamma_t})$,
(i') $E_n \rightharpoonup E$ *in*[1] $D(\mathrm{curl}_{\Gamma_t})$,
(ii) $H_n, H \in D(\mathrm{div}_{\Gamma_n})$,
(ii') $H_n \rightharpoonup H$ *in*[2] $D(\mathrm{div}_{\Gamma_n})$.

Then

(iii) $\langle E_n, H_n \rangle_{L^2(\Omega)} \to \langle E, H \rangle_{L^2(\Omega)}$.

[1]In particular, $E_n \rightharpoonup E$ in $L^2(\Omega)$ and curl $E_n \rightharpoonup$ curl E in $L^2(\Omega)$.
[2]In particular, $H_n \rightharpoonup H$ in $L^2(\Omega)$ and div $H_n \rightharpoonup$ div H in $L^2(\Omega)$.

D. Pauly (✉)
Fakultät für Mathematik, Universität Duisburg-Essen, Essen, Germany
e-mail: dirk.pauly@uni-due.de

W. Dörfler et al. (eds.), *Mathematics of Wave Phenomena*, Trends in Mathematics,
https://doi.org/10.1007/978-3-030-47174-3_15

Here, we introduce the densely defined and closed linear operators ∇_{Γ_t}, $\operatorname{curl}_{\Gamma_t}$, $\operatorname{div}_{\Gamma_t}$ with domains of definition[3] $D(\nabla_{\Gamma_t})$, $D(\operatorname{curl}_{\Gamma_t})$, $D(\operatorname{div}_{\Gamma_t})$ as closures of the classical differential operators from vector analysis acting on $\mathsf{L}^2(\Omega)$ and defined on smooth test functions resp. test vector fields bounded away from the boundary part Γ_t given by

$$\mathsf{C}^\infty_{\Gamma_t}(\Omega) := \big\{\varphi|_\Omega \,:\, \varphi \in \mathsf{C}^\infty(\mathbb{R}^3),\ \operatorname{supp}\varphi \text{ compact},\ \operatorname{dist}(\operatorname{supp}\varphi, \Gamma_t) > 0\big\}.$$

As shown in [3, Theorem 4.5] (weak equals strong in terms of definitions of boundary conditions) their adjoints are given by $-\operatorname{div}_{\Gamma_n}$, $\operatorname{curl}_{\Gamma_n}$, $-\nabla_{\Gamma_n}$ defined in the same way. Note that these operators are unbounded and that the domains of definition are Hilbert spaces equipped with the respective proper graph inner products.

Corollary 1 (Local div-curl-**Lemma)** *Let $\Omega \subset \mathbb{R}^3$ be an open set and let*

(i) $E_n, E \in D(\operatorname{curl})$,
(i') $E_n \rightharpoonup E$ *in* $D(\operatorname{curl})$,
(ii) $H_n, H \in D(\operatorname{div})$,
(ii') $H_n \rightharpoonup H$ *in* $D(\operatorname{div})$.

Then

(iii) $\forall\, \varphi \in \mathsf{C}^\infty_\Gamma(\Omega) \qquad \langle \varphi\, E_n, H_n\rangle_{\mathsf{L}^2(\Omega)} \to \langle \varphi\, E, H\rangle_{\mathsf{L}^2(\Omega)}.$

$\mathsf{C}^\infty_\Gamma(\Omega)$ *may be replaced by* $\mathsf{C}^1_\Gamma(\Omega)$ *or even* $\mathsf{C}^{0,1}_\Gamma(\Omega)$*, the space of Lipschitz continuous functions vanishing in a neighbourhood of* Γ*. Moreover, the boundedness of* (E_n) *and* (H_n) *in local spaces is sufficient for the assertion to hold.*

The div-curl-lemma, or compensated compactness, see the original papers by Murat [13] and Tartar [23] or [6, 22], and its variants and extensions have plenty important applications. It is widely used in the theory of homogenization of (nonlinear) partial differential equations, see, e.g., [22]. Moreover, it is crucial in establishing compactness and regularity results for nonlinear partial differential equations such as harmonic maps, see, e.g., [7, 8, 19]. Numerical applications can be found, e.g., in [2]. The div-curl-lemma is further a crucial tool in the

---

[3] Various notations like

$$D(\nabla_{\Gamma_t}) = \mathsf{H}(\nabla_{\Gamma_t}, \Omega) = \mathsf{H}_{\Gamma_t}(\nabla, \Omega) = \mathsf{H}^1_{\Gamma_t}(\Omega) = \mathring{\mathsf{H}}^1_{\Gamma_t}(\Omega),$$

$$D(\operatorname{curl}_{\Gamma_t}) = \mathsf{H}(\operatorname{curl}_{\Gamma_t}, \Omega) = \mathsf{H}_{\Gamma_t}(\operatorname{curl}, \Omega) = \mathsf{R}_{\Gamma_t}(\Omega) = \mathring{\mathsf{R}}_{\Gamma_t}(\Omega),$$

$$D(\operatorname{div}_{\Gamma_t}) = \mathsf{H}(\operatorname{div}_{\Gamma_t}, \Omega) = \mathsf{H}_{\Gamma_t}(\operatorname{div}, \Omega) = \mathsf{D}_{\Gamma_t}(\Omega) = \mathring{\mathsf{D}}_{\Gamma_t}(\Omega)$$

can be found frequently in the literature, where also curl = rot is used.

homogenization of stochastic partial differential equations, especially with certain random coefficients, see, e.g., the survey [1] and the literature cited therein, e.g., [9].

For an extensive discussion and a historical overview of the div-curl-lemma see [24]. More recent discussions can be found, e.g., in [4, 25]. Recently, in [17, 26] the div-curl-lemma has been proved in a general Hilbert space setting which allows for various applications in mathematical physics. Interesting and new applications to homogenization of partial differential equations can be found in [27].

Let us also mention that the div-curl-lemma is particularly useful to treat homogenization of problems arising in plasticity, see, e.g., a recent preprint on this topic [21], for which the preprint [20] provides the important key div-curl-lemma. Unfortunately, in [20, 21] a $\mathsf{H}^1(\Omega)$-detour is used as the core argument for the proofs. The same detour is utilized in the recent contribution [11] where div-curl-type lemmas are presented which also allow for inhomogeneous boundary conditions. This unnecessarily high regularity assumption of $\mathsf{H}^1(\Omega)$-fields excludes results like [11, 20, 21] to be applied to important applications which are stated, e.g., in Lipschitz domains.

2 Notations, Preliminaries, and Proofs

Definition 1 (Admissible Domains) We call a pair $(\Omega, \Gamma_{\mathsf{t}})$ admissible, if

(i) $\Omega \subset \mathbb{R}^3$ is a bounded weak Lipschitz domain in the sense of [3, Definition 2.3]
(ii) with boundary $\Gamma := \partial\,\Omega$, which is divided into two relatively open weak Lipschitz subsets $\Gamma_{\mathsf{t}} \subset \Gamma$ and its complement $\Gamma_{\mathsf{n}} := \Gamma \setminus \overline{\Gamma_{\mathsf{t}}}$ in the sense of [3, Definition 2.5].

Note that strong Lipschitz domains (locally below a graph of a Lipschitz function) are weak Lipschitz domains (the boundary is a Lipschitz manifold) which holds for the boundary as well as for the interface. The reverse implication is not true due to the failure of the implicit function theorem for Lipschitz mappings. Throughout this paper we shall assume the latter regularity of Ω, and Γ, Γ_{t}, Γ_{n}.

Recently, in [3], Weck's selection theorem [29], also known as the Maxwell compactness property, has been shown to hold for such bounded weak Lipschitz domains and mixed boundary conditions. More precisely, the following holds:

Lemma 1 (Weck's Selection Theorem) *Let* $(\Omega, \Gamma_{\mathsf{t}})$ *be admissible. Then the embedding*

$$D(\operatorname{curl}_{\Gamma_{\mathsf{t}}}) \cap D(\operatorname{div}_{\Gamma_{\mathsf{n}}}) \hookrightarrow\!\!\!\rightarrow \mathsf{L}^2(\Omega)$$

is compact.

For a proof see [3, Theorem 4.7]. A short historical overview of Weck's selection theorem is given in the introduction of [3], see also the original paper [29] and [5, 10, 12, 18, 28, 30] for simpler proofs and generalizations.

Let us emphasize that our assumptions also allow for Rellich's selection theorem, i.e., the embedding

$$D(\nabla_{\Gamma_t}) \hookrightarrow\!\!\!\rightarrow \mathsf{L}^2(\Omega) \tag{1}$$

is compact, see, e.g., [3, Theorem 4.8]. By density we have the two rules of integration by parts

$$\forall\, u \in D(\nabla_{\Gamma_t}) \qquad \forall\, H \in D(\operatorname{div}_{\Gamma_n}) \qquad \langle \nabla u, H\rangle_{\mathsf{L}^2(\Omega)} = -\langle u, \operatorname{div} H\rangle_{\mathsf{L}^2(\Omega)}, \tag{2}$$

$$\forall\, E \in D(\operatorname{curl}_{\Gamma_t}) \quad \forall\, H \in D(\operatorname{curl}_{\Gamma_n}) \quad \langle \operatorname{curl} E, H\rangle_{\mathsf{L}^2(\Omega)} = \langle E, \operatorname{curl} H\rangle_{\mathsf{L}^2(\Omega)}. \tag{3}$$

A direct consequence of Lemma 1 is the compactness of the unit ball in

$$\mathcal{H}(\Omega) := N(\operatorname{curl}_{\Gamma_t}) \cap N(\operatorname{div}_{\Gamma_n}),$$

the space of so-called Dirichlet-Neumann fields. Hence $\mathcal{H}(\Omega)$ is finite-dimensional. Here and in the following we denote the kernels and the ranges of our operators ∇_{Γ_t}, $\operatorname{curl}_{\Gamma_t}$, $\operatorname{div}_{\Gamma_t}$ by

$$N(\nabla_{\Gamma_t}), \quad N(\operatorname{curl}_{\Gamma_t}), \quad N(\operatorname{div}_{\Gamma_t}), \qquad R(\nabla_{\Gamma_t}), \quad R(\operatorname{curl}_{\Gamma_t}), \quad R(\operatorname{div}_{\Gamma_t}).$$

Another immediate consequence of Weck's selection theorem, stated in Lemma 1, using a standard indirect argument, is the so-called Maxwell estimate, i.e., there exists $c_{\mathsf{m}} > 0$ such that for all $E \in D(\operatorname{curl}_{\Gamma_t}) \cap D(\operatorname{div}_{\Gamma_n}) \cap \mathcal{H}(\Omega)^{\perp_{\mathsf{L}^2(\Omega)}}$

$$|E|_{\mathsf{L}^2(\Omega)} \le c_{\mathsf{m}}\big(|\operatorname{curl} E|_{\mathsf{L}^2(\Omega)} + |\operatorname{div} E|_{\mathsf{L}^2(\Omega)}\big), \tag{4}$$

see [3, Theorem 5.1]. Recent estimates for the Maxwell constant c_{m} can be found in [14–16]. Analogously, Rellich's selection theorem (1) shows the Friedrichs/Poincaré estimate, i.e., there exists $c_{\mathsf{f,p}} > 0$ such that for all $u \in D(\nabla_{\Gamma_t})$

$$|u|_{\mathsf{L}^2(\Omega)} \le c_{\mathsf{f,p}} |\nabla u|_{\mathsf{L}^2(\Omega)}, \tag{5}$$

see [3, Theorem 4.8]. To avoid case studies due to the one-dimensional kernel $\mathbb{R}$ of ∇ when using the Friedrichs/Poincaré estimate in the case $\Gamma_t = \emptyset$, we also define

$$D(\nabla_{\emptyset}) := D(\nabla) \cap \mathbb{R}^{\perp_{\mathsf{L}^2(\Omega)}} = \big\{u \in \mathsf{H}^1(\Omega) : \int_\Omega u = 0\big\}.$$

By the projection theorem, applied to our densely defined and closed (unbounded) linear operator

$$\nabla_{\Gamma_t} : D(\nabla_{\Gamma_t}) \subset \mathsf{L}^2(\Omega) \longrightarrow \mathsf{L}^2(\Omega)$$

with (Hilbert space) adjoint

$$\nabla_{\Gamma_t}^* = -\operatorname{div}_{\Gamma_n} : D(\operatorname{div}_{\Gamma_n}) \subset \mathsf{L}^2(\Omega) \longrightarrow \mathsf{L}^2(\Omega),$$

where we have used [3, Theorem 4.5] (weak equals strong), we get the simple (orthogonal) Helmholtz decomposition

$$\mathsf{L}^2(\Omega) = R(\nabla_{\Gamma_t}) \oplus_{\mathsf{L}^2(\Omega)} N(\operatorname{div}_{\Gamma_n}), \tag{6}$$

see [3, Theorem 5.3 or (13)], which immediately implies the orthogonal decomposition

$$D(\operatorname{curl}_{\Gamma_t}) = R(\nabla_{\Gamma_t}) \oplus_{\mathsf{L}^2(\Omega)} \big(D(\operatorname{curl}_{\Gamma_t}) \cap N(\operatorname{div}_{\Gamma_n})\big) \tag{7}$$

as the complex property $R(\nabla_{\Gamma_t}) \subset N(\operatorname{curl}_{\Gamma_t})$ holds. Here $\oplus_{\mathsf{L}^2(\Omega)}$ in the decompositions (6) and (7) denotes the orthogonal sum in the Hilbert space $\mathsf{L}^2(\Omega)$. By (5) the range $R(\nabla_{\Gamma_t})$ is closed in $\mathsf{L}^2(\Omega)$, see also [3, Lemma 5.2]. Note that we call (6) a simple Helmholtz decomposition, since the refined Helmholtz decomposition

$$\mathsf{L}^2(\Omega) = R(\nabla_{\Gamma_t}) \oplus_{\mathsf{L}^2(\Omega)} \mathcal{H}(\Omega) \oplus_{\mathsf{L}^2(\Omega)} R(\operatorname{curl}_{\Gamma_n})$$

holds as well, see [3, Theorem 5.3], where also $R(\operatorname{curl}_{\Gamma_n})$ is closed in $\mathsf{L}^2(\Omega)$ as a consequence of (4), see [3, Lemma 5.2].

Proof of Theorem 1 By (7) we have $D(\operatorname{curl}_{\Gamma_t}) \ni E_n = \nabla u_n + \tilde{E}_n$ with some $u_n \in D(\nabla_{\Gamma_t})$ and $\tilde{E}_n \in D(\operatorname{curl}_{\Gamma_t}) \cap N(\operatorname{div}_{\Gamma_n})$. Then (u_n) is bounded in $\mathsf{H}^1(\Omega)$ by orthogonality and the Friedrichs/Poincaré estimate (5). By orthogonality $(\tilde{E}_n)$ is bounded in $D(\operatorname{curl}_{\Gamma_t}) \cap N(\operatorname{div}_{\Gamma_n})$ and $\operatorname{curl} \tilde{E}_n = \operatorname{curl} E_n$. Hence, using Rellich's and Weck's selection theorems there exist $u \in D(\nabla_{\Gamma_t})$ and $\tilde{E} \in D(\operatorname{curl}_{\Gamma_t}) \cap N(\operatorname{div}_{\Gamma_n})$ and we can extract two subsequences, again denoted by (u_n) and $(\tilde{E}_n)$ such that $u_n \rightharpoonup u$ in $D(\nabla_{\Gamma_t})$ and $u_n \to u$ in $\mathsf{L}^2(\Omega)$ as well as $\tilde{E}_n \rightharpoonup \tilde{E}$ in $D(\operatorname{curl}_{\Gamma_t}) \cap N(\operatorname{div}_{\Gamma_n})$ and $\tilde{E}_n \to \tilde{E}$ in $\mathsf{L}^2(\Omega)$. We observe $E = \nabla u + \tilde{E}$, giving the simple Helmholtz decomposition for E. Finally, by (2)

$$\begin{aligned}\langle E_n, H_n\rangle_{\mathsf{L}^2(\Omega)} &= \langle \nabla u_n, H_n\rangle_{\mathsf{L}^2(\Omega)} + \langle \tilde{E}_n, H_n\rangle_{\mathsf{L}^2(\Omega)} \\ &= -\langle u_n, \operatorname{div} H_n\rangle_{\mathsf{L}^2(\Omega)} + \langle \tilde{E}_n, H_n\rangle_{\mathsf{L}^2(\Omega)}\end{aligned}$$

$$\begin{aligned}&\to -\langle u, \operatorname{div} H\rangle_{\mathsf{L}^2(\Omega)} + \langle \tilde{E}, H\rangle_{\mathsf{L}^2(\Omega)}\\&= \langle \nabla u, H\rangle_{\mathsf{L}^2(\Omega)} + \langle \tilde{E}, H\rangle_{\mathsf{L}^2(\Omega)}\\&= \langle E, H\rangle_{\mathsf{L}^2(\Omega)}.\end{aligned}$$

As the limit is unique, the original sequence $\big(\langle E_n, H_n\rangle_{\mathsf{L}^2(\Omega)}\big)$ already converges to the limit $\langle E, H\rangle_{\mathsf{L}^2(\Omega)}$. ∎

Proof of Corollary 1 Let $\Gamma_{\mathsf{t}} := \Gamma$ and hence $\Gamma_{\mathsf{n}} = \emptyset$. The sequence (φE_n) is bounded in $D(\operatorname{curl}_\Gamma)$ and (H_n) is bounded in $D(\operatorname{div})$. Theorem 1 shows the assertion. ∎

3 Generalizations and the Classical Div-curl-Lemma

In [17, 26] more general div-curl-lemmas have been presented. In particular in [17] we can find the following generalization to distributions.

Theorem 2 (Alternative Global div-curl-**Lemma)** *Let $\Omega \subset \mathbb{R}^3$ be a bounded strong Lipschitz domain with trivial topology (all Betti numbers vanish) and let*

(i) $E_n, H_n, E, H \in \mathsf{L}^2(\Omega)$,
(i') $E_n \rightharpoonup E$ *and* $H_n \rightharpoonup H$ *in* $\mathsf{L}^2(\Omega)$.

Moreover, let either

(ii) $(\widehat{\operatorname{curl}} E_n)$ *be relatively compact in* $\mathring{\mathsf{H}}^{-1}(\Omega)$,
(iii) $(\widetilde{\operatorname{div}} H_n)$ *be relatively compact in* $\mathsf{H}^{-1}(\Omega)$,

or

(ii') $(\widetilde{\operatorname{curl}} E_n)$ *be relatively compact in* $\mathsf{H}^{-1}(\Omega)$,
(iii') $(\widehat{\operatorname{div}} H_n)$ *be relatively compact in* $\mathring{\mathsf{H}}^{-1}(\Omega)$.

Then

(iv) $\langle E_n, H_n\rangle_{\mathsf{L}^2(\Omega)} \to \langle E, H\rangle_{\mathsf{L}^2(\Omega)}$.

Here, $\mathsf{H}^{-1}(\Omega) := \mathring{\mathsf{H}}^1(\Omega)'$ and $\mathring{\mathsf{H}}^{-1}(\Omega) := \mathsf{H}^1(\Omega)'$ and the distributional extensions

$$\widetilde{\operatorname{curl}} : \mathsf{L}^2(\Omega) \to \mathsf{H}^{-1}(\Omega), \qquad \widetilde{\operatorname{div}} : \mathsf{L}^2(\Omega) \to \mathsf{H}^{-1}(\Omega),$$

$$\widehat{\operatorname{curl}} : \mathsf{L}^2(\Omega) \to \mathring{\mathsf{H}}^{-1}(\Omega), \qquad \widehat{\operatorname{div}} : \mathsf{L}^2(\Omega) \to \mathring{\mathsf{H}}^{-1}(\Omega)$$

of curl and div, respectively, are defined for $E \in \mathsf{L}^2(\Omega)$ by

$$\widetilde{\operatorname{curl}}\, E\,(\Phi) := \langle \operatorname{curl} \Phi, E \rangle_{\mathsf{L}^2(\Omega)}, \qquad \Phi \in \mathring{\mathsf{H}}^1(\Omega),$$

$$\widehat{\operatorname{curl}}\, E\,(\Phi) := \langle \operatorname{curl} \Phi, E \rangle_{\mathsf{L}^2(\Omega)}, \qquad \Phi \in \mathsf{H}^1(\Omega),$$

$$\widetilde{\operatorname{div}}\, E\,(\varphi) := -\langle \nabla \varphi, E \rangle_{\mathsf{L}^2(\Omega)}, \qquad \varphi \in \mathring{\mathsf{H}}^1(\Omega),$$

$$\widehat{\operatorname{div}}\, E\,(\varphi) := -\langle \nabla \varphi, E \rangle_{\mathsf{L}^2(\Omega)}, \qquad \varphi \in \mathsf{H}^1(\Omega).$$

Finally, we emphasize that Theorem 2 provides a global result while the classical div-curl-lemma by Murat [13] and Tartar [23] is a local one. The latter may be formulated as follows:

Theorem 3 (Classical div-curl-**Lemma)** *Let $\Omega \subset \mathbb{R}^3$ be an open set and let*

(i) $E_n, H_n, E, H \in \mathsf{L}^2(\Omega)$,
(i') $E_n \rightharpoonup E$ *and* $H_n \rightharpoonup H$ *in* $\mathsf{L}^2(\Omega)$,
(ii) $(\widetilde{\operatorname{curl}} E_n)$ *and* $(\widetilde{\operatorname{div}} H_n)$ *be relatively compact in* $\mathsf{H}^{-1}(\Omega)$.

Then

(iii) $\forall\, \varphi \in \mathsf{C}^\infty_\Gamma(\Omega) \qquad \langle \varphi\, E_n, H_n \rangle_{\mathsf{L}^2(\Omega)} \to \langle \varphi\, E, H \rangle_{\mathsf{L}^2(\Omega)}.$

Acknowledgments The author is grateful to Sören Bartels for bringing up the topic of the div-curl-lemma, and especially to Marcus Waurick for lots of inspiring discussions on the div-curl-lemma and for his substantial contributions to the Special Semester at RICAM in Linz late 2016.

References

1. Alexanderian, A.: Expository paper: a primer on homogenization of elliptic pdes with stationary and ergodic random coefficient functions. Rocky Mountain J. Math. **45**(3), 703–735 (2015)
2. Bartels, S.: Numerical analysis of a finite element scheme for the approximation of harmonic maps into surfaces. Math. Comp. **79**(271), 1263–1301 (2010)
3. Bauer, S., Pauly, D., Schomburg, M.: The Maxwell compactness property in bounded weak Lipschitz domains with mixed boundary conditions. SIAM J. Math. Anal. **48**(4), 2912–2943 (2016)
4. Briane, M., Casado-Dáz, J., Murat, F.: The div-curl lemma "trente ans après": an extension and an application to the g-convergence of unbounded monotone operators. J. Math. Pures Appl. **91**(5), 476–494 (2009)
5. Costabel, M.: A remark on the regularity of solutions of Maxwell's equations on Lipschitz domains. Math. Methods Appl. Sci. **12**(4), 365–368 (1990)
6. Evans, L.: Weak Convergence Methods for Nonlinear Partial Differential Equations. American Mathematical Society, Providence (1990)

7. Evans, L.: Partial regularity for stationary harmonic maps into spheres. Arch. Rational Mech. Anal. **116**(2), 101–113 (1991)
8. Freire, A., Müller, S., Struwe, M.: Weak compactness of wave maps and harmonic maps. Ann. Inst. H. Poincaré Anal., Non Linaire **15**(6), 725–754 (1998)
9. Gloria, A., Neukamm, S., Otto, F.: Quantification of ergodicity in stochastic homogenization: optimal bounds via spectral gap on glauber dynamics. Invent. Math. **199**(2), 455–515 (2015)
10. Jochmann, F.: A compactness result for vector fields with divergence and curl in $L^q(\Omega)$ involving mixed boundary conditions. Appl. Anal. **66**, 189–203 (1997)
11. Kozono, H., Yanagisawa, T.: Global compensated compactness theorem for general differential operators of first order. Arch. Ration. Mech. Anal. **207**(3), 879–905 (2013)
12. Leis, R.: Initial Boundary Value Problems in Mathematical Physics. Teubner, Stuttgart (1986)
13. Murat, F.: Compacité par compensation. Annali della Scuola Normale Superiore di Pisa-Classe di Scienze **5**(3), 489–507 (1978)
14. Pauly, D.: On constants in Maxwell inequalities for bounded and convex domains. Zapiski POMI **435**, 46–54 (2014). J. Math. Sci. (N.Y.) **210**(6), 787–792 (2015)
15. Pauly, D.: On Maxwell's and Poincaré's constants. Discrete Contin. Dyn. Syst. Ser. S **8**(3), 607–618 (2015)
16. Pauly, D.: On the Maxwell constants in 3D. Math. Methods Appl. Sci. **40**(2), 435–447 (2017)
17. Pauly, D.: A Global div-curl-Lemma for Mixed Boundary Conditions in Weak Lipschitz Domains and a Corresponding Generalized A_0^*-A_1-Lemma in Hilbert Spaces. Analysis (Munich) **39**(2), 33–58 (2019)
18. Picard, R.: An elementary proof for a compact imbedding result in generalized electromagnetic theory. Math. Z. **187**, 151–164 (1984)
19. Rivière, T.: Conservation laws for conformally invariant variational problems. Invent. Math. **168**(1), 1–22 (2007)
20. Schweizer, B.: On Friedrichs Inequality, Helmholtz Decomposition, Vector Potentials, and the div-curl Lemma (2017). Preprint, http://www.mathematik.uni-dortmund.de/lsi/schweizer/Preprints/publist.html
21. Schweizer, B., Röger, M.: Strain Gradient Visco-Plasticity with Dislocation Densities Contributing to the Energy (2017). Preprint, http://www.mathematik.uni-dortmund.de/lsi/schweizer/Preprints/publist.html, https://arxiv.org/abs/1704.05326
22. Struwe, M.: Variational methods. Applications to Nonlinear Partial Differential Equations and Hamiltonian Systems. Springer, Berlin (2008)
23. Tartar, L.: Compensated Compactness and Applications to Partial Differential Equations. Nonlinear Anal. Mech. Heriot-Watt Symp. **4**, 136–211 (1979)
24. Tartar, L.: The General Theory of Homogenization: A Personalized Introduction. Springer, Berlin (2009)
25. Tartar, L.: Compensated compactness with more geometry. Springer Proc. Math. Stat. **137**, 74–101 (2015)
26. Waurick, M.: A functional analytic perspective to the div-curl lemma. J. Oper. Theory **80**(1), 95–111 (2018)
27. Waurick, M.: Nonlocal H-convergence. Calc. Var. Partial Differ. Equ. **57**(6), Art. 159, 46 pp. (2018)
28. Weber, C.: A local compactness theorem for Maxwell's equations. Math. Methods Appl. Sci. **2**, 12–25 (1980)
29. Weck, N.: Maxwell's boundary value problems on Riemannian manifolds with nonsmooth boundaries. J. Math. Anal. Appl. **46**, 410–437 (1974)
30. Witsch, K.J.: A remark on a compactness result in electromagnetic theory. Math. Methods Appl. Sci. **16**, 123–129 (1993)

Existence and Stability of Klein–Gordon Breathers in the Small-Amplitude Limit

Dmitry E. Pelinovsky, Tiziano Penati, and Simone Paleari

Abstract We consider a discrete Klein–Gordon (dKG) equation on $\mathbb{Z}^d$ in the limit of the discrete nonlinear Schrödinger (dNLS) equation, for which small-amplitude breathers have precise scaling with respect to the small coupling strength ϵ. By using the classical Lyapunov–Schmidt method, we show existence and linear stability of the KG breather from existence and linear stability of the corresponding dNLS soliton. Nonlinear stability, for an exponentially long time scale of the order $\mathcal{O}(\exp(\epsilon^{-1}))$, is obtained via the normal form technique, together with higher order approximations of the KG breather through perturbations of the corresponding dNLS soliton.

1 Introduction

Nonlinear oscillators with weak linear couplings on the d-dimensional cubic lattice are described by the discrete Klein–Gordon (dKG) equation

$$\ddot{u}_n + V'(u_n) = \epsilon(\Delta u)_n, \quad n \in \mathbb{Z}^d, \tag{1}$$

where $\epsilon > 0$ is the small coupling strength, Δ is the discrete Laplacian operator on $\ell^2(\mathbb{Z}^d)$, and $V(u)$ is a nonlinear potential for each oscillator. The total energy of the nonlinear oscillators conserves in time t and is given by the Hamiltonian function

$$H(u) = \frac{1}{2} \sum_{n\in\mathbb{Z}^d} \dot{u}_n^2 + \frac{\epsilon}{2} \sum_{n\in\mathbb{Z}^d} \sum_{|k-n|=1} (u_k - u_n)^2 + \sum_{n\in\mathbb{Z}^d} V(u_n). \tag{2}$$

D. E. Pelinovsky (✉)
Department of Mathematics, McMaster University, Hamilton, ON, Canada
e-mail: dmpeli@math.mcmaster.ca

T. Penati · S. Paleari
Department of Mathematics "F. Enriques", Milano University, Milano, Italy
e-mail: tiziano.penati@unimi.it; simone.paleari@unimi.it

W. Dörfler et al. (eds.), *Mathematics of Wave Phenomena*, Trends in Mathematics,
https://doi.org/10.1007/978-3-030-47174-3_16

For illustrative purposes, we deal with a hard anharmonic potential in the form

$$V(u) = \frac{1}{2}u^2 + \frac{1}{2+2p}u^{2+2p}, \tag{3}$$

where $p \in \mathbb{N}$ is assumed for analyticity of the vector field. There exists the unique global solution to the Cauchy problem for the dKG equation (1) with (3) equipped with the initial datum $(u, \dot{u}) \in \ell^2(\mathbb{Z}^d) \times \ell^2(\mathbb{Z}^d)$, where u stands for $\{u_n\}_{n\in\mathbb{Z}^d}$. Because our main results are formulated for small initial datum $(u, \dot{u})$, most of the results are applicable for general anharmonic potentials expanded as

$$V(u) = \frac{1}{2}u^2 + \alpha_p u^{2+2p} + O(u^{4+2p}) \quad \text{as} \quad u \to 0, \tag{4}$$

if $\alpha_p \neq 0$. The general anharmonic potential V is classified as *soft* if $\alpha_p < 0$ and *hard* if $\alpha_p > 0$.

Discrete breathers are time-periodic solutions localized on the lattice. Such solutions can be constructed asymptotically by exploring the two opposite limit: the anti-continuum limit $\epsilon \to 0$ of weak coupling between the oscillators [18] and the continuum limit $\epsilon \to \infty$ of strong coupling [5]. Compared to these asymptotic approximations, we explore here a different limit of the dKG equation to the discrete nonlinear Schrödinger (dNLS) equation, where the weak coupling between the oscillators is combined together with small amplitudes of each oscillator. To be precise, we assume the scaling

$$u_n = \epsilon^{1/2p}\tilde{u}_n, \quad n \in \mathbb{Z}^d \tag{5}$$

and rewrite the dKG equation (1) with the potential (3) in the perturbed form:

$$\ddot{u}_n + u_n + \epsilon u_n^{1+2p} = \epsilon(\Delta u)_n, \quad n \in \mathbb{Z}^d, \tag{6}$$

where the tilde notations have been dropped. By using a formal expansion

$$u_n(t) = a_n(\epsilon t)e^{it} + \bar{a}_n(\epsilon t)e^{-it} + O(\epsilon),$$

the following dNLS equation for the complex amplitudes is derived from the requirement that the correction term $O(\epsilon)$ remains bounded in $\ell^2(\mathbb{Z}^d)$ on the time scale of $O(\epsilon^{-1})$:

$$2ia_n' + \gamma_p|a_n|^{2p}a_n = (\Delta a)_n, \quad n \in \mathbb{Z}^d, \tag{7}$$

where the prime denotes the derivative with respect to the slow time variable $\tau = \epsilon t$ and the numerical coefficient γ_p is given by

$$\gamma_p = \binom{1+2p}{1+p} = \frac{(2p+1)!}{p!(p+1)!}. \tag{8}$$

The asymptotic relation between the dKG equation (6) and the dNLS equation (7) was observed first in [24] and was made rigorous by using two equivalent analytical methods in our previous work [26].

Discrete breathers of the dKG equation (6) are approximated by discrete solitons (standing localized waves) of the dNLS equation (7) in the form $a_n(\tau) = A_n e^{-\frac{i}{2}\Omega\tau}$, where the time-independent amplitudes satisfies the stationary dNLS equation

$$\Omega A_n + \gamma_p |A_n|^{2p} A_n = (\Delta A)_n, \quad n \in \mathbb{Z}^d. \tag{9}$$

The elementary staggering transformation

$$A_n = (-1)^n \tilde{A}_n, \quad \Omega = -4d - \tilde{\Omega} \tag{10}$$

relates the defocusing version (9) to the focusing version

$$(\Delta \tilde{A})_n + \gamma_p |\tilde{A}_n|^{2p} \tilde{A}_n = \tilde{\Omega} \tilde{A}_n, \quad n \in \mathbb{Z}^d. \tag{11}$$

In the recent past, existence and stability of discrete solitons in the focusing version (11) has been studied in many details depending on the exponent p and the dimension d. Various approximations of discrete solitons of the dNLS equation (11) are described in [9]. Let us review some relevant results on this subject.

The stationary dNLS equation (11) is the Euler–Lagrange equation of the constrained variational problem

$$\mathcal{E}_\nu = \inf_{a \in \ell^2(\mathbb{Z}^d)} \{E(a) : \ N(a) = \nu\}, \tag{12}$$

where

$$E(a) = \sum_{n \in \mathbb{Z}^d} \sum_{|k-n|=1} |a_k - a_n|^2 - \frac{1}{p+1} \sum_{n \in \mathbb{Z}^d} |a_n|^{2p+2} \tag{13}$$

is the conserved energy, $N(a) = \sum_{n \in \mathbb{Z}^d} |a_n|^2$ is the conserved mass, and $\nu > 0$ is fixed. The existence of a ground state as a minimizer of the constrained variational problem (12) was proven in Theorem 2.1 in [27] for every $\mathcal{E}_\nu < 0$. By Theorem 3.1 in [27], if $p < \frac{2}{d}$, the ground state exists for every $\nu > 0$, however, if $p \geqslant \frac{2}{d}$, there exists an excitation threshold $\nu_d > 0$ and the ground state only exists for $\nu > \nu_d$.

Variational and numerical approximations for $d = 1$ were employed to analyze the structure of discrete solitons of the stationary dNLS equation (11) near the critical case $p = 2$ [11, 19]. It was shown for single-pulse solitons that although the dependence $\Omega \mapsto \nu$ is monotone for $p = 1$, it becomes non-monotone for $p \gtrsim 1.5$ covering the whole range $\nu > 0$ for $p < 2$ and featuring the excitation threshold for $p \geqslant 2$. Further analytical estimates on the excitation threshold in the stationary dNLS equation were developed in [8, 10, 12].

Spectral stability of discrete solitons in the dNLS equation (7) was analyzed in the limit $\Omega \to \infty$, which can be recast as the anti-continuum limit of the dNLS equation. It was shown for $d = 1$ in [23, 25] (see Section 4.3.3 in [22]) that the single-pulse solitons are stable in the limit $\Omega \to \infty$ for every $p \in \mathbb{N}$. Asymptotic stability of single-pulse solitons for $d = 1$ and $p \geqslant 3$ was also proven in the same limit in [3] after similar asymptotical stability results were obtained for small solitons of the dNLS equation in the presence of a localized potential [7, 16].

Spectral and orbital stability of single-pulse discrete solitons in the dNLS equation (7) is determined by the monotonicity of the dependence $\Omega \mapsto \nu$ according to the Vakhitov–Kolokolov criterion [14, 22]. It was shown in [17] that this criterion is related to a similar energy criterion for spectral stability of discrete breathers in the dKG equation (6). If ω is a frequency of the discrete breathers and H is the value of their energy, then the monotonicity of the dependence $\omega \mapsto H$ is related to the monotonicity of the dependence $\Omega \mapsto \nu$ in the dNLS limit. Further results on the energy criterion for spectral stability of discrete breathers in the dKG equation (1) are given in [13, 28]. In spite of many convincing numerical evidences, the orbital stability of single-pulse discrete breathers is still out of reach in the energy methods.

The purpose of this paper is to make precise the correspondence between existence and linear stability of discrete breathers in the dKG equation (6) and discrete solitons in the dNLS equation (7). This work clarifies applications mentioned in Section 4 of our previous paper [26]. We show how the Lyapunov–Schmidt reduction method can be employed equally well to study existence and linear stability of small-amplitude discrete breathers near the point of their bifurcation from the dNLS limit under reasonable assumptions on existence and linear stability of the discrete solitons of the dNLS equation. We also show how normal form methods (see [2, 4, 6, 20, 21]), combined with the Lyapunov-Schmidt reduction, are implemented to provide higher order approximation of the discrete breathers, in the same dNLS limit. These results represent a considerable improvement with respect to the corresponding results in [20]. Long-time nonlinear stability of small-amplitude discrete breathers then follows, assuming the discrete soliton of the stationary dNLS equation (11) is a ground state of the variational problem (12).

The remainder of this paper consists of three sections. Section 2 proves the existence of discrete breathers obtained via the Lyapunov–Schmidt decomposition. Section 3 describes the linear stability results obtained by the extension of the same technique. Section 4 gives the normal form arguments towards the long-time nonlinear stability of small-amplitude breathers.

2 Existence via Lyapunov–Schmidt Decomposition

Breathers are T-periodic solutions of the dKG equation (1) localized on the lattice. One can consider such strong solutions of the dKG equation (1) in the space $u \in H^2_{\text{per}}([0, T]; \ell^2(\mathbb{Z}^d))$. By scaling the time variable as $\tau = \omega t$ with $\omega = 2\pi/T$, it is convenient to consider 2π-periodic solutions $U \in H^2_{\rm per}([-\pi,\pi];\ell^2(\mathbb{Z}^d))$ with parameter ω, such that $u(t) = U(\omega t)$. Breather solutions can be equivalently represented by the Fourier series

$$U(\tau) = \sum_{m\in\mathbb{Z}} A^{(m)} e^{im\tau}. \tag{14}$$

Since U is real, the complex-valued Fourier coefficients satisfy the constraints:

$$A^{(m)} = \overline{A^{(-m)}}, \quad m \in \mathbb{Z}. \tag{15}$$

If the periodic solution has zero initial velocity, i.e., $U'(0) = 0$, then it follows from reversibility of the dKG equation (1) in time[1] that the periodic solution is even in time, which implies

$$A^{(m)} = A^{(-m)}, \quad m \in \mathbb{Z}. \tag{16}$$

As a consequence of the two symmetries, the Fourier coefficients are real, hence the representation (14) becomes Fourier cosine series with real-valued coefficients:

$$U(\tau) = A^{(0)} + 2\sum_{m\in\mathbb{N}} A^{(m)} \cos(m\tau). \tag{17}$$

After the scaling transformation (5), breather solutions to the scaled dKG equation (6) satisfy the following boundary-value problem:

$$\omega^2 U'' + U + \epsilon U^{1+2p} = \epsilon \Delta U, \quad U \in H^2_{\rm per}([-\pi,\pi];\ell^2(\mathbb{Z}^d)), \tag{18}$$

where we use componentwise multiplication of sequences with the following convention: $(U^{1+2p})_n = (U_n)^{1+2p}$. The existence problem (18) can be rewritten in real-valued Fourier coefficients as

$$(1 - m^2\omega^2)A^{(m)} + \frac{\epsilon}{2\pi}\int_{-\pi}^{\pi} U^{1+2p}(\tau)e^{-im\tau}d\tau = \epsilon\Delta A^{(m)}, \quad m \in \mathbb{N}_0. \tag{19}$$

At $\epsilon = 0$, bifurcation of breathers is expected at $\omega_m = 1/m$, $m \in \mathbb{N}$, from which the lowest bifurcation value $\omega_1 = 1$ gives a branch of fundamental breathers with one oscillation on the period $[-\pi,\pi]$. If the solution branch $\omega(\epsilon)$ and $\{A^{(m)}(\epsilon)\}_{m\in\mathbb{N}} \in \ell^{2,2}(\mathbb{Z};\ell^2(\mathbb{Z}^d))$ is parameterized by ϵ, then we are looking for

---

[1]Given a solution $\gamma := \{u(t), \dot{u}(t)\}$ to the dKG equation (1), another solution is $\tilde{\gamma} = \{\tilde{u}(t) = u(-t),\ \dot{\tilde{u}}(t) = -\dot{u}(-t)\}$. If γ is a periodic solution with initial zero velocity, then the same is true for $\tilde{\gamma}$, and since the two solutions have the same initial configuration $u(0) = \tilde{u}(0)$, they are solutions of the same Cauchy problem, hence they coincide.

the branch of fundamental breathers to satisfy the limiting conditions:

$$\lim_{\epsilon\to 0}\omega(\epsilon)=1,\quad \lim_{\epsilon\to 0}A^{(1)}(\epsilon)\neq 0,\quad \text{and}\quad \lim_{\epsilon\to 0}A^{(m)}(\epsilon)=0,\quad m\neq 1. \tag{20}$$

The limiting conditions (20) are not sufficient for persistence argument. In order to define uniquely a continuation of the solution branch in ϵ, we consider the stationary dNLS equation in the form:

$$\Omega\mathcal{A}+\gamma_p|\mathcal{A}|^{2p}\mathcal{A}=\Delta\mathcal{A},\quad \mathcal{A}\in\ell^2(\mathbb{Z}^d), \tag{21}$$

where Ω is parameter and γ_p is a numerical coefficient given by (8). We restrict consideration to the case of dNLS solitons given by real $\mathcal{A}$, for which we introduce the Jacobian operator for the stationary dNLS equation (21) at $\mathcal{A}$:

$$J_\Omega := \Omega+(1+2p)\gamma_p|\mathcal{A}|^{2p}-\Delta. \tag{22}$$

Since $\sigma(\Delta)=[-4d,0]$ in $\ell^2(\mathbb{Z}^d)$ and $\mathcal{A}\in\ell^2(\mathbb{Z}^d)$ is expected to decay exponentially at infinity, we need to consider Ω in $\mathbb{R}\backslash[-4d,0]$.

Remark 1 Since the discrete solitons in the focusing stationary dNLS equation (11) exist for $\tilde{\Omega}>0$ [27], the staggering transformation (10) suggests that the discrete solitons in the defocusing stationary dNLS equation (21) exist for $\Omega<-4d$.

Assuming existence of a dNLS soliton $\mathcal{A}$ in the stationary dNLS equation (21) for some $\Omega\in\mathbb{R}\backslash[-4d,0]$ and invertibility of J_Ω at this $\mathcal{A}$ in (22), we will prove existence and uniqueness of the branch $\omega(\epsilon)$ and $\{A^{(m)}(\epsilon)\}_{m\in\mathbb{N}}\in\ell^{2,2}(\mathbb{Z};\ell^2(\mathbb{Z}^d))$ of fundamental breathers satisfying the limiting conditions:

$$\lim_{\epsilon\to 0}\frac{\omega(\epsilon)-1}{\epsilon}=-\frac{1}{2}\Omega,\qquad \lim_{\epsilon\to 0}A^{(m)}(\epsilon)=\begin{cases}\mathcal{A}, & m=1,\\ 0, & m\neq 1.\end{cases} \tag{23}$$

The following theorem gives the existence and uniqueness result for breathers.

Theorem 1 *Fix $p\in\mathbb{N}$. Assume the existence of real $\mathcal{A}\in\ell^2(\mathbb{Z}^d)$ in the stationary dNLS equation (21) for some $\Omega\in\mathbb{R}\backslash[-4d,0]$ such that the Jacobian operator J_Ω at this $\mathcal{A}$ in (22) has trivial null space in $\ell^2(\mathbb{Z}^d)$. There exists $\epsilon_0>0$ and $C_0>0$ such that the breather equation (19) for every $\epsilon\in(0,\epsilon_0)$ admits a unique C^ω (analytic) solution branch $\omega(\epsilon)\in\mathbb{R}$ and $\{A^{(m)}(\epsilon)\}_{m\in\mathbb{N}}\in\ell^{2,2}(\mathbb{Z};\ell^2(\mathbb{Z}^d))$ satisfying the bounds*

$$\left|\omega(\epsilon)-1+\frac{1}{2}\epsilon\Omega\right|\leqslant C_0\epsilon^2 \tag{24}$$

and

$$\|A^{(0)}\|_{\ell^2(\mathbb{Z}^d)}+\|A^{(1)}-\mathcal{A}\|_{\ell^2(\mathbb{Z}^d)}+\sum_{m\geqslant 2}\|A^{(m)}\|_{\ell^2(\mathbb{Z}^d)}\leqslant C_0\epsilon, \tag{25}$$

for every $\epsilon\in(0,\epsilon_0)$.

Remark 2 In order to explain the relevance of the stationary dNLS equation (21), we set $\omega^2 = 1 - \epsilon\Omega$, where Ω is fixed independently of ϵ, and rewrite Eq. (19) for $m = 1$ after dividing it by ϵ. This procedure yields the bifurcation equation:

$$\Omega A^{(1)}(\epsilon) + \frac{1}{2\pi}\int_{-\pi}^{\pi} U^{1+2p}(\tau,\epsilon)e^{-i\tau}d\tau = \Delta A^{(1)}(\epsilon),$$

where $U(\tau,\epsilon)$ is given by the Fourier series (14) with amplitudes $\{A^{(m)}(\epsilon)\}_{m\in\mathbb{Z}}$ satisfying symmetries (15) and (16). Formally, at the leading order (20), we have:

$$\Omega A^{(1)}(0) + \frac{1}{2\pi}\int_{-\pi}^{\pi}\left[A^{(1)}(0)e^{i\tau} + \overline{A^{(1)}(0)}e^{-i\tau}\right]^{1+2p} e^{-i\tau}d\tau = \Delta A^{(1)}(0). \quad (26)$$

By expanding

$$\left[A^{(1)}(0)e^{i\tau} + \overline{A^{(1)}(0)}e^{-i\tau}\right]^{1+2p} = \sum_{k=0}^{1+2p}\binom{1+2p}{k}\left(A^{(1)}(0)\right)^k\left(\overline{A^{(1)}(0)}\right)^{1+2p-k} e^{i(2k-2p-1)\tau}$$

and evaluating the integral in (26), we get the only nonzero term for $k = p+1$. As a result, Eq. (26) becomes the limiting dNLS equation (21) with $\mathcal{A} := A^{(1)}(0)$.

Proof In order to solve the breather equation (19) as $\epsilon \to 0$ near the limiting solution (20), we proceed with the classical Lyapunov-Schmidt decomposition [1] (see applications of this method in a similar context in [5, 24]). We introduce the Hilbert spaces

$$X_2 := H^2_{\text{per}}([-\pi,\pi];\ell^2(\mathbb{Z}^d)), \qquad X_0 := L^2_{\text{per}}([-\pi,\pi];\ell^2(\mathbb{Z}^d))$$

and the dual spaces under the Fourier series (14):

$$\hat{X}_2 := \ell^{2,2}(\mathbb{Z};\ell^2(\mathbb{Z}^d)), \qquad \hat{X}_0 := \ell^2(\mathbb{Z};\ell^2(\mathbb{Z}^d)).$$

The breather solution U is an element of X_2, which is uniquely identified by the sequence A in $\hat{X}_2$. In other words, a solution is given by a sequence of Fourier coefficients $\{A^{(m)}\}_{m\in\mathbb{Z}}$ in $\ell^{2,2}(\mathbb{Z})$, where each Fourier coefficient $A^{(m)}$ is a complex sequence $A^{(m)} = \{A_n^{(m)}\}_{n\in\mathbb{Z}^d}$ in $\ell^2(\mathbb{Z}^d)$. The Sobolev norm in space $\hat{X}_2$ is given by

$$\|A\|_{\hat{X}_2} = \left(\sum_{m\in\mathbb{Z}}(1+|m|^2)^2\left\|A^{(m)}\right\|^2_{\ell^2(\mathbb{Z}^d)}\right)^{1/2}.$$

Let us introduce also the linear operator $L_\omega : X_2 \to X_0$, which is given in Fourier space by $\hat{L}_\omega : \hat{X}_2 \to \hat{X}_0$:

$$\left(\hat{L}_\omega A\right)^{(m)} = (1 - m^2\omega^2)A^{(m)}, \quad m \in \mathbb{Z}. \quad (27)$$

Let $\{e_n\}_{n\in\mathbb{Z}^d}$ be a set of unit vectors in $\ell^2(\mathbb{Z}^d)$. We define the linear subspace $V_2 = \text{span}(\cup_{n\in\mathbb{Z}^d} e_n e^{i\tau}, \cup_{n\in\mathbb{Z}^d} e_n e^{-i\tau})$ as the kernel of $L_{\omega=1}$ in X_2 and the linear subspace W_2 as its orthogonal complement in $X_2 = V_2 \oplus W_2$. In the Fourier space, we set $\hat{V}_2$ as a kernel of $\hat{L}_{\omega=1}$ in $\hat{X}_2$ and $\hat{W}_2$ as its orthogonal complement in $\hat{X}_2 = \hat{V}_2 \oplus \hat{W}_2$. In a similar way, we introduce the range subspace W_0 for the operator $L_{\omega=1}$, which is a subspace of X_0 and is orthogonal to V_2, so that $X_0 = V_2 \oplus W_0$, and similarly $\hat{X}_0 = \hat{V}_2 \oplus \hat{W}_0$. Any element of $\hat{X}_2$ can be decomposed into

$$A = A^\sharp + A^\flat \,, \qquad A^\sharp \in \hat{V}_2 \,, \qquad A^\flat \in \hat{W}_2 \,. \tag{28}$$

The breather equation (19) can be written in the abstract form:

$$F(A, \omega, \epsilon) := \hat{L}_\omega A + \epsilon N(A) - \epsilon \Delta A = 0 \,, \tag{29}$$

where $N(A)$ is the nonlinear term. If $p \in \mathbb{N}$, then the nonlinear map $F(A, \omega, \epsilon)$: $\hat{X}_2 \times \mathbb{R} \times \mathbb{R} \to \hat{X}_0$ is C^ω in its variables. The nonlinear equation (29) is projected onto $\hat{V}_0$ and $\hat{W}_0$, thus yields the following two equations:

$$\Pi_{\hat{V}_0} F(A^\sharp + A^\flat, \omega, \epsilon) = 0 \,, \qquad \Pi_{\hat{W}_0} F(A^\sharp + A^\flat, \omega, \epsilon) = 0 \,. \tag{30}$$

The former one is known as the kernel equation and the latter one is known as the range equation. We shall solve the range equation for small ϵ assuming that $|\omega - 1| = O(\epsilon)$ by using the implicit function theorem.

Exploiting the fact that $\hat{V}_0$ and $\hat{W}_0$ are invariant under Δ and that $\hat{L}_{\omega=1} A^\sharp = 0$ by definition, the range equation in (30) takes the form

$$\left(\hat{L}_\omega - \epsilon \Delta\right) A^\flat + \epsilon \Pi_{\hat{W}_0} N(A^\sharp + A^\flat) = 0 \,. \tag{31}$$

The perturbed linear operator $\hat{L}_\omega - \epsilon\Delta$ can be inverted on $\hat{W}_0$ for ϵ small enough if $|\omega - 1| = O(\epsilon)$. Indeed, first write using Neumann series

$$\left(\hat{L}_\omega - \epsilon\Delta\right)^{-1} = \left[\sum_{k=0}^{\infty} \left(\epsilon \hat{L}_\omega^{-1} \Delta\right)^k\right] \hat{L}_\omega^{-1} \,, \tag{32}$$

where $\hat{L}_\omega^{-1}$ is well defined on $\hat{W}_0$ thanks to the diagonal form:

$$(\hat{L}_\omega^{-1} A)^{(m)} = \frac{1}{1 - m^2\omega^2} A^{(m)} \,, \qquad m \neq \pm 1 \,.$$

Let us introduce a parametrization of ω by

$$\omega^2 = 1 - \epsilon\Omega, \tag{33}$$

where Ω is fixed independently of ϵ. It follows by elementary computation that there exists $\epsilon_*(\Omega)$ that only depends on Ω such that for every $\epsilon \in (0, \epsilon_*(\Omega))$,

$$|1 - m^2\omega^2| = |1 - m^2(1 - \epsilon\Omega)| > \frac{1}{2}(1 + m^2), \quad \forall m \in \mathbb{Z}\backslash\{-1, 1\},$$

thus obtaining the estimate

$$\|\hat{L}_\omega^{-1}\|_{\hat{W}_0 \to \hat{W}_2} \leqslant 2,$$

and consequently

$$\|\epsilon\hat{L}_\omega^{-1}\Delta\|_{\hat{W}_0 \to \hat{W}_2} \leqslant 8d\epsilon.$$

By Neumann formula (32) there exists $\epsilon_0 > 0$ and $C_0 > 0$ such that for every $\epsilon \in (0, \epsilon_0)$,

$$\|(\hat{L}_\omega - \epsilon\Delta)^{-1}\|_{\hat{W}_0 \to \hat{W}_2} \leqslant C_0. \tag{34}$$

For example, we can take $\epsilon_0 := \min\{\epsilon^*(\Omega), (8d)^{-1}\}$.

Since X_2 is a Banach algebra with respect to componentwise multiplication of sequence of functions and $\hat{X}_2$ is a Banach algebra with respect to convolution, the nonlinear term $N(A)$ in (31) is closed in $\hat{X}_2$. By writing the range equation as the fixed-point equation for $A^\flat$:

$$A^\flat = -\epsilon\left(\hat{L}_\omega - \epsilon\Delta\right)^{-1}\Pi_{\hat{W}_0} N(A^\sharp + A^\flat) \tag{35}$$

and using the implicit function theorem thanks to the parametrization (33) and the uniform bound (34), we conclude that for every $\epsilon \in (0, \epsilon_0)$, $\Omega \in \mathbb{R}$, and $A^\sharp \in \hat{V}_2 \subset \hat{X}_2$, there exists a unique solution $A^\flat \in \hat{W}_2 \subset \hat{X}_2$ to the fixed-point equation (35) such that the mapping $(A^\sharp, \Omega, \epsilon) \to A^\flat$ is C^ω and the solution is as small as $O(\epsilon)$ thanks to the leading order approximation

$$A^\flat = -\epsilon\hat{L}_\omega^{-1}\Pi_{\hat{W}_0} N(A^\sharp) + O(\epsilon^2), \tag{36}$$

which provides the bound

$$\|A^\flat\|_{\hat{X}_2} \leqslant C\epsilon, \tag{37}$$

for some ϵ-independent C.

Inserting the parametrization (33) and the mapping $(A^\sharp, \Omega, \epsilon) \to A^\flat$ into the kernel equation in (30) and dividing by ϵ, we obtain

$$\Omega A^\sharp - \Delta A^\sharp + \Pi_{\hat{V}_0} N(A^\sharp + A^\flat(A^\sharp, \Omega, \epsilon)) = 0.$$

Thanks to the computations in Remark 2 and the bound (37), one can rewrite the kernel equation explicitly in terms of the real-valued amplitude $A^{(1)}$ as follows:

$$f(A^{(1)}, \Omega, \epsilon) := \Omega A^{(1)} - \Delta A^{(1)} + \gamma_p A^{(1)} |A^{(1)}|^{2p} + \epsilon R(A^{(1)}, \Omega, \epsilon) = 0, \quad (38)$$

where $R(A^{(1)}, \Omega, \epsilon) : \ell^2(\mathbb{Z}^d) \times \mathbb{R} \times \mathbb{R} \to \ell^2(\mathbb{Z}^d)$ is C^ω and bounded as $\epsilon \to 0$ thanks to the bound (37). Thanks to the assumptions of the theorem, $\mathcal{A} \in \ell^2(\mathbb{Z}^d)$ is a root of

$$f(\mathcal{A}, \Omega, 0) = 0 \quad (39)$$

and

$$D_{A^{(1)}} f(\mathcal{A}, \Omega, 0) = J_\Omega \quad (40)$$

is a bounded and invertible operator on $\ell^2(\mathbb{Z}^d)$. By the implicit function theorem, there exists $\epsilon_1 < \epsilon_0$ such that for every $\epsilon \in (0, \epsilon_1)$ and $\Omega \in \mathbb{R}$ for which $\mathcal{A} \in \ell^2(\mathbb{Z}^d)$ exists in (39) and J_Ω is invertible in (40), there exists a unique solution $A^{(1)} \in \ell^2(\mathbb{Z}^d)$ to the kernel equation (38) such that the mapping $(\Omega, \epsilon) \to A^{(1)}$ is C^ω and the solution satisfies the bound

$$\|A^{(1)} - \mathcal{A}\|_{\ell^2(\mathbb{Z}^d)} \leqslant C\epsilon, \quad (41)$$

for some ϵ-independent C. Combining (37) and (41) with the decompositions (28) and (33) yields bounds (24) and (25). □

3 Stability via Lyapunov–Schmidt Decomposition

Linearizing $u(t) = U(\tau) + w(t)$ of the dKG equation (6) at the breather solution $U \in H^2_{\text{per}}([-\pi, \pi]; \ell^2(\mathbb{Z}^d))$ with $\tau = \omega t$ yields the linearized dKG equation:

$$\ddot{w} + w + \epsilon(1 + 2p)U^{2p} w = \epsilon \Delta w. \quad (42)$$

By Floquet theorem, every solution of the 2π-periodic linear equation (42) can be represented in the form $w(t) = W(\tau)e^{\lambda t}$, where $\lambda \in \mathbb{C}$ is the spectral parameter and $W \in H^2_{\text{per}}([-\pi, \pi]; \ell^2(\mathbb{Z}^d))$ is an eigenfunction of the spectral problem:

$$\omega^2 W'' + 2\lambda\omega W' + \lambda^2 W + W + \epsilon(1 + 2p)U^{2p} W = \epsilon \Delta W. \quad (43)$$

Let us represent $W \in H^2_{\text{per}}([-\pi, \pi]; \ell^2(\mathbb{Z}^d))$ by the Fourier series:

$$W(\tau) = \sum_{m \in \mathbb{Z}} B^{(m)} e^{im\tau}. \quad (44)$$

With the help of (14) and (44), the spectral problem (43) is rewritten in Fourier coefficients as

$$\left[1+(\lambda+im\omega)^2\right]B^{(m)}+\frac{\epsilon(1+2p)}{2\pi}\int_{-\pi}^{\pi}U^{2p}(\tau)W(\tau)e^{-im\tau}d\tau=\epsilon\Delta B^{(m)}. \tag{45}$$

No symmetry reductions exist generally for the Fourier coefficients $\{B^{(m)}\}_{m\in\mathbb{Z}}$.

At $\epsilon=0$ and $\omega=1$, the spectral problem (45) admits a double set of eigenvalues λ defined by

$$\Sigma_{\pm}:=\{i(\pm 1-m),\quad m\in\mathbb{Z}\}, \tag{46}$$

where $\Sigma_+=\Sigma_-$ and each eigenvalue has infinite multiplicity due to the lattice $\mathbb{Z}^d$. In terms of the Floquet multipliers

$$\mu:=e^{\lambda T}=e^{2\pi\lambda/\omega}, \tag{47}$$

all eigenvalues at $\epsilon=0$ and $\omega=1$ correspond to the same Floquet multiplier $\mu=1$.

Remark 3 The degeneracy of the Floquet multiplier μ in (47) is understood in terms of the following symmetry for the spectral problem (45). Fix $k\in\mathbb{Z}$ and apply transformation

$$\lambda=ik+\tilde{\lambda},\quad m=-k+\tilde{m},\quad B^{(m)}=\tilde{B}^{(\tilde{m})}.$$

The eigenvalue-eigenvector pair $\left(\tilde{\lambda},\{\tilde{B}^{(\tilde{m})}\}_{\tilde{m}\in\mathbb{Z}}\right)$ satisfies the same spectral problem (45) but in tilde variables. Therefore, the spectral problem (45) near every nonzero point $\lambda\in\Sigma_{\pm}$ repeats its behavior near $\lambda=0$. It is hence sufficient to consider the spectral problem (45) near $\lambda=0$.

Let us review the spectral stability problem for the dNLS equation (7). The dNLS soliton $a(\tau)=e^{-\frac{i}{2}\Omega\tau}\mathcal{A}$ is defined by solutions of the stationary dNLS equation (21) with real $\mathcal{A}\in\ell^2(\mathbb{Z}^d)$. Linearizing with the expansion $a(\tau)=e^{-\frac{i}{2}\Omega\tau}\left[\mathcal{A}+b(\tau)\right]$ yields the linearized dNLS equation:

$$2ib'+\left(\Omega-\Delta+\gamma_p(p+1)\mathcal{A}^{2p}\right)b+\gamma_p p\mathcal{A}^{2p}\bar{b}=0. \tag{48}$$

Separating variables by

$$b(\tau)=(b_++ib_-)e^{\Lambda\tau}+(\bar{b}_++i\bar{b}_-)e^{\bar{\Lambda}\tau}$$

and

$$\bar{b}(\tau)=(b_+-ib_-)e^{\Lambda\tau}+(\bar{b}_+-i\bar{b}_-)e^{\bar{\Lambda}\tau},$$

where $\Lambda \in \mathbb{C}$ is the spectral parameter and $(b_+, b_-) \in \ell^2(\mathbb{Z}^d) \times \ell^2(\mathbb{Z}^d)$ is a complex-valued eigenfunction, yields the spectral problem:

$$\begin{bmatrix} 0 & -(\Omega - \Delta + \gamma_p \mathcal{A}^{2p}) \\ \Omega - \Delta + \gamma_p(1+2p)\mathcal{A}^{2p} & 0 \end{bmatrix} \begin{bmatrix} b_+ \\ b_- \end{bmatrix} = 2\Lambda \begin{bmatrix} b_+ \\ b_- \end{bmatrix}. \tag{49}$$

The spectral problem (49) can be written in the Hamiltonian form $\mathcal{J}\mathcal{H}''(\mathcal{A})\mathbf{f} = 2\Lambda\mathbf{f}$, where $\mathbf{f} = (b_+, b_-)^T$,

$$\mathcal{J} = \begin{bmatrix} 0 & -1 \\ 1 & 0 \end{bmatrix}, \quad \mathcal{H}''(\mathcal{A}) = \begin{bmatrix} \Omega - \Delta + \gamma_p(1+2p)\mathcal{A}^{2p} & 0 \\ 0 & \Omega - \Delta + \gamma_p \mathcal{A}^{2p} \end{bmatrix}.$$

The first diagonal entry in $\mathcal{H}''(\mathcal{A})$ coincides with the Jacobian operator (22) for the stationary dNLS equation (21).

Remark 4 Since $\mathcal{H}''(\mathcal{A})$ and $\Omega - \Delta$ are bounded operators in $\ell^2(\mathbb{Z}^d)$, whereas $\Omega \in \mathbb{R}\backslash[-4d, 0]$ and $\mathcal{A}^{2p}$ decays exponentially at infinity, the operator $(\Omega - \Delta)^{-1}\mathcal{A}^{2p}$ is a compact (Hilbert–Schmidt) operator. As a result, $\sigma_c(\mathcal{H}''(\mathcal{A})) = [\Omega, \Omega + 4d]$ and $\sigma_d(\mathcal{H}''(\mathcal{A}))$ consists of finitely many eigenvalues of finite multiplicities, where σ_c and σ_d denotes the absolutely continuous and discrete spectra of the self-adjoint operator $\mathcal{H}''(\mathcal{A})$ in the Hilbert space $\ell^2(\mathbb{Z}^d)$.

It follows from Remark 4 that if $\Omega < -4d$ (see Remark 1), there exist finitely many positive eigenvalues of $\sigma_d(\mathcal{H}''(\mathcal{A}))$, whereas if $\Omega > 0$, there exist finitely many negative eigenvalues of $\sigma_d(\mathcal{H}''(\mathcal{A}))$. In either case, the stability theory in linear Hamiltonian systems [14, 22] is applied to conclude that there exist finitely many eigenvalues Λ with $\mathrm{Re}(\Lambda) \neq 0$ in the spectral problem (49). The continuous spectrum of $\mathcal{J}\mathcal{H}''(\mathcal{A})$ coincides with the purely continuous spectrum of $\mathcal{J}\mathcal{H}''(0)$ and is located on

$$\sigma_c(\mathcal{J}\mathcal{H}''(\mathcal{A})) = \{i[\Omega, \Omega + 4d]\} \cup \{-i[\Omega, \Omega + 4d]\}. \tag{50}$$

The following theorem guarantees the persistence of simple isolated eigenvalues of the spectral problem (49) in the spectral problem (45) near $\lambda = 0$.

Theorem 2 *Under the assumption of Theorem 1, assume that $\Lambda \in \mathbb{C}$ is a simple isolated eigenvalue of the spectral problem (49) such that $2\Lambda \notin \sigma_c(\mathcal{J}\mathcal{H}''(\mathcal{A}))$ and $(b_+, b_-) \in \ell^2(\mathbb{Z}^d) \times \ell^2(\mathbb{Z}^d)$. There exists $\epsilon_0 > 0$ and $C_0 > 0$ such that the spectral problem (45) for every $\epsilon \in (0, \epsilon_0)$ admits a unique C^ω branch of the eigenvalue–eigenvector pair with $\lambda(\epsilon) \in \mathbb{C}$ and $\{B^{(m)}(\epsilon)\}_{m\in\mathbb{N}} \in \ell^{2,2}(\mathbb{Z}; \ell^2(\mathbb{Z}^d))$ satisfying*

$$|\lambda(\epsilon) - \epsilon\Lambda| \leqslant C_0\epsilon^2, \tag{51}$$

$$\|B^{(1)} - b_+ - ib_-\|_{\ell^2(\mathbb{Z}^d)} + \|B^{(-1)} - b_+ + ib_-\|_{\ell^2(\mathbb{Z}^d)} \leqslant C_0\epsilon, \tag{52}$$

and

$$\|B^{(0)}\|_{\ell^2(\mathbb{Z}^d)} + \sum_{m\geqslant 2} \|B^{(m)}\|_{\ell^2(\mathbb{Z}^d)} \leqslant C_0\epsilon, \tag{53}$$

for every $\epsilon \in (0, \epsilon_0)$.

Proof We adopt the same Hilbert spaces as those used in the proof of Theorem 1. Any element of $\hat{X}_2$ can be decomposed into

$$B = B^\sharp + B^\flat, \qquad B^\sharp \in \hat{V}_2, \qquad B^\flat \in \hat{W}_2. \tag{54}$$

We assume that $\omega(\epsilon)$ and $\{A^{(m)}(\epsilon)\}_{m\in\mathbb{Z}}$ are given by Theorem 1 with the error bounds (24) and (25). Let us introduce the linear operator $\hat{M}_{\lambda,\omega} : \hat{X}_2 \to \hat{X}_0$:

$$\left(\hat{M}_{\lambda,\omega} B\right)^{(m)} = \left[1 + (\lambda + im\omega)^2\right] B^{(m)}, \quad m \in \mathbb{Z}. \tag{55}$$

The spectral problem (45) for Fourier coefficients can be written in the abstract form:

$$F(B, \lambda, \epsilon) := \hat{M}_{\lambda,\omega(\epsilon)} B + \epsilon S(A(\epsilon), B) - \epsilon\Delta B = 0, \tag{56}$$

where $S(A(\epsilon), B)$ is the linear map on B obtained from the nonlinear term $N(A)$. Since $p \in \mathbb{N}$, the map $F(B, \lambda, \epsilon) : \hat{X}_2 \times \mathbb{C} \times \mathbb{R} \to \hat{X}_0$ is C^ω in its arguments. Projecting equation (56) onto $\hat{V}_0$ and $\hat{W}_0$ yields the following kernel and range equations:

$$\Pi_{\hat{V}_0} F(B^\sharp + B^\flat, \lambda, \epsilon) = 0, \qquad \Pi_{\hat{W}_0} F(B^\sharp + B^\flat, \lambda, \epsilon) = 0. \tag{57}$$

The range equation in system (57) can be solved in the same way as the range equation in system (30). By using the implicit function theorem, for every $\epsilon \in (0, \epsilon_0)$, $\Lambda \in \mathbb{C}$, and $B^\sharp \in \hat{V}_2 \subset \hat{X}_2$, there exists a unique solution $B^\flat \in \hat{W}_2 \subset \hat{X}_2$ of the range equation $\Pi_{\hat{W}_0} F(B^\sharp + B^\flat, \epsilon\Lambda, \epsilon) = 0$ such that the mapping $(B^\sharp, \Lambda, \epsilon) \to B^\flat$ is C^ω and the solution is as small as $O(\epsilon)$ thanks to the bound

$$\|B^\flat\|_{\hat{X}_2} \leqslant C\epsilon, \tag{58}$$

for some ϵ-independent C.

Inserting $\omega = 1 - \frac{1}{2}\epsilon\Omega + O(\epsilon^2)$, $\lambda = \epsilon\Lambda$, and the C^ω mapping $(B^\sharp, \Lambda, \epsilon) \to B^\flat$ into the kernel equation in system (57) and dividing by ϵ, we obtain the following system of two equations on the two amplitudes $(B^{(1)}, B^{(-1)})$:

$$(\Omega \pm 2i\Lambda)B^{(\pm 1)} - \Delta B^{(\pm 1)} + \gamma_p \mathcal{A}^{2p}\left[(p+1)B^{(\pm 1)} + pB^{(-1)}\right] + \epsilon R^{(\pm 1)}(B^{(1)}, B^{(-1)}, \Lambda, \epsilon) = 0, \tag{59}$$

where $R^{(\pm 1)}(B^{(1)}, B^{(-1)}, \Lambda, \epsilon) : \ell^2(\mathbb{Z}^d) \times \ell^2(\mathbb{Z}^d) \times \mathbb{R} \times \mathbb{R} \to \ell^2(\mathbb{Z}^d)$ is a linear map on $(B^{(1)}, B^{(-1)})$ with C^ω coefficients which are bounded as $\epsilon \to 0$ thanks to the bound (58). In the derivation of numerical coefficients in (59), we have used the following explicit computation:

$$\frac{1}{2\pi}\int_{-\pi}^{\pi}\left[\mathcal{A}e^{i\tau} + \mathcal{A}e^{-i\tau}\right]^{2p}\left[B^{(1)}e^{i\tau} + B^{(-1)}e^{-i\tau}\right]e^{\mp i\tau}d\tau$$

$$= \sum_{k=0}^{2p}\binom{2p}{k}\mathcal{A}^{2p}\frac{1}{2\pi}\int_{-\pi}^{\pi}\left[B^{(1)}e^{i(2k-2p+1\mp 1)\tau} + B^{(-1)}e^{i(2k-2p-1\mp 1)\tau}\right]d\tau$$

$$= \frac{p+1}{2p+1}\gamma_p\mathcal{A}^{2p}B^{(\pm 1)} + \frac{p}{2p+1}\gamma_p\mathcal{A}^{2p}B^{(-1)}.$$

At $\epsilon = 0$, the system (59) becomes the spectral problem (49) in variables $B^{(\pm 1)} = b_+ \pm ib_-$. It is assume that Λ is a simple isolated eigenvalue in the spectral problem (49) with $2\Lambda \notin \sigma_c(\mathcal{J}\mathcal{H}''(\mathcal{A}))$ and a related eigenvector $(b_+, b_-) \in \ell^2(\mathbb{Z}^d) \times \ell^2(\mathbb{Z}^d)$. For $\epsilon \neq 0$, the eigenvalue Λ becomes the characteristic root of the linear system (59). By the analytic perturbation theory for closed linear operators (see Theorem 1.7 in Chapter VII on p. 368 in [15]), simple characteristic roots and the associated eigenvectors are continued in ϵ as C^ω functions. This completes justification of the bounds (51) and (52). □

Remark 5 If $2\Lambda \in i\mathbb{R}\backslash\sigma_c(\mathcal{J}\mathcal{H}''(\mathcal{A}))$, the bound (51) is not sufficient to guarantee that the eigenvalue λ remains on $i\mathbb{R}$.

In order to obtain a definite prediction that the simple isolated eigenvalue $\Lambda \in i\mathbb{R}$ of the spectral problem (49) persist as a simple isolated eigenvalue $\lambda \in i\mathbb{R}$ of the spectral problem (45), we use the Krein signature theory for linearized Hamiltonian systems. Consider the linearized dKG equation (42) and define

$$k(w) := i\sum_{n\in\mathbb{Z}^d} w_n\dot{\bar{w}}_n - \bar{w}_n\dot{w}_n. \tag{60}$$

It is straightforward to verify that $k(w)$ is independent of t. Let us represent the eigenvalue-eigenvector pair by $w(t) = W(\tau)e^{\lambda t}$ with $\lambda \in \mathbb{C}$ and $W \in H^2_{\text{per}}([-\pi, \pi]; \ell^2(\mathbb{Z}^d))$. Then, $k(w) = K(W, \lambda)e^{(\lambda+\bar{\lambda})t}$ with

$$K(W, \lambda) := i\omega\sum_{n\in\mathbb{Z}^d}\left(W_n\bar{W}_n' - \bar{W}_n W_n'\right) - i(\lambda - \bar{\lambda})\sum_{n\in\mathbb{Z}^d}|W_n|^2. \tag{61}$$

The following lemma reproduces the main result of the Krein theory.

Lemma 1 *Let $\lambda \in \mathbb{C}$ be a simple isolated eigenvalue in the spectral problem (43) with the eigenvector $W \in H^2_{\text{per}}([-\pi, \pi]; \ell^2(\mathbb{Z}^d))$. Then, $K(W, \lambda) = 0$ if* $\text{Re}(\lambda) \neq 0$ *and $K(W, \lambda) \neq 0$ if $\lambda \in i\mathbb{R}\backslash\{0\}$.*

Proof Let $Q := \lambda W + \omega W'$, where $W \in H^2_{\text{per}}([-\pi, \pi]; \ell^2(\mathbb{Z}^d))$ is an eigenvector of the spectral problem (43). Then, system (43) can be formulated in the Hamiltonian form $JH''(U)\mathbf{f} = \lambda \mathbf{f}$, where $\mathbf{f} = (W, Q)$, $J^* = -J = J^{-1}$, and $H''(U)$ is self-adjoint in $L^2_{\text{per}}([-\pi, \pi]; \ell^2(\mathbb{Z}^d))$. By using (61), we obtain that

$$\lambda K(W, \lambda) = i\langle H''(U)\mathbf{f}, \mathbf{f}\rangle = i\langle \mathbf{f}, H''(U)\mathbf{f}\rangle = -\bar{\lambda} K(W, \lambda),$$

so that if $\text{Re}(\lambda) \neq 0$ then $K(W, \lambda) = 0$. If $\lambda \in i\mathbb{R}\backslash\{0\}$ is a simple isolated eigenvalue, then we claim that $K(W, \lambda) \neq 0$. Indeed, if we assume $K(W, \lambda) = 0$, then there exists a generalized eigenvector from solution of the nonhomogeneous equation

$$JH''(U)\mathbf{g} = \lambda \mathbf{g} + \mathbf{f},$$

since the condition of the Fredholm alternative theorem is satisfied:

$$\langle J^{-1}\mathbf{f}, \mathbf{f}\rangle = \lambda^{-1}\langle H''(U)\mathbf{f}, \mathbf{f}\rangle = -iK(W, \lambda) = 0.$$

Therefore, λ is at least a double eigenvalue in contradiction with the assumption that λ is simple. Therefore, $K(W, \lambda) \neq 0$. □

Equipped with Lemma 1, we can now prove an analogue of Theorem 2 about persistence of simple isolated eigenvalues on $i\mathbb{R}$.

Theorem 3 *Under the assumption of Theorem 1, assume that $\Lambda \in i\mathbb{R}\backslash\{0\}$ is a simple isolated eigenvalue of the spectral problem (49) such that $2\Lambda \notin \sigma_c(\mathcal{J}\mathcal{H}''(\mathcal{A}))$ and $(b_+, b_-) \subset \ell^2(\mathbb{Z}^d) \times \ell^2(\mathbb{Z}^d)$. There exists $\epsilon_0 > 0$ and $C_0 > 0$ such that the spectral problem (45) for every $\epsilon \in (0, \epsilon_0)$ admits a unique C^ω branch of the eigenvalue–eigenvector pair with $\lambda(\epsilon) \in i\mathbb{R}$ and $\{B^{(m)}(\epsilon)\}_{m\in\mathbb{N}} \in \ell^{2,2}(\mathbb{Z}; \ell^2(\mathbb{Z}^d))$ satisfying (51), (52), and (53).*

Proof By Remark 5, we only need to prove that $\lambda(\epsilon) = \epsilon\Lambda + O(\epsilon^2)$ remains on $i\mathbb{R}$. By smoothness of the branch of eigenvalue-eigenvectors in ϵ, we can compute the limit $\epsilon \to 0$ for the Krein quantity $K(W, \lambda)$ in (61). We obtain

$$\lim_{\epsilon\to 0} K(W, \lambda) = 2\|B^{(1)}\|^2_{\ell^2(\mathbb{Z}^d)} - 2\|B^{(-1)}\|^2_{\ell^2(\mathbb{Z}^d)} = 4i\langle b_-, b_+\rangle_{\ell^2(\mathbb{Z}^d)} - 4i\langle b_+, b_-\rangle_{\ell^2(\mathbb{Z}^d)},$$

which is the Krein quantity for the spectral problem (49). Since $\Lambda \in i\mathbb{R}\backslash\{0\}$ is simple and isolated, the Krein quantity for the spectral problem (49) enjoys the same properties as in Lemma 1. In particular, it is real and nonzero. By continuity in ϵ, $K(W, \lambda)$ is nonzero for every $\epsilon \in (0, \epsilon_0)$, so that by Lemma 1, the eigenvalue $\lambda(\epsilon) = \epsilon\Lambda + O(\epsilon^2)$ of the spectral problem (43) satisfies $\text{Re}(\lambda) = 0$. □

Remark 6 Assume that $\sigma_d(\mathcal{J}\mathcal{H}''(\mathcal{A}))$ contains no eigenvalues 2Λ with $\text{Re}(\Lambda) \neq 0$, hence the dNLS solitons are spectrally stable. Theorem 3 implies that the spectral stability of dNLS solitons is transferred to the spectral stability of dKG breathers if bifurcations of new isolated eigenvalues from the continuous spectrum in (50) do not result in the appearance of new eigenvalues with $\mathrm{Re}(\lambda) \neq 0$ in the spectral problem (49). Such arguments generally follow from the Krein theory [22]. In the anti-continuum limit of the dNLS equation (7), one can find precise conditions excluding bifurcations of new isolated eigenvalues from the continuous spectrum of the spectral problem (49) [23].

4 Long-Time Nonlinear Stability via Resonant Normal Forms

The resonant normal form we consider here is based on the scheme already illustrated in [4, 6], which is suitable for infinite dimensional Hamiltonian systems and can be implemented by working at the level of either the Hamiltonian fields (as we decide to do, following [4]) or the Hamiltonian function (as in [6]).

In what follows we first present a result according to which the Hamiltonian of our problem can be put into a resonant normal form up to an exponentially small remainder. The truncated normal form represents a generalized dNLS equation in the same spirit as in [21]. We then give a theorem about the existence of a breather for the dKG equation, exponentially close to discrete soliton of the normal form; we stress here that such an estimate is a significant improvement with respect to the one obtained in [20] where the two objects were proven to be only order one close in the small parameter. As a last step, under additional hypothesis that the dNLS soliton is a minimizer in the variational problem (12), we state a stability result for the discrete breathers on an exponentially long time scale. The proofs of the above mentioned results are illustrated respectively in Sects. 4.3–4.5. Stability over longer, even infinite, time scales cannot be excluded a priori, but may require different ingredients than the present normal form construction. Indeed, although a standard control of the remainder does not generally prevent the orbit to leave a small neighbourhood of the breather beyond the time scale here considered, the stability could in principle persist even if we are not able to ensure it.

4.1 Setting, Preliminaries and Normal Form Result

We consider the Hamiltonian corresponding to the scaled model (6)

$$H = \frac{1}{2}\sum_{j\in\mathbb{Z}^d}\left(u_j^2 + v_j^2\right) + \frac{\epsilon}{2p+2}\sum_{j\in\mathbb{Z}^d} u_j^{2p+2} + \frac{\epsilon}{2}\sum_{j\in\mathbb{Z}^d}\sum_{|j-h|=1}\left(u_j - u_h\right)^2, \tag{62}$$

where $v_j = \dot{u}_j$. The Hamiltonian (62) can be obtained scaling both the variables $(u_n, \dot{u}_n)$ according to (5), and the original Hamiltonian original energy (2) by $\epsilon^{-\frac{1}{p}}$.

In the following, (62) will be considered as a nearly integrable Hamiltonian system

$$H = G + F\,, \qquad G := \frac{1}{2}\sum_{j\in\mathbb{Z}^d}\left(u_j^2 + v_j^2\right), \qquad F := H - G = O(\epsilon)\,, \tag{63}$$

where G is an integrable Hamiltonian and F is a perturbation of order $O(\epsilon)$.

We need some notations (we refer to Section 5 of [4] for further details). We consider $z := (u, v)$ in the complexified phase space $\mathcal{P} = \ell^2(\mathbb{C}) \times \ell^2(\mathbb{C})$ with the usual ℓ^2 norm, which makes it Hilbert with the usual inner product. Given $0 < R < 1$ and $0 < \mathfrak{d} \leqslant \frac{1}{4}$, we restrict to a ball around the origin $B_{R,\mathfrak{d}} := \{z \in \mathcal{P} \;\; s.t. \;\; \|z\| < R(1 - \mathfrak{d})\}$. To deal with complex valued functions g and Hamiltonian vector fields X_g on such a generic ball, we make use of the supremum norm

$$N_{\mathfrak{d}}(g) := \sup_{z\in B_{R,\mathfrak{d}}} |g(z)|\,, \qquad N_{\mathfrak{d}}^{\nabla}(g) := \frac{1}{R}\sup_{z\in B_{R,\mathfrak{d}}} \left\|X_g(z)\right\|\,. \tag{64}$$

Our aim is to construct a normal form K admitting a second conserved quantity G

$$H = K + \mathcal{P}\,, \qquad \{K, G\} = 0\,;$$

this additional conserved quantity, which corresponds to the ℓ^2 norm, is related to the invariance under the rotation symmetry, given by the periodic flow Φ_G^t of the Hamiltonian field X_G. The normal form K is thus a generalized dNLS model (see also [21]); given the smallness of $\mathcal{P}$, G turns out to be an approximated conserved quantity for H, whose variation can be kept bounded on exponentially long times.

Theorem 4 *For any positive $\mathfrak{d} \leqslant 1/4$, any dimension $d \geqslant 1$ and any $R < 1$, there exists $\epsilon^*(\mathfrak{d}, d, R)$ such that, for $\epsilon < \epsilon^*$ there exists a canonical change of coordinates $T_{\mathcal{X}}$ mapping*

$$B_{R,2\mathfrak{d}} \subset T_{\mathcal{X}}\big(B_{R,\mathfrak{d}}\big) \subset B_{R,0} \qquad B_{R,3\mathfrak{d}} \subset T_{\mathcal{X}}\big(B_{R,2\mathfrak{d}}\big) \subset B_{R,\mathfrak{d}} \tag{65}$$

which puts the Hamiltonian (62) *into the resonant normal form*

$$H = G + Z + \mathcal{P}\,, \qquad \{G, Z\} = 0\,, \qquad N_{\mathfrak{d}}^{\nabla}(\mathcal{P}) \leqslant \mu \exp\left(-\frac{1}{\mu}\right), \tag{66}$$

where $\mu := \frac{12e\pi\epsilon}{\mathfrak{d}} = O(\epsilon)$. Moreover, for any initial datum $z_0 \in B_{R,3\mathfrak{d}}$, there exists a positive constant C such that the variations of G and Z are bounded as follows

$$\begin{aligned} |G(z(t)) - G(z_0)| &< C\mu N_0(G)\,, \\ |Z(z(t)) - Z(z_0)| &< C\mu N_0(F)\,, \end{aligned} \qquad |t| \leqslant T^* := \exp\left(\frac{1}{\mu}\right). \tag{67}$$

The construction is based on the linear operator $T_{\mathcal{X}}$ associated to a generating sequence $\{\mathcal{X}_s\}_{s=1}^r$, where $\mathcal{X}_s = \mathcal{O}(\epsilon^s)$, which acts recursively on G and F as follows

$$T_{\mathcal{X}} G = \sum_{r\geqslant 0} G_r\,, \qquad G_0 := G\,, \qquad G_r := \sum_{l=1}^{r} \frac{l}{r}\{\mathcal{X}_l, G_{r-l}\}\,,$$

$$T_{\mathcal{X}} F = \sum_{r\geqslant 0} F_r\,, \qquad F_0 := 0\,, \quad F_1 := F\,, \qquad F_r := \sum_{l=1}^{r-1} \frac{l}{r-1}\{\mathcal{X}_l, F_{r-l}\}\,. \tag{68}$$

In the above recursive definition, it coherently turns out that $F_r = \mathcal{O}(\epsilon^r)$. Such a linear operator also provides the close-to-the-identity nonlinear transformation

$$T_{\mathcal{X}} z = z + \sum_{r\geqslant 1} z_r\,, \qquad z_r = \sum_{l=1}^{r} \frac{l}{r}\{\mathcal{X}_l, z\}_{r-l}\,. \tag{69}$$

The generating sequence $\mathcal{X}$, and the corresponding transformation $T_{\mathcal{X}}$, will be determined in order to put the Hamiltonian in resonant normal form up to order $\mathcal{O}(\epsilon^r)$

$$H^{(r)} = T_{\mathcal{X}} H = G + Z + \mathcal{R}^{(r+1)}\,, \qquad \{G, Z\} = 0\,, \qquad \mathcal{R}^{(r+1)} = \mathcal{O}(\epsilon^{r+1})\,. \tag{70}$$

Thus $\mathcal{X} = \{\mathcal{X}_s\}$ and the normal form terms $Z = \sum_{s=1}^r Z_s$ have to satisfy

$$\{G, \mathcal{X}_s\} + Z_s = \Psi_s\,, \qquad 1 \leqslant s \leqslant r\,, \tag{71}$$

where $\mathcal{X}_s$, Z_s and Ψ_s are all homogeneous terms of order ϵ^s, with

$$\Psi_1 = F_1 = F\,, \qquad \Psi_s := \frac{1}{s} F_s + \sum_{l=1}^{s-1} \frac{l}{s}\{\mathcal{X}_l, Z_{s-l}\}\,. \tag{72}$$

At first order $r = 1$, we obtain again Eq. (21) as leading order approximation of the dKG breather. Indeed we have to put into normal form the initial perturbation $\Psi_1 := F_1$. The first normal form term Z_1 represents its average, and it turns out that at first order the Hamiltonian $K^{(1)}$ can be given by the corresponding dNLS model

$$K^{(1)} = \sum_j |\psi_j|^2 + \frac{\epsilon}{p+1} \sum_j |\psi_j|^{2p+2} + \epsilon \sum_{|j-h|=1} |\psi_j - \psi_h|^2\,, \tag{73}$$

once complex coordinates are introduced

$$u_j = \psi_j + i\overline{\psi_j} = \frac{1}{\sqrt{2}}(\zeta_j + i\eta_j)\,, \quad \Rightarrow \quad \psi_j = \zeta_j/\sqrt{2}\,, \qquad i\eta_j = \overline{\zeta_j}\,, \tag{74}$$

so that the quadratic part of $K^{(1)}$ reads $\sum_j |\zeta_j|^2 + \epsilon \sum_{|j-h|=1} |\zeta_j - \zeta_h|^2$. To average the nonlinearity one follows the same calculations already used in the Remark 2

$$\frac{1}{2\pi}\int_0^{2\pi} u_j^{2p+2} \circ \Phi_g^t dt = \frac{1}{2\pi}\int_0^{2\pi} \frac{1}{2^{p+1}}\left(\zeta_j e^{it} + i\eta_j e^{-it}\right)^{2p+2} dt = \Gamma_p |\zeta_j|^{2p+2}\,,$$

with $\Gamma_p := \frac{1}{2^p}\gamma_p$; thus the nonlinear term reads $\frac{\Gamma_p}{2(p+1)} \sum_j |\zeta_j|^{2p+2}$, and its standard shape is recovered introducing the complex variable (74) which allows to rescale the prefactor 2^{-p}. Discrete solitons of (73) with frequency close to one

$$\psi = \mathcal{A} e^{i(1-\frac{\epsilon}{2}\Omega)t} \tag{75}$$

are then extremizer of $\mathcal{Z}_1 := \epsilon^{-1} Z_1$ constrained to constant values of the norm $G = \nu$; this again provides (21) with $\Omega = \Omega(\nu)$.

4.2 High Order Approximation and Nonlinear Stability Results

Let us consider $K := H - \mathcal{P}$ in (66) and its equations

$$\dot{z} = X_K(z)\,, \qquad K = G + Z\,, \qquad \{Z, G\} = 0\,. \tag{76}$$

To generalize the discrete soliton approximation, we rewrite the ansatz (75) as

$$\zeta_{\rm ds} = \mathfrak{A} e^{i(1-\frac{1}{2}\epsilon\Omega)t} \tag{77}$$

where $\mathfrak{A}$ is the real amplitude of the soliton,[2] which is assumed to be small enough to belong to the domain of validity of the normal form (66); once inserted in (76), it provides the equation for $\mathfrak{A}$

$$f := f_0 + \epsilon f_1 = 0\,, \qquad \begin{cases} f_0 & := \Omega\mathfrak{A} + \gamma_p |\mathfrak{A}|^{2p}\mathfrak{A} - \Delta\mathfrak{A}\,, \\ f_1 & := \epsilon^{-2} X_Z(\mathfrak{A})\,; \end{cases} \tag{78}$$

[2] Notice the use of the gothic font instead of the calligraphic one to distinguish between the objects of the generalized dNLS—given by the higher order normal form—to those of the standard dNLS

where f_0 gives the standard dNLS equation (21), while f_1 is the perturbation due to the normal form steps $r \geqslant 2$; we recall that, due to $\{G, Z\} = 0$, X_Z is equivariant under the action of the symmetry $e^{i\theta}$

$$X_Z\left(\mathfrak{A}e^{i(1-\frac{1}{2}\epsilon\Omega)t}\right) = X_Z(\mathfrak{A})e^{i(1-\frac{1}{2}\epsilon\Omega)t} .$$

The next statement represents the higher order version of Theorem 1, under the same assumption on $\mathcal{A}$ and J_Ω: it claims the existence of a Klein–Gordon breather for each discrete soliton of the normal form K.

Theorem 5 *Let $\mathcal{A}$ be a solution of* (21) *with J_Ω of* (22) *invertible in $\ell^2(\mathbb{Z}^d, \mathbb{R})$. Then:*

1. *there exists $\epsilon_1^* < \epsilon^*$ such that for any $0 < \epsilon < \epsilon_1^*$ there exists a unique solution $\mathfrak{A}(\Omega, \epsilon)$ of* (78)*, analytic in ϵ. Moreover, the following estimates hold true*

$$\|\mathfrak{A} - \mathcal{A}\|_{\ell^2} \leqslant C\epsilon , \qquad \sup \frac{\|(J_{\Omega,\epsilon} - J_\Omega)(z)\|_{\ell^2(\mathbb{Z}^d;\mathbb{R})}}{\|z\|_{\ell^2(\mathbb{Z}^d;\mathbb{R})}} \leqslant C\epsilon , \tag{79}$$

where $J_{\Omega,\epsilon} := D_{\mathfrak{A}} f(\mathfrak{A}(\Omega, \epsilon), \Omega, \epsilon)$ is the differential of f evaluated at $\mathfrak{A}(\Omega, \epsilon)$.

2. *Let*

$$\zeta_{\rm br}(\tau) = \sum_m \mathfrak{A}^{(m)} e^{im\tau} , \qquad \tau := \omega t , \tag{80}$$

be the Fourier expansion of the breather of $\dot{z} = X_H(z)$. Then, there exists positive $\epsilon_2^* < \epsilon_1^*$ such that for every $0 < \epsilon < \epsilon_2^*$ the breather (80) *admits a unique analytic solution branch $\omega(\epsilon)$ and $\{\mathfrak{A}^{(m)}(\epsilon)\} \in \ell^{2,2}(\mathbb{Z}; \ell^2(\mathbb{Z}^d))$ satisfying the bounds*

$$|\omega(\epsilon) - 1 + \frac{\epsilon}{2}\Omega| \leqslant C\epsilon^2 , \tag{81}$$

$$\|\mathfrak{A}^{(0)}\|_{\ell^2(\mathbb{Z}^d)} + \|\mathfrak{A}^{(1)} - \mathfrak{A}\|_{\ell^2(\mathbb{Z}^d)} + \sum_{m\geqslant 2} \|\mathfrak{A}^{(m)}\|_{\ell^2(\mathbb{Z}^d)} \leqslant C \exp\left(-\frac{c}{\epsilon}\right) . \tag{82}$$

3. *Let $z_{\rm ds}(t) = T_{\mathcal{X}}^{-1}(\zeta_{\rm ds})(t)$ and $z_{\rm br}(t) = T_{\mathcal{X}}^{-1}(\zeta_{\rm br})(t)$ be the discrete soliton and the discrete breather solutions in the original coordinates, and $T_{\rm ds}$ and $T_{\rm br}$ the corresponding periods; then it holds true*

$$\sup_{|t|\leqslant \max\{T_{\rm ds}, T_{\rm br}\}} \|z_{\rm ds}(t) - z_{\rm br}(t)\| \leqslant C \exp\left(-\frac{c}{\epsilon}\right) . \tag{83}$$

We now assume a stronger condition than the invertibility of the Jacobian operator J_Ω; we require $\mathcal{A}$ to be a nondegenerate extremizer for $\mathcal{Z}_1$ constrained to constant values of the norm G. Under this assumption, which implies invertibility of J_Ω, it follows that for ϵ sufficiently small also the discrete soliton $\mathfrak{A}$ obtained in Proposition 4.1 is a nondegenerate extremizer for $\mathcal{Z} := \epsilon^{-1}Z$ constrained to the sphere $\mathcal{S} := \{G(z) = \nu\}$, with ν sufficiently small (as required by the normal form construction). As a consequence, $\mathfrak{A}$ is an orbitally stable periodic orbits (see [2, 20, 27]) for the normal form $K = G + Z$. Once we add the remainder P to K we cannot guarantee anymore that such a geometry is preserved, since P is not known to Poisson commute with the ℓ^2 norm G. However, the smallness of P in the norm (64) ensures that the orbit remains in a prescribed region for long times. We are going to show that $\mathfrak{A}$ is an approximate periodic orbit for the full system $H = G + Z + P$ which is orbitally stable for exponentially long times and that the same kind of stability holds true for the Klein–Gordon breathers.

Let us introduce with $\bar{\mathfrak{A}} := \{\mathfrak{A}^{(m)}\}$ and denote with $\mathsf{O}(\bar{\mathfrak{A}})$ the closed curve described by the Klein–Gordon breather

$$\mathsf{O}(\bar{\mathfrak{A}}) := \{\zeta_{\rm br}(t),\ t \in [0, T]\} \qquad\qquad \mathsf{O} := T_{\mathcal{X}}^{-1}\mathsf{O}(\bar{\mathfrak{A}})\,.$$

The next Theorem provides the orbital stability of $\mathsf{O}(\bar{\mathfrak{A}})$.

Theorem 6 *Assume $\mathcal{A}$ to be a nondegenerate extremizer for $\mathcal{Z}_1$ constrained to constant values of the norm G. Let $z_0 \in B_{R,3\mathfrak{d}}$ with $R < 1$. Then $\forall\, 0 < \mu \ll 1$, $\exists\, 0 < \delta \ll 1$ such that*

$$\inf_{w\in\mathcal{O}} \|z_0 - w\| < \delta \quad \Rightarrow \quad \inf_{w\in\mathcal{O}} \left\|\Phi_H^t(z_0) - w\right\| < \mu\,, \quad |t| < \exp\left(\frac{c}{\epsilon}\right). \tag{84}$$

Remark 7 In Theorems 5 and 6, c and C are suitable constants independent of ϵ.

4.3 *Proof of Theorem 4 (Normal Form Theorem)*

We give a sketch of the proof, which would be long and technical if all the details were included. The estimates here included can be obtained by following [4, 6].

Recursive estimates, which are the most technical aspect of the whole construction, need estimates on the initial size of the perturbation F and its vector field X_F. We thus introduce the main quantities E and ω_1 providing the initial estimates

$$N_0(F) \leqslant E := \epsilon\left[4dR^2 + \frac{1}{2p+2}R^{2p+2}\right], \qquad\qquad N_0^{\nabla}(F) \leqslant \omega_1 := 2\epsilon\left[C_d + R^{2p}\right].$$

Remark 8 The magnitudes of E and ω_1 are coherent: since $E = O(\epsilon R^2)$, then its differential, divided by R according to the definition of $N_{\mathfrak{d}}^{\nabla}(\cdot)$ in (64), has to be $O(\epsilon)$.

In order to solve (71) we average along the periodic flow Φ_G^t of period 2π, as claimed by the following Lemma (for the proof, see [2]):

Lemma 2 *The homological equation* (71), *i.e.* $\{G, \mathcal{X}\} = \Psi(z) - Z(z)$, *is solved by*

$$Z(z) = \frac{1}{2\pi}\int_0^{2\pi} \Psi \circ \Phi_G^t(z)dt\ , \qquad \mathcal{X}(z) = \frac{1}{2\pi}\int_0^{2\pi} t\big[(\Psi - Z)\circ \Phi_G^t\big](z)dt\ ;$$

for any $\eth$ *it satisfies the following estimates*

$$\begin{aligned} N_\eth(Z) &\leqslant N_\eth(\Psi)\ , & N_\eth^\nabla(Z) &\leqslant N_\eth^\nabla(\Psi)\ , \\ N_\eth(\mathcal{X}) &\leqslant 2\pi N_\eth(\Psi)\ , & N_\eth^\nabla(\mathcal{X}) &\leqslant 2\pi N_\eth^\nabla(\Psi)\ . \end{aligned} \tag{85}$$

At first order, Lemma 2 immediately provides the estimates

$$N_\eth^\nabla(Z_1) \leqslant \omega_1\ , \qquad N_\eth^\nabla(\mathcal{X}_1) \leqslant \phi := 2\pi\omega_1\ , \tag{86}$$

which introduce the main perturbation parameter $\phi = O(\epsilon)$ of the normal form scheme. Let the arbitrary integer $r \geqslant 1$ be the order of the normal form construction, i.e. the number of generating functions $\mathcal{X}_s$ in the generating sequence $\mathcal{X} = \{\mathcal{X}_s\}_{s=1}^r$ and thus the number of homological equations (71) to be solved. The first important result gives the bounds for the quantities involved in (71)

Lemma 3 *Let* $\eth_s = \frac{s\eth}{r}$, *with* $\eth \leqslant \frac{1}{4}$. *Then, for any* $1 \leqslant s \leqslant r$ *it holds true*

$$N_{\eth_{s-1}}^\nabla(\Theta) \leqslant \frac{\omega_1}{s}\left(\frac{6r\phi}{\eth}\right)^{s-1} \quad \forall\Theta \in \{F_s, \Psi_s, Z_s\}\ , \quad N_{\eth_{s-1}}^\nabla(\mathcal{X}_s) \leqslant \frac{\phi}{s}\left(\frac{6r\phi}{\eth}\right)^{s-1}. \tag{87}$$

The above Lemma, and in particular the last of (87), allows to control the deformation of functions and vector fields under the canonical transformation; indeed, let

$$\begin{aligned} T_{\mathcal{X}} f &= \sum_{r\geqslant 0} f_r\ , & f_r &:= \sum_{j=1}^{r} \frac{j}{r} L_{\mathcal{X}_j} f_{r-j}\ , & f_0 &= f\ , \\ T_{\mathcal{X}} g &= \sum_{r\geqslant 1} g_r\ , & g_r &:= \sum_{j=1}^{r-1} \frac{j}{r-1} L_{\mathcal{X}_j} g_{r-j}\ , & g_1 &= g\ , \end{aligned} \tag{88}$$

then the following holds true

Lemma 4 *Let us introduce* $M_1 < M_2 < 1$

$$M_1(\phi, r) := \frac{\phi(e+3r)}{\eth}\,, \qquad M_2(\phi, r) := \frac{\phi(2e+3r)}{\eth}\,. \tag{89}$$

Then, for any $\eth \leqslant \frac{1}{4}$ *and any* $r \geqslant 1$ *the following bounds hold true*

$$\begin{aligned} &N_\eth(z_r) \leqslant M_1^{r-1}\tilde{G}_1\,, \qquad &&N_\eth(f_r) \leqslant M_1^{r-1}\tilde{B}_1\,, \qquad &&N_\eth(g_r) \leqslant M_1^{r-2}\tilde{\Gamma}_2\,,\\ & &&N_\eth^\nabla(f_r) \leqslant M_2^{r-1}\bar{B}_1\,, &&N_\eth^\nabla(g_r) \leqslant M_1^{r-2}\bar{\Gamma}_2\,, \end{aligned} \tag{90}$$

together with

$$\tilde{G}_1 := \frac{R\phi}{\eth}, \quad \tilde{B}_1 := \frac{\phi}{\eth}N_0(f), \quad \tilde{\Gamma}_2 := \frac{\phi}{\eth}N_0(g), \quad \bar{B}_1 := \frac{2\phi}{\eth}N_0^\nabla(f), \quad \bar{\Gamma}_2 := \frac{2\phi}{\eth}N_0^\nabla(g),$$

Given the above estimates, we can obtain the inclusions (65); indeed, we have

$$N_\eth(T_\chi z - z) \leqslant \sum_{r\geqslant 1} N_\eth(z_r) \leqslant \frac{\tilde{G}_1}{1-M_1} < 2\tilde{G}_1 = \phi\left(\frac{2R}{\eth}\right),$$

provided we ask for $M_1 < \frac{1}{2}$; thus the deformation is $O(R\epsilon)$. The remainder $\mathcal{R}^{(r+1)}$ in (70), at an arbitrary step r, is given by

$$\mathcal{R}^{(r+1)} = \sum_{s\geqslant r+1} G_s + \sum_{s\geqslant r+1} F_s\,;$$

by exploiting (90) and the initial estimates, the following bounds hold true

$$N_\eth^\nabla(\Theta) \leqslant \left(\frac{2\phi}{\eth}\right)M_2^{s-1}, \ \forall \Theta \in \{G_s, F_s\} \quad \Rightarrow \quad N_\eth^\nabla(\mathcal{R}^{(r+1)}) \leqslant \left(\frac{2\phi}{\eth}\right)\frac{M_2^r}{1-M_2}\,. \tag{91}$$

The exponential estimate (66) is derived from (91) by expanding M_2^r as

$$M_2^r = \left(\frac{6\phi}{\eth}\right)^r r^r \left(1+\frac{1}{r}\right)^r < e\left(\frac{6\phi}{\eth}\right)^r r^r\,, \tag{92}$$

and optimizing the number of normal form steps $r = r_{opt} := \lfloor \frac{\eth}{6e\phi} \rfloor$; in this way r is related ϵ. Finally, the variation of G along the generic orbit in the neighbourhood of the origin is obtained combining the estimate of the Poisson bracket

$$|G(t) - G(0)| \leqslant \int_0^t |\{G, \mathcal{R}^{(r+1)}\}(z(s))|ds \leqslant |t|N_\eth(G)\left(\frac{4e\phi}{\eth^2}\right)\exp\left[-\left(\frac{\eth}{6e\phi}\right)\right],$$

with the bound on the deformation $|G(z) - G(T_\chi(z))| = |(T_\chi G - G)(z)|$ and exploiting the fact that G coincides with the norm on the phase space. Similarly one can control the variation of Z.

4.4 Proof of Theorem 5 (High Order Approximation)

The proof of point *(1)* is an easy application of the Implicit Function Theorem (I.F.T.), based on the main assumption that J_Ω is invertible in the subspace of real square-summable sequences. Indeed $\mathfrak{A} = \mathcal{A}$ is the "unperturbed" solution of Eq. (21) $f(\mathcal{A}, \Omega, 0) = f_0(\mathcal{A}, \Omega) = 0$, and the Jacobian of f evaluated at $(\mathcal{A}, \epsilon = 0)$ is given by $D_{\mathfrak{A}} f(\mathcal{A}, \Omega, 0) = J_\Omega$, which is invertible. The first of (79) is standard, while the second can be easily obtained from

$$D_{\mathfrak{A}} f(\mathfrak{A}) = D_{\mathfrak{A}} f_0(\mathfrak{A}) + O(\epsilon) = D_{\mathfrak{A}} f_0(\mathcal{A}) + D^2_{\mathfrak{A}} f_0(\mathcal{A})(\mathfrak{A} - \mathcal{A}) + O(\epsilon) = J_\Omega + O(\epsilon) \, .$$

Once proved the existence of $\mathfrak{A}$, discrete soliton of K, we follow the same strategy used for Theorem (1) in order to prove the existence of an analytic branch of Klein–Gordon breathers[3] $\bar{\mathfrak{A}}$ exponentially close to $\mathfrak{A}(\Omega, \epsilon)$. Let us consider the Hamilton equation for (70) restricting to the variable ζ only (recall $i\eta = -\bar{\zeta}$)

$$\omega \partial_\tau \zeta = i\zeta + \partial_\eta Z + \partial_\eta \mathcal{R}^{(r+1)} \, ;$$

by inserting the Fourier expansion (80) with real coefficients $\mathfrak{A}^{(m)}$, we get the equation

$$\hat{L}_\omega \bar{\mathfrak{A}} + \partial_\eta Z(\bar{\mathfrak{A}}) + \partial_\eta \mathcal{R}^{(r+1)}(\bar{\mathfrak{A}}) = 0 \, , \qquad \left(\hat{L}_\omega \bar{\mathfrak{A}}\right)^{(m)} := i(1 - m\omega)\mathfrak{A}^{(m)} \, ,$$

since the Hamilton equations are first order in time. The Kernel $\hat{V}_2$ of the linear operator $\hat{L}_{\omega=1}$ is given only by $e^{i\tau}$, all the other harmonics belonging to the Range $\hat{W}_2$. We decompose $\bar{\mathfrak{A}} = \bar{\mathfrak{A}}^\sharp + \bar{\mathfrak{A}}^\flat$ and project the equations on $\hat{V}_0$ and $\hat{W}_0$. The great difference with respect to the proof of Theorem 1, is that the resonant normal form construction, performed averaging with respect to the periodic flow $e^{i\tau}$, provides a natural decomposition of the term $\partial_\eta Z(\bar{\mathfrak{A}}^\sharp)$, so that one has for free

$$\Pi_{\hat{W}_0} \partial_\eta Z(\bar{\mathfrak{A}}^\sharp) = 0 \, , \qquad \Pi_{\hat{V}_0} \partial_\eta Z(\bar{\mathfrak{A}}^\sharp) = \partial_\eta Z(\bar{\mathfrak{A}}^\sharp) \, . \tag{93}$$

This remarkable property allows to write the Range equation as

[3]We again "identify" solitons and breathers with their amplitudes.

$$\hat{L}_\omega \bar{\mathfrak{A}}^\flat + \left[\Pi_{\hat{W}_0} \partial_\eta Z(\bar{\mathfrak{A}}^\flat + \bar{\mathfrak{A}}^\sharp) - \Pi_{\hat{W}_0} \partial_\eta Z(\bar{\mathfrak{A}}^\sharp)\right] + \Pi_{\hat{W}_0} \partial_\eta \mathcal{R}^{(r+1)}(\bar{\mathfrak{A}}^\flat + \bar{\mathfrak{A}}^\sharp) = 0\,,$$

where the term in square brackets is at least linear in $\bar{\mathfrak{A}}^\flat$, $O(\epsilon\bar{\mathfrak{A}}^\flat)$, hence a small perturbation with respect to $\hat{L}_{\omega=1}$. The usual leading order approximation provided by the I.F.T. gives, for some constant C independent of ϵ,

$$\bar{\mathfrak{A}}^\flat \approx -\left(\hat{L}_\omega\right)^{-1} \Pi_{\hat{W}_0} \partial_\eta \mathcal{R}^{(r+1)}(\bar{\mathfrak{A}}^\flat) \qquad \Rightarrow \qquad \left\|\bar{\mathfrak{A}}^\flat\right\| \leqslant C\epsilon^{r+1}\,. \tag{94}$$

The Kernel equation, after dividing by ϵ, takes the form

$$\Omega\bar{\mathfrak{A}}^\sharp - \Delta\bar{\mathfrak{A}}^\sharp + \Gamma_p|\bar{\mathfrak{A}}^\sharp|^{2p}\bar{\mathfrak{A}}^\sharp + \sum_{s=2}^{r} \epsilon^{s-1}\Pi_{\hat{V}_0}\partial_\eta \mathcal{Z}_s(\bar{\mathfrak{A}}^\sharp)) + O(\epsilon^r) = 0\,,$$

where we have introduced the scaled functions $\mathcal{Z}_s := \epsilon^{-s} Z_s$. Notice that in the small remainder we have included not only the smallness of the vector field $\partial_\eta \mathcal{R}^{(r+1)}$, but also

$$\Pi_{\hat{V}_0}\partial_\eta \mathcal{Z}_s(\bar{\mathfrak{A}}^\flat + \bar{\mathfrak{A}}^\sharp) - \Pi_{\hat{V}_0}\partial_\eta \mathcal{Z}_s(\bar{\mathfrak{A}}^\sharp) = \Pi_{\hat{V}_0}\partial_\eta \mathcal{Z}_s(\bar{\mathfrak{A}}^\flat + \bar{\mathfrak{A}}^\sharp) - \partial_\eta \mathcal{Z}_s(\bar{\mathfrak{A}}^\sharp) = O(\bar{\mathfrak{A}}^\flat)\,,$$

which is of order $O\big(\epsilon^{r+1}\big)$ for any $1 \leqslant s \leqslant r$, because of (94). The Kernel equation now turns out to be a perturbation of order $O(\epsilon^r)$ of

$$f(\bar{\mathfrak{A}}^\sharp, \Omega, \epsilon) := \Omega\bar{\mathfrak{A}}^\sharp - \Delta\bar{\mathfrak{A}}^\sharp + \Gamma_p|\bar{\mathfrak{A}}^\sharp|^{2p}\bar{\mathfrak{A}}^\sharp + \sum_{s=2}^{r} \epsilon^{s-1}\Pi_{\hat{V}_0}\partial_\eta \mathcal{Z}_s(\bar{\mathfrak{A}}^\sharp)) = 0\,, \tag{95}$$

since the last term in the sum of $\mathcal{Z}_s$ is of order $O(\epsilon^{r-1})$. Equation (95) admits the solution $\bar{\mathfrak{A}} = \bar{\mathfrak{A}}(\Omega, \epsilon)$ given by Proposition 4.1., and since $J_{\Omega,\epsilon}$ is invertible for ϵ sufficiently small, a fixed point argument (see for example Appendix of [5]) can be used to conclude the proof; estimate (81) is standard, while estimate (82) follows once the generic step r is replaced with the optimal choice $r = r_{opt}(\epsilon)$.

In order to prove point *(3)* we exploit the fact that the two periods, T_{ds} and T_{br}, are both approximately equal to 2π, because of (81). Hence

$$\|\zeta_{\mathrm{ds}}(t) - \zeta_{\mathrm{br}}(t)\| \leqslant \sum_{m\neq 1} \left\|\mathfrak{A}^{(m)}\right\| + \left\|\mathfrak{A}^{(1)}e^{i\omega t} - \mathfrak{A}e^{i(1-\frac{\epsilon}{2}\Omega)t}\right\|\,;$$

moreover, on a time interval of order $O(1) = \max\{T_{\mathrm{ds}}, T_{\mathrm{br}}\}$, we have

$$\left\|\mathfrak{A}^{(1)}e^{i\omega t} - \mathfrak{A}e^{i(1-\frac{\epsilon}{2}\Omega)t}\right\| \leqslant \left\|\mathfrak{A}^{(1)}e^{O(\epsilon^2)t} - \mathfrak{A}\right\| \leqslant C\left\|\mathfrak{A}^{(1)} - \mathfrak{A}\right\|\,.$$

The estimates holds true also in the original coordinates, exploiting the Lipschitz continuity of the canonical transformation $T_{\mathcal{X}}^{-1}$, with a Lipschitz constant $L = \mathcal{O}(1)$.

4.5 Proof of Theorem 6 (Exponentially Long Time Stability)

We collect the main geometrical ideas already exploited in [2, 20], omitting most of the details that the interested reader can find in the quoted papers.

We first have to prove the orbital stability of the discrete soliton $\mathfrak{A}$, interpreted as approximate breather solution of the Klein–Gordon model. We denote with $\mathsf{O}(\mathfrak{A})$ the closed orbit described by the discrete soliton profile $\mathfrak{A}$ during its periodic evolution

$$\mathsf{O}(\mathfrak{A}) := \{e^{i(1-\frac{\xi}{2}\Omega)t}\mathfrak{A},\ t \in [0, T_{\mathrm{ds}}]\} .$$

We consider a tubolar neighbourhood $\mathcal{W}_0$ of $\mathsf{O}(\mathfrak{A})$, in the transformed coordinates which give the original Hamiltonian the normal form (66). Any point $z \in \mathcal{W}_0$ can be represented with a local set of coordinates

$$z = (\varphi, E, v) \in \mathbb{R} \times \mathbb{R} \times V_\xi ,$$

where ξ is the projection of z on $\mathsf{O}(\mathfrak{A})$ and V_ξ is the orthogonal complement to the symmetry field $X_G(\xi)$ in the tangent space $T_\xi\mathcal{S} = V_\xi \oplus X_G(\xi)$. The three coordinates represent, respectively: the scalar coordinate φ is the tangential displacement along the field X_G, the scalar coordinate E is the displacement in the direction ∇G orthogonal to the surface $\mathcal{S}$ and the vector valued coordinate v is the tangential displacement in the directions of V_ξ. In order to measure the orbital distance of a generic point z from $\mathsf{O}(\mathfrak{A})$ we need to control only the directions transversal to the orbits, hence E and v. The main point is that E is related to the variation of G while v is related to the variation of $\mathcal{Z}$: the first is obvious, while the second holds because we have asked $\mathfrak{A}$ to be a nondegenerate extremizer of $\mathcal{Z}$ on $\mathcal{S}$, hence locally we have

$$\|v\|^2 = \|z - \xi\|^2 \leqslant \frac{1}{C}|\mathcal{Z}(z) - \mathcal{Z}(\xi)| ,$$

where C is a constant depending on $\mathcal{Z}''(\xi)$. Let us consider an initial datum $z_0 \in \mathcal{W}_0$ and its piece of orbit $\Phi_H^t(z_0) \cap \mathcal{W}_0$; for any point on this curve we have

$$\inf_{w\in\mathsf{O}(\mathfrak{A})\cap\mathcal{W}_0} \left\|\Phi_H^t(z_0) - w\right\| \leqslant c_1|E(t)| + c_2\,\|v(t)\| \leqslant$$

$$\leqslant C\sqrt{|G(\Phi_H^t(z_0)) - G(\mathfrak{A})| + |\mathcal{Z}(\Phi_H^t(z_0)) - \mathcal{Z}(\mathfrak{A})|} .$$

If z_0 is taken in a suitable domain where (67) hold true, then the two terms in the square root can be bounded by

$$|G(\Phi_H^t(z_0)) - G(\mathfrak{A})| \leqslant |G(\Phi_H^t(z_0)) - G(z_0)| + |G(z_0) - G(\mathfrak{A})| \tag{96}$$

$$|\mathcal{Z}(\Phi_H^t(z_0)) - \mathcal{Z}(\mathfrak{A})| \leqslant |\mathcal{Z}(\Phi_H^t(z_0)) - |\mathcal{Z}(z_0)| + |\mathcal{Z}(z_0) - \mathcal{Z}(\mathfrak{A})| \,; \tag{97}$$

the first right hand terms are exactly controlled by (67) on exponentially long times, while the second right hand terms are controlled by the initial distance from the orbit. This allows to get (84) for the discrete soliton $\mathfrak{A}$ in the normal form coordinates. The stability result then is transferred to the orbit $\mathsf{O}(\bar{\mathfrak{A}})$ of the Klein–Gordon breather exploiting the exponentially small distance between the two orbits, as stated by (83). Finally, the same stability result holds also in the original coordinates, since $T_{\mathcal{X}}^{-1}$ is Lipschitz with a Lipschitz constant $L = \mathcal{O}(1)$.

References

1. Ambrosetti, A., Prodi, G.: A primer of nonlinear analysis. Cambridge Studies in Advanced Mathematics, vol. 34. Cambridge University Press, Cambridge (1995)
2. Bambusi, D.:Exponential stability of breathers in Hamiltonian networks of weakly coupled oscillators. Nonlinearity **9**, 433–457 (1996)
3. Bambusi, D.: Asymptotic stability of breathers in some Hamiltonian networks of weakly coupled oscillators. Comm. Math. Phys. **324**, 515–547 (2013)
4. Bambusi, D., Giorgilli, A.: Exponential stability of states close to resonance in infinite-dimensional Hamiltonian systems. J. Statist. Phys. **71**, 569–606 (1993)
5. Bambusi, D., Paleari, S., Penati, T.: Existence and continuous approximation of small amplitude breathers in 1D and 2D Klein-Gordon lattices. Appl. Anal. **89**, 1313–1334 (2010)
6. Benettin, G., Galgani, L., Giorgilli, A.: Realization of holonomic constraints and freezing of high frequency degrees of freedom in the light of classical perturbation theory. II. Comm. Math. Phys. **121**, 557–601 (1989)
7. Cuccagna, S., Tarulli, M.: On asymptotic stability of standing waves of discrete Schrödinger equation in $\mathbb{Z}$. SIAM J. Math. Anal. **41**, 861–885 (2009)
8. Cuevas, J., Eilbeck, J.C., Karachalios, N.I.: Thresholds for breather solutions of the discrete nonlinear Schrödinger equation with saturable and power nonlinearity. Discr. Contin. Dynam. Syst. **21**, 445–475 (2008)
9. Cuevas, J., James, G., Kevrekidis, P.G., Malomed, B.A., Sánchez–Rey, B.: Approximation of solitons in the discrete NLS equation. J. Nonlin. Math. Phys. **15**, 124–136 (2008)
10. Cuevas, J., Karachalios, N.I., Palmero, F.: Lower and upper estimates on the excitation threshold for breathers in discrete nonlinear Schrödinger lattices. J. Math. Phys. **50**, 112705 (10p) (2009)
11. Cuevas, J., Kevrekidis, P.G., Franzeskakis, D.J., Malomed, B.A.: Discrete solitons in nonlinear Schrödinger lattices with a power-law nonlinearity. Physica D **238**, 67–76 (2009)
12. Cuevas, J., Karachalios, N.I., Palmero, F.: Energy thresholds for the existence of breather solutions and travelling waves on lattices. Appl. Anal. **89**, 1351–1385 (2010)
13. Cuevas-Maraver, J., Kevrekidis, P.G., Vainchtein, A., Xu, H.: Unifying perspective: Solitary traveling waves as discrete breathers in Hamiltonian lattices and energy criteria for their stability. Phys. Rev. E **96**, 032214 (9p) (2017)
14. Kapitula, T., Promislow, K.: Spectral and dynamical stability of nonlinear waves. Applied Mathematical Sciences, vol. 185. Springer, Berlin (2013)

15. Kato, T.: Perturbation Theory for Linear Operators. Springer, Berlin (1995)
16. Kevrekidis, P.G., Pelinovsky, D.E., Stefanov, A.: Asymptotic stability of small bound states in the discrete nonlinear Schrödinger equation in one dimension. SIAM J. Math. Anal. **41**, 2010–2030 (2009)
17. Kevrekidis, P.G., Cuevas-Maraver, J., Pelinovsky, D.E.: Energy criterion for the spectral stability of discrete breathers. Phys. Rev. Lett. **117**, 094101 (5p) (2016)
18. MacKay, R.S., Aubry, S.: Proof of existence of breathers for time-reversible or Hamiltonian networks of weakly coupled oscillators. Nonlinearity **7**, 1623–1643 (1994)
19. Malomed, B.A., Weinstein, M.I.: Soliton dynamics in the discrete nonlinear Schrödinger equation. Phys. Lett. A **220**, 91–96 (1996)
20. Paleari, S., Penati, T.: Long time stability of small amplitude Breathers in a mixed FPU-KG model. Z. Angew. Math. Phys. **67**(6), 148 (2016)
21. Paleari, S., Penati, T.: An extensive resonant normal form for an arbitrary large KG model. Annali di Matematica **195**, 133–165 (2016)
22. Pelinovsky, D.E.: Localization in periodic potentials: from Schrödinger operators to the Gross–Pitaevskii equation. London Mathematical Society Lecture Note Series, vol. 390. Cambridge University Press, Cambridge (2011)
23. Pelinovsky, D., Sakovich, A.: Internal modes of discrete solitons near the anti-continuum limit of the dNLS equation. Physica D **240**, 265–281 (2011)
24. Pelinovsky, D., Sakovich, A.: Multi-site breathers in Klein-Gordon lattices: stability, resonances and bifurcations. Nonlinearity **25**, 3423–3451 (2012)
25. Pelinovsky, D.E., Kevrekidis, P.G., Frantzeskakis, D.: Stability of discrete solitons in nonlinear Schrodinger lattices. Physica D **212**, 1–19 (2005)
26. Pelinovsky, D.E., Penati, T., Paleari, S.: Approximation of small-amplitude weakly coupled oscillators by discrete nonlinear Schrödinger equations. Rev. Math. Phys. **28**, 1650015 (25p) (2016)
27. Weinstein, M.I.: Excitation thresholds for nonlinear localized modes on lattices. Nonlinearity **12**, 673–691 (1999)
28. Xu, H., Cuevas-Maraver, J., Kevrekidis, P.G., Vainchtein, A.: An energy-based stability criterion for solitary travelling waves in Hamiltonian lattices. Phil. Trans. R. Soc. A **376**, 20170192 (26p) (2018)

On Strichartz Estimates from ℓ^2-Decoupling and Applications

Robert Schippa

Abstract Strichartz estimates are derived from ℓ^2-decoupling for phase functions satisfying a curvature condition. Bilinear refinements without loss in the high frequency are discussed. Estimates are established from uniform curvature generalizing Galilean invariance or from transversality in one dimension. The bilinear refinements are utilized to prove local well-posedness for generalized cubic nonlinear Schrödinger equations.

1 Introduction

We point out how ℓ^2-decoupling implies Strichartz estimates for non-degenerate phase functions on tori $\mathbb{T}^n = (\mathbb{R}/2\pi\mathbb{Z})^n$. These estimates apply to solutions to linear dispersive equations

$$\begin{cases} i\partial_t u + \varphi(\nabla/i)u = 0, & (t,x) \in \mathbb{R} \times \mathbb{T}^n, \\ u(0) = u_0, \end{cases} \tag{1}$$

where $\varphi \in C^2(\mathbb{R}^n, \mathbb{R})$.
The eigenvalues of $D^2\varphi(\xi)$ are denoted by $\{\gamma_1(\xi), \ldots, \gamma_n(\xi)\}$ and we set

$$\sigma_\varphi(\xi) = \min(\{\#neg.\gamma_i(\xi), \#pos.\gamma_i(\xi)\}).$$

The non-degeneracy hypothesis we assume reads as follows:

$$\exists \psi : 2^{\mathbb{N}_0} \to \mathbb{R}^{>0} \text{ such that } \min(|\gamma_i(\xi)|) \sim \max(|\gamma_i(\xi)|) \sim \psi(N), \; |\xi| \in [N, 2N) \tag{$\mathcal{E}^{\sigma_\varphi}(\psi)$}$$

and $\sigma_\varphi(\xi)$ is independent of ξ.

R. Schippa (✉)
Fakultät für Mathematik, Karlsruher Institut für Technologie, Karlsruhe, Germany
e-mail: robert.schippa@kit.edu

W. Dörfler et al. (eds.), *Mathematics of Wave Phenomena*, Trends in Mathematics,
https://doi.org/10.1007/978-3-030-47174-3_17

By P_N we denote the frequency projector

$$(P_N f)\widehat{(\xi)} = \begin{cases} 1_{[N,2N)}(|\xi|)\hat{f}(\xi), & N \in 2^{\mathbb{N}_0}, \\ 1_{[0,1)}(|\xi|)\hat{f}(\xi), & N = 0. \end{cases}$$

The Strichartz estimates we will prove read

$$\|P_N e^{it\varphi(\nabla/i)} u_0\|_{L^p(I\times\mathbb{T}^n)} \lesssim |I|^{1/p} N^{s(\varphi)} \|P_N u_0\|_{L^2}. \tag{2}$$

To prove (2), we will use ℓ^2-decoupling (cf. [2, 3]), more precisely, (variants of) the discrete L^2-restriction theorem. This was carried out in [2, 3] in the special cases of $\varphi(\xi) = \sum_{i=1}^n \alpha_i \xi_i^2$, $\alpha_i \in \mathbb{R}\backslash 0$. The modest generalization will clarify the role of the asymptotic behaviour of the eigenvalues of $D^2\varphi$, i.e., the curvature of the characteristic surface of (1).

The following proposition is proved:

Proposition 1 *Suppose that φ satisfies $(\mathcal{E}^k(\psi))$ and let $I \subseteq \mathbb{R}$ be a compact interval. Then, we find the following estimates to hold for any $\varepsilon > 0$:*

$$\|P_N e^{it\varphi(\nabla/i)} u_0\|_{L^p(I\times\mathbb{T}^n)} \lesssim_\varepsilon |I|^{1/p} \frac{N^{\left(\frac{n}{2}-\frac{n+2}{p}\right)+\varepsilon}}{(\min(\psi(N),1))^{1/p}}, \qquad \frac{2(n+2-k)}{n-k} \leq p < \infty. \tag{3}$$

Recall that certain Strichartz estimates from [1–3] are known to be sharp up to endpoints. With the above proposition being a generalization, the Strichartz estimates proved above are also sharp in this sense. We shall also consider the example $\varphi(\xi) = |\xi|^a$, $1 < a < 2$, where the proposition gives an additional loss of derivatives due to decreased curvature compared to the Schrödinger case.

When we consider the associated nonlinear Schrödinger equation, we shall see why this additional loss does probably not admit relaxation. Moreover, as in [2, 3] there are estimates for $2 \leq p \leq \frac{2(n+2-k)}{n-k}$, which follow from interpolation. In fact, as $p \neq 2$, Proposition 1 does not yield Strichartz estimates without loss of derivatives.

When we aim to apply these estimates to prove well-posedness of generalized cubic nonlinear Schrödinger equations

$$\begin{cases} i\partial_t u + \varphi(\nabla/i)u &= \pm|u|^2 u, \ (t,x) \in \mathbb{R}\times\mathbb{T}^n, \\ \quad u(0) &= u_0 \in H^s(\mathbb{T}^n), \end{cases} \tag{4}$$

we will use orthogonality considerations to prove bilinear $L^2_{t,x}$-estimates for $High \times Low \to High$-interaction without loss of derivatives in the high frequency. In [4, Theorem 3, p. 193] was proved the following proposition to derive well-posedness to cubic Schrödinger equations on compact manifolds:

Proposition 2 *Let $u_0, v_0 \in L^2(\mathbb{T}^n)$, $K, N \in 2^{\mathbb{N}}$. If there exists $s_0 > 0$ such that*

$$\|P_N e^{\pm it\varphi(\nabla/i)} u_0 P_K e^{\pm it\varphi(\nabla/i)} v_0\|_{L^2_{t,x}(I\times\mathbb{T}^n)} \lesssim |I|^{1/2} \min(N,K)^{s_0} \|P_N u_0\|_{L^2} \|P_K v_0\|_{L^2}, \tag{5}$$

where $I \subseteq \mathbb{R}$ is a compact time interval with $|I| \gtrsim 1$, then the Cauchy problem (4) *is locally well-posed in H^s for $s > s_0$.*

For $\varphi = \sum_{i=1}^n \alpha_i \xi^2$, (5) follows from almost orthogonality and the Galilean transformation (cf. [1, 14]). It turns out that it is enough to require ($\mathcal{E}^{\sigma_\varphi}(\psi)$) to hold for some uniform constant:

$$\exists C_\varphi > 0 : \ \forall \xi \in \mathbb{R}^n : \quad \min(|\gamma_i(\xi)|) \sim \max(|\gamma_i(\xi)|) \sim C_\varphi. \tag{$\mathcal{E}^{\sigma_\varphi}(C_\varphi)$}$$

This will be sufficient to generalize the Galilean transformation and prove the following:

Proposition 3 *Suppose that $\varphi \in C^2(\mathbb{R}^n, \mathbb{R})$ satisfies ($\mathcal{E}^k(C_\varphi)$). Then, there is $s(n,k)$ such that we find the estimate*

$$\|P_N e^{\pm it\varphi(\nabla/i)} u_0 P_K e^{\pm it\varphi(\nabla/i)} v_0\|_{L^2_{t,x}(I\times\mathbb{T}^n)} \lesssim_{C_\varphi,s} K^{2s} |I|^{1/2} \|P_N u_0\|_{L^2} \|P_K v_0\|_{L^2} \tag{6}$$

to hold for $s > s(n,k)$, where $I \subseteq \mathbb{R}$ denotes a compact time interval, $|I| \gtrsim 1$.

This bilinear improvement can also stem from transversality: In [10, 12, 13] short-time bilinear Strichartz estimates were discussed and the following transversality condition played a crucial role in the derivation of the estimates:

$$\text{There is } \alpha > 0 \text{ so that } |\nabla\varphi(\xi_1) \pm \nabla\varphi(\xi_2)| \sim N^\alpha \text{ whenever } |\xi_1| \sim K,\ |\xi_2| \sim N, 3 \tag{$\mathcal{T}_\alpha$}$$

$K \ll N, K, N \in 2^{\mathbb{N}}$.

The corresponding short-time estimate reads

$$\|P_N e^{\pm it\varphi(\nabla/i)} u_0 P_K e^{\pm it\varphi(\nabla/i)} v_0\|_{L^2_{t,x}([0,N^{-\alpha}],L^2_x(\mathbb{T}))} \lesssim_\varphi N^{-\alpha/2} \|P_N u_0\|_{L^2} \|P_K v_0\|_{L^2}. \tag{7}$$

This is sufficient to derive an $L^2_{t,x}$-estimate for finite times by gluing together the short time intervals:

Proposition 4 *Suppose that φ satisfies $(\mathcal{T}_\alpha)$ and let $K \ll N$, $K, N \in 2^{\mathbb{N}}$. Then, we find the following estimate to hold:*

$$\| P_N e^{\pm it\varphi(\nabla/i)} u_0 P_K e^{\pm it\varphi(\nabla/i)} v_0 \|_{L^2_{t,x}(I\times\mathbb{T})} \lesssim_\varphi |I|^{1/2} \| P_N u_0 \|_{L^2} \| P_K v_0 \|_{L^2}, \qquad (8)$$

whenever $I \subseteq \mathbb{R}$ is a compact time interval with $|I| \gtrsim N^{-\alpha}$.

Proof Let $I = \bigcup_j I_j$, $|I_j| \sim N^{-\alpha}$, where the I_j are disjoint. Then,

$$\begin{aligned} \text{lhs}(8)^2 &\lesssim \sum_{I_j} \| P_N e^{\pm it\varphi(\nabla/i)} u_0 P_K e^{\pm it\varphi(\nabla/i)} v_0 \|^2_{L^2_{t,x}(I_j\times\mathbb{T})} \\ &\lesssim (\# I_j) N^{-\alpha} \| P_N u_0 \|^2_{L^2} \| P_K v_0 \|^2_{L^2} \end{aligned}$$

and the claim follows from $\# I_j \sim |I| N^\alpha$. □

Invoking Proposition 2 together with Propositions 3 or 4, the below theorem follows:

Theorem 1 *Suppose that $\varphi \in C^2(\mathbb{R}^n, \mathbb{R})$ satisfies $(\mathcal{E}^k(C_\varphi))$. Then, there is $s_0(n,k)$ such that* (4) *is locally well-posed for $s > s_0(n,k)$.*

Let $n = 1$ and suppose that φ satisfies $(\mathcal{T}_\alpha)$. Then, there is $s_0 = s_0(\varphi)$ such that (4) *is locally well-posed for $s > s_0(\varphi)$.*

Finally, we give examples: In one dimension we treat the fractional Schrödinger equation

$$\begin{cases} i\partial_t u + D^a u = \pm |u|^2 u, \ (t,x) \in \mathbb{R}\times\mathbb{T}, \\ \quad u(0) \quad = u_0 \in H^s(\mathbb{T}), \end{cases} \qquad (9)$$

where $D = (-\Delta)^{1/2}$.

Theorem 1 yields uniform local well-posedness for $s > \frac{2-a}{4}$, $1 < a < 2$, which is presumably sharp up to endpoints as discussed in [5], where the endpoint $s = \frac{2-a}{4}$ was covered by resonance considerations.

For $0 < a < 1$, by varying the above arguments, we can also prove local well-posedness for $s > \frac{2-a}{4}$, which was previously proved in [8] in the context of Strichartz estimates for fractional Schrödinger equations on compact manifolds. In [8] short-time arguments were used to derive Strichartz estimates on arbitrary compact manifolds. These estimates we can improve on tori for $1 < a < 2$ because we do not have to sum up frequency dependent time intervals. Since the derived estimates essentially resemble the estimates on Euclidean space (cf. [11]), we conjecture the estimates to be sharp up to endpoints.

We also discuss hyperbolic Schrödinger equations. The well-posedness result from [9, 14] is recovered for the hyperbolic nonlinear Schrödinger equation in two dimensions, which is known to be sharp up to endpoints. Generalizing the example, which probes sharpness in higher dimensions, indicates that there is only a significant difference between hyperbolic and elliptic Schrödinger equations in low dimensions.

This note is structured as follows: In Sect. 2 we prove linear Strichartz estimates utilizing ℓ^2-decoupling, in Sect. 3 we discuss bilinear Strichartz estimates without loss in the high frequency, and in Sect. 4 the implied well-posedness results for generalized cubic nonlinear Schrödinger equations are discussed.

2 Linear Strichartz Estimates

We prove Proposition 1 generalizing the proofs from [2, 3]:

Proof (Proposition 1) Without loss of generality let $I = [0,T]$. First, let $p \geq \frac{2(n+2-k)}{n-k}$ and compute

$$\text{lhs}(3)^p = \int_{\substack{0\leq x_1,\ldots,x_n\leq 2\pi,\\ 0\leq t\leq T}} \Big| \sum_{|\xi|\sim N} e^{i(x.\xi + t\varphi(\xi))} \hat{u}_0(\xi) \Big|^p dxdt$$

$$\sim \frac{N^{-(n+2)}}{\psi(N)} \int_{\substack{0\leq x_1,\ldots,x_n\leq N,\\ 0\leq t\leq TN^2\psi(N)}} \Big| \sum_{|\xi|\sim 1, \xi\in\mathbb{Z}^n/N} e^{i(x.\xi + \frac{t}{N^2\psi(N)}\varphi(N\xi))} \hat{u}_0(N\xi) \Big|^p dxdt.$$

We distinguish between $\psi(N) \ll 1$ and $\psi(N) \gtrsim 1$. In the latter case, we use periodicity in space to find for the above display

$$\sim \frac{N^{-(n+2)}}{(TN\psi(N))^n \psi(N)} \int_{\substack{0\leq x_1,\ldots,x_n\leq TN^2\psi(N),\\ 0\leq t\leq TN^2\psi(N)}} \Big| \sum_{\substack{|\xi|\sim 1,\\ \xi\in\mathbb{Z}^n/N}} \hat{u}_0(N\xi) e^{i(x.\xi + \frac{t}{N^2\psi(N)}\varphi(N\xi))} \Big|^p dxdt.$$

This expression is amenable to the discrete L^2-restriction theorem [2, Theorem 2.1, p. 354] or the variant for hyperboloids because $TN^2\psi(N) \gtrsim N^2$, and the frequency points are separated of size $\frac{1}{N}$ and the eigenvalues of $\frac{\varphi(N\cdot)}{N^2\psi(N)}$ are approximately one.

Hence, we have the following estimate uniform in φ (the dependence is encoded in $\psi(N)$, which drops out in the ultimate estimate):

$$\begin{aligned}\text{lhs}(3)^p &\lesssim_\varepsilon \frac{N^{-(n+2)}}{(TN\psi(N))^n\psi(N)} (TN^2\psi(N))^{n+1} N^{\left(\frac{n}{2}-\frac{n+2}{p}\right)p+\varepsilon} \|P_N u_0\|_2^p \\ &\lesssim TN^{\left(\frac{n}{2}-\frac{n+2}{p}\right)p+\varepsilon} \|P_N u_0\|_2^p.\end{aligned}$$

Next, suppose that $\psi(N) \ll 1$. In this case we can not avoid loss of derivatives in general. Following along the above lines, we find for $p \geq \frac{2(n+2-k)}{n-k}$

$$\text{lhs}(3)^p \sim \frac{N^{-(n+2)}}{\psi(N)} \int_{\substack{0 \leq x_1,\ldots,x_n \leq N, \\ 0 \leq t \leq TN^2\psi(N)}} \Bigg| \sum_{\substack{|\xi|\sim 1, \\ \xi \in \mathbb{Z}^n/N}} e^{i(x.\xi + t\frac{\varphi(N\xi)}{N^2\psi(N)})} \hat{u}_0(N\xi) \Bigg|^p dxdt$$

$$\lesssim \frac{N^{-(n+2)}}{(NT)^n \psi(N)} \int_{\substack{0 \leq x_1,\ldots,x_n \leq TN^2, \\ 0 \leq t \leq TN^2}} \Big| \sum e^{i(x.\xi + \frac{t\varphi(N\xi)}{N^2\psi(N)})} \hat{u}_0(N\xi) \Big|^p dxdt$$

$$\lesssim_\varepsilon \frac{T}{\psi(N)} N^{\left(\frac{n}{2} - \frac{n+2}{p}\right)p + \varepsilon} \|P_N u_0\|_2^p,$$

which yields the claim. □

As an example consider Strichartz estimates for the free fractional Schrödinger equation

$$\begin{cases} i\partial_t u + D^a u = 0, \ (t, x) \in \mathbb{R} \times \mathbb{T}^n, \\ \quad u(0) \quad = u_0. \end{cases} \tag{10}$$

The phase function $\varphi(\xi) = |\xi|^a, \ 1 < a < 2$ is elliptic and the lack of higher differentiability in the origin is not an issue because low frequencies can always be treated with Bernstein's inequality. φ satisfies $(\mathcal{E}^0(\psi))$ with $\psi(N) = N^{a-2}$, hence we find by virtue of Proposition 1

$$\|e^{itD^a} u_0\|_{L^4_{t,x}(I \times \mathbb{T}^n)} \lesssim_{n,a,s} |I|^{1/4} \|u_0\|_{H^s}, \quad s > s_0 = \begin{cases} \frac{2-a}{8}, & n = 1, \\ \frac{2-a}{4} + \left(\frac{n}{2} - \frac{n+2}{4}\right), & \text{else.} \end{cases} \tag{11}$$

To find the $L^4_{t,x}$-estimate in one dimension, we interpolate the $L^6_{t,x}$-endpoint estimate with the trivial $L^2_{t,x}$-estimate.

In case $n = 1, \ 1 < a < 2$, this recovers the Strichartz estimates from [7], and for $0 < a < 1$, corresponding estimates were proved in [8]. For $n \geq 2, \ 1 < a < 2$ this estimate seems to be new. Comparing with the results in Euclidean space (cf. [11]), the estimates are presumably sharp up to endpoints.

3 Bilinear Strichartz Estimates and Transversality

The argument from Sect. 2 admits bilinearization provided that the dispersion relation satisfies $(\mathcal{E}^{\sigma_\varphi}(C_\varphi))$. This generalizes Galilean invariance, which was previously used to infer a bilinear estimate with no loss in the high frequency (cf. [1, 14]).

Proof (Proposition 3) Let $P_N = \sum_{K_1} R_{K_1}$, where R_K projects to cubes of sidelength K. Then, by means of almost orthogonality

$$\begin{aligned}
&\| P_N e^{it\varphi(\nabla/i)} u_0 P_K e^{it\varphi(\nabla/i)} v_0 \|^2_{L^2_{t,x}(I\times\mathbb{T}^n)} \\
&\lesssim \sum_{K_1} \| R_{K_1} e^{it\varphi(\nabla/i)} u_0 P_K e^{it\varphi(\nabla/i)} v_0 \|^2_{L^2_{t,x}(I\times\mathbb{T}^n)}.
\end{aligned}$$

After applying Hölder's inequality, we are left with estimating two $L^4_{t,x}$-norms. Clearly, by Proposition 1,

$$\| P_K e^{it\varphi(\nabla/i)} v_0 \|_{L^4_{t,x}(I\times\mathbb{T}^n)} \lesssim_{\varphi,s} K^s \| P_K v_0 \|_{L^2}$$

provided that $s > s(n, \sigma_\varphi)$.

To treat the other term, let ξ_0 denote the center of the cube Q_{K_1} onto which R_{K_1} is projecting in frequency space. Following along the above lines, we write

$$\begin{aligned}
&\| R_{K_1} e^{it\varphi(\nabla/i)} u_0 \|^4_{L^4_{t,x}(I\times\mathbb{T}^n)} \\
&= \int_{\substack{0\le x_1,\dots,x_n\le 2\pi,\\ 0\le t\le T}} \Big| \sum_{\xi\in Q_{K_1}} e^{i(x.\xi + t\varphi(\xi))} \hat{u}_0(\xi) \Big|^4 dxdt \\
&= \int_{\substack{0\le x_1,\dots,x_n\le 2\pi,\\ 0\le t\le T}} \Big| \sum_{|\xi'|\le K} \hat{u}_0(\xi + \xi') e^{i(x.(\xi_0+\xi') + t\varphi(\xi_0+\xi'))} \Big|^4 dxdt \\
&= \int_{\substack{0\le x_1,\dots,x_n\le 2\pi,\\ 0\le t\le T}} \Big| \sum_{|\xi'|\le K} e^{i((x+t\nabla\varphi(\xi_0)).\xi' + t\psi_{\xi_0}(\xi'))} \hat{w}_0(\xi') \Big|^4 dxdt \\
&= \| P_{\le K_1} e^{it\psi_{\xi_0}(\nabla/i)} w_0(x + t\nabla\varphi(\xi_0)) \|^4_{L^4(I\times\mathbb{T}^n)},
\end{aligned}$$

where $\psi_{\xi_0}(\xi') = \varphi(\xi_0 + \xi') - \varphi(\xi_0) - \nabla\varphi(\xi_0).\xi'$.

After breaking $\| P_{\le K} e^{it\psi_{\xi_0}(\nabla/i)} w_0 \|_{L^4_{t,x}(I\times\mathbb{T}^n)} \le \sum_{1\le L\le K} \| P_L e^{it\psi_{\xi_0}(\nabla/i)} w_0 \|_{L^4}$, the single expressions are amenable to Proposition 1. Indeed, the size of the moduli of the eigenvalues of $D^2\psi_{\xi_0}$ are approximately independent of the frequencies. Hence,

$$\| P_L e^{it\psi_{\xi_0}(\nabla/i)} w_0 \|_{L^4_{t,x}(I\times\mathbb{T}^n)} \lesssim_{\varepsilon, C_\varphi} L^{s(n,k)+\varepsilon} \| P_L w_0 \|_{L^2}.$$

Carrying out the sum and the relation of u_0 and w_0, we find

$$\| P_{\le K} e^{it\psi_{\xi_0}(\nabla/i)} w_0 \|_{L^4(I\times\mathbb{T}^n)} \lesssim_{\varepsilon,\varphi} K^{s(n,k)+\varepsilon} \| R_{K_1} u_0 \|_{L^2}$$

By almost orthogonality,

$$\left(\sum_{K_1} \| R_{K_1} u_0 \|_{L^2}^2 \right)^{1/2} \lesssim \| P_N u_0 \|_{L^2}^2,$$

and the proof is complete. □

In one dimension (and for certain phase functions also in higher dimensions, see [13]) transversality $(\mathcal{T}_\alpha)$ of the phase function allows us to derive Proposition 4, which improves the above estimate.

4 Local Well-Posedness of Generalized Cubic Schrödinger Equation

Deploying Proposition 2 by use of the estimates from Sects. 2 and 3, we can conclude the proof of Theorem 1:

Proof First, suppose that φ satisfies $(\mathcal{E}^{\sigma_\varphi}(C_\varphi))$. In case $K \ll N$, Proposition 3 yields the estimate

$$\begin{aligned} &\| P_N e^{\pm it\varphi(\nabla/i)} u_0 P_K e^{\pm it\varphi(\nabla/i)} v_0 \|_{L^2_{t,x}(I \times \mathbb{T}^n)} \\ &\lesssim_{\varepsilon,\varphi} |I|^{1/2} K^{2s(n,\sigma_\varphi)+\varepsilon} \| P_N u_0 \|_{L^2} \| P_K v_0 \|_{L^2}. \end{aligned} \tag{12}$$

For $K \sim N$, the claim follows after applying Hölder's inequality and Proposition 1. From Proposition 2 we find (4) to be locally well-posed provided that $s > 2s_0(n, \sigma_\varphi)$. In case φ satisfies $(\mathcal{E}^0(\psi(N)))$ and $(\mathcal{T}_\alpha)$, we have the improved bilinear bound

$$\| P_N e^{\pm it\varphi(\nabla/i)} u_0 P_K e^{\pm it\varphi(\nabla/i)} v_0 \|_{L^2_{t,x}(I \times \mathbb{T})} \lesssim_\varphi |I|^{1/2} \| P_N u_0 \|_{L^2} \| P_K v_0 \|_{L^2}$$

due to Proposition 4. Hence, only loss stems from $High \times High \to High$-interaction, where $K \sim N$: By means of Proposition 1 and Hölder's inequality, we derive

$$\| P_N e^{\pm it\varphi(\nabla/i)} u_0 P_K e^{\pm it\varphi(\nabla/i)} v_0 \|_{L^2_{t,x}(I \times \mathbb{T})} \lesssim_\varphi K^{2s} |I|^{1/2} \| P_N u_0 \|_{L^2} \| P_K v_0 \|_{L^2}$$

and by Proposition 2 we find (4) to be locally well-posed for $s > 2s_0(\varphi)$. □

We turn to examples: As discussed in Sect. 2, the phase functions $\varphi(\xi) = |\xi|^a$ do not satisfy $(\mathcal{E}^0(C_\varphi))$, but $(\mathcal{T}_{a-1})$ for $1 < a < 2$. For $0 < a < 1$, $K \ll N$, $|I| \gtrsim 1$, we have the following bilinear Strichartz estimate

$$\| P_N e^{\pm it D^a} u_0 P_K e^{\pm it D^a} v_0 \|_{L^2_{t,x}(I \times \mathbb{T})} \lesssim |I|^{1/2} K^{\frac{1-a}{2}} \| P_N u_0 \|_{L^2} \| P_K v_0 \|_{L^2},$$

which can be proved like in [12, 13].

Consequently, by (the proof of) Theorem 1 we find (9) to be locally well-posed for $s > \frac{2-a}{4}$. As discussed in [5], this is likely to be the threshold of uniform local well-posedness, which indicates that the linear Strichartz estimates from Sect. 2 are in this case sharp up to endpoints.

In Euclidean space fractional Schrödinger equations were considered in [11]. Key ingredient to well-posedness are linear and bilinear Strichartz estimates, which hold globally in time due to dispersive effects. On the circle we can reach the same regularity like in [11] up to the endpoint. Although the linear Strichartz estimates might well be sharp in higher dimensions, satisfactory bilinear $L^2_{t,x}$-Strichartz estimates appear to be beyond the methods of this paper so that we can not prove non-trivial well-posedness results in higher dimensions. Progress presumably requires an additional angular decomposition (cp. [6]) to improve control of the resonance function.

For hyperbolic phase functions, Theorem 1 recovers the results from [9, 14], where essentially sharp local well-posedness of

$$\begin{cases} i\partial_t u + (\partial_{x_1}^2 - \partial_{x_2}^2) u = \pm |u|^2 u, \; (t,x) \in \mathbb{R} \times \mathbb{T}^2, \\ u(0) \qquad\qquad\qquad = u_0 \end{cases} \tag{13}$$

was proved for $s > 1/2$. Notably, due to subcriticality of the $L^4_{t,x}$-Strichartz estimate already for the hyperbolic equations

$$\begin{cases} i\partial_t u + (\partial_{x_1}^2 - \partial_{x_2}^2 + \partial_{x_3}^2) u = \pm |u|^2 u, \; (t,x) \in \mathbb{R} \times \mathbb{T}^3, \\ u(0) \qquad\qquad\qquad = u_0, \end{cases} \tag{14}$$

and

$$\begin{cases} i\partial_t u + (\partial_{x_1}^2 - \partial_{x_2}^2 + \partial_{x_3}^2 - \partial_{x_4}^2) u = \pm |u|^2 u, \; (t,x) \in \mathbb{R} \times \mathbb{T}^4, \\ u(0) \qquad\qquad\qquad = u_0, \end{cases} \tag{15}$$

the (essentially sharp) Strichartz estimates yield the same well-posedness results as for the elliptic counterparts.

Firstly, recall the counterexample from [14], which showed C^3-ill-posedness of (13) for $s < 1/2$. As initial data consider

$$\phi_N(x) = N^{-1/2} \sum_{|k| \leq N} e^{ikx_1} e^{-ikx_2},$$

which satisfies $\|\phi_N\|_{H^s} \sim N^s$ and $S[\phi_N](t) := e^{it(\partial_{x_1}^2 - \partial_{x_2}^2)}\phi_N = \phi_N$. This implies

$$\left\| \int_0^T |S[\phi_N](s)|^2 S[\phi_N](s) ds \right\|_{H^s} = T \| |\phi_N|^2 \phi_N \|_{H^s} \gtrsim T N^{1+s}$$

For details on this estimate see [14].

The veracity of the estimate

$$\left\| \int_0^T |S[\phi_N](s)|^2 S[\phi_N](s) ds \right\|_{H^s} \lesssim \|\phi_N\|_{H^s}^3 \quad (T \lesssim 1)$$

requires $s \geq 1/2$.

The same counterexample shows that $s \geq 1/2$ is required for C^3-well-posedness of (14). This regularity is reached up to the endpoint by Theorem 1.
When considering (15), we modify the above example to

$$\phi_N(x) = N^{-1} \sum_{|k_1|,|k_2| \leq N} e^{ik_1 x_1} e^{-ik_1 x_2} e^{ik_2 x_3} e^{-ik_2 x_4},$$

which again satisfies $\|\phi_N\|_{H^s} \sim N^s$.
Carrying out the estimate for the first Picard iterate with the necessary modifications yields

$$\left\| \int_0^T |S[\phi_N](s)|^2 S[\phi_N](s) ds \right\|_{H^s} = T \| |\phi_N|^2 \phi_N \|_{H^s} \gtrsim T N^{2+s},$$

which implies C^3-ill-posedness unless $s \geq 1$. This regularity is again obtained up to the endpoint by Theorem 1.

Apparently, for other hyperbolic Schrödinger equations the $L^4_{t,x}$-Strichartz estimate also coincides with the elliptic $L^4_{t,x}$-estimate and modifications of the above counterexample yield lower thresholds than in the elliptic case. This indicates that the difference between elliptic and hyperbolic Schrödinger equations is only significant for lower dimensions.

Acknowledgments Financial support by the German Science Foundation (IRTG 2235) is gratefully acknowledged.

References

1. Bourgain, J.: Fourier transform restriction phenomena for certain lattice subsets and applications to nonlinear evolution equations. I. Schrödinger equations. Geom. Funct. Anal. **3**(2), 107–156 (1993). https://doi.org/10.1007/BF01896020

2. Bourgain, J., Demeter, C.: The proof of the l^2 decoupling conjecture. Ann. Math. (2) **182**(1), 351–389 (2015). https://doi.org/10.4007/annals.2015.182.1.9
3. Bourgain, J., Demeter, C.: Decouplings for curves and hypersurfaces with nonzero Gaussian curvature. J. Anal. Math. **133**, 279–311 (2017). https://doi.org/10.1007/s11854-017-0034-3
4. Burq, N., Gérard, P., Tzvetkov, N.: Bilinear eigenfunction estimates and the nonlinear Schrödinger equation on surfaces. Invent. Math. **159**(1), 187–223 (2005). https://doi.org/10.1007/s00222-004-0388-x
5. Cho, Y., Hwang, G., Kwon, S., Lee, S.: Well-posedness and ill-posedness for the cubic fractional Schrödinger equations. Discrete Contin. Dynam. Syst. **35**(7), 2863–2880 (2015). https://doi.org/10.3934/dcds.2015.35.2863
6. Colliander, J., Keel, M., Staffilani, G., Takaoka, H., Tao, T.: Resonant decompositions and the I-method for the cubic nonlinear Schrödinger equation on $\mathbb{R}^2$. Discrete Contin. Dynam. Syst. **21**(3), 665–686 (2008). https://doi.org/10.3934/dcds.2008.21.665
7. Demirbas, S., Erdoğan, M.B., Tzirakis, N.: Existence and uniqueness theory for the fractional Schrödinger equation on the torus. In: Some Topics in Harmonic Analysis and Applications. Adv. Lect. Math. (ALM), vol. 34, pp. 145–162. Int. Press, Somerville (2016)
8. Dinh, V.D.: Strichartz estimates for the fractional Schrödinger and wave equations on compact manifolds without boundary. J. Differ. Equ. **263**(12), 8804–8837 (2017). https://doi.org/10.1016/j.jde.2017.08.045
9. Godet, N., Tzvetkov, N.: Strichartz estimates for the periodic non-elliptic Schrödinger equation. C. R. Math. Acad. Sci. Paris **350**(21-22), 955–958 (2012). https://doi.org/10.1016/j.crma.2012.10.029
10. Hani, Z.: A bilinear oscillatory integral estimate and bilinear refinements to Strichartz estimates on closed manifolds. Anal. PDE **5**(2), 339–363 (2012). https://doi.org/10.2140/apde.2012.5.339
11. Hong, Y., Sire, Y.: On fractional Schrödinger equations in Sobolev spaces. Commun. Pure Appl. Anal. **14**(6), 2265–2282 (2015). https://doi.org/10.3934/cpaa.2015.14.2265
12. Moyua, A., Vega, L.: Bounds for the maximal function associated to periodic solutions of one-dimensional dispersive equations. Bull. Lond. Math. Soc. **40**(1), 117–128 (2008). https://doi.org/10.1112/blms/bdm096
13. Schippa, R.: On short-time bilinear Strichartz estimates and applications to the Shrira equation. Nonlinear Anal. **198**, 111910, p 22 (2020)
14. Wang, Y.: Periodic cubic hyperbolic Schrödinger equation on $\mathbb{T}^2$. J. Funct. Anal. **265**(3), 424–434 (2013). https://doi.org/10.1016/j.jfa.2013.05.016

On a Limiting Absorption Principle for Sesquilinear Forms with an Application to the Helmholtz Equation in a Waveguide

Ben Schweizer and Maik Urban

Abstract We prove a limiting absorption principle for sesquilinear forms on Hilbert spaces and apply the abstract result to a Helmholtz equation with radiation condition. The limiting absorption principle is based on a Fredholm alternative. It is applied to Helmholtz-type equations in a truncated waveguide geometry. We analyse a problem with radiation conditions on truncated domains, recently introduced in [4]. We improve the previous results by treating the limit $\delta \to 0$.

1 Introduction

This article is devoted to the analysis of time-harmonic versions of wave equations, the most prominent example being the Helmholtz equation $\mathcal{L}u := -\nabla \cdot (a\nabla u) - \omega^2 u = f$, where $\omega > 0$ is the prescribed frequency of the problem, a a positive coefficient, and f a source term. The equation is posed in a domain $\Omega \subset \mathbb{R}^n$ and it is complemented by boundary conditions. For applications, it is interesting to solve the equation in an unbounded domain Ω. In order to make the equation well-posed, radiation conditions must be imposed in those directions in which Ω is unbounded.

A vast body of literature is devoted to radiation conditions such as the classical Sommerfeld condition and their analysis, we mention [2, 12, 15] and the overview of [13]. Recently, the interest turned to radiation conditions in periodic media. This requires new methods, see [5, 6, 10]. Related to radiation conditions is the question of a numerical treatment of the equation; the domain must be truncated (replaced by a bounded domain), and the radiation condition must be replaced appropriately. Methods like non-reflecting boundary conditions or perfectly matched layers are successfully applied to problems with constant coefficients, but these methods cannot be used in a periodic medium. The approach of [4] is to introduce radiation boxes and to demand that the solution of the truncated problem is an outgoing wave

B. Schweizer (✉) · M. Urban
Technische Universität Dortmund, Fakultät für Mathematik, Dortmund, Germany
e-mail: ben.schweizer@tu-dortmund.de; maik.urban@tu-dortmund.de

W. Dörfler et al. (eds.), *Mathematics of Wave Phenomena*, Trends in Mathematics,
https://doi.org/10.1007/978-3-030-47174-3_18

in the radiation boxes. The problem can be formulated with the help of a sesquilinear form.

Limiting Absorption Principle A tool to establish the existence of solutions to $\mathcal{L}u = f$ in an unbounded domain is the limiting absorption principle. The idea of that approach is to add a small absorption term. Denoting the imaginary unit by i, we study for each $\delta > 0$ the equation

$$\mathcal{L}u_\delta + \delta \mathrm{i} u_\delta = f\,, \tag{1}$$

such that the original problem is recovered for $\delta = 0$. Usually, it is easy to establish the existence of a solution to (1). Given a family of solutions $(u_\delta)_\delta$, we can study the limit $\delta \to 0$. Any limit of the sequence $(u_\delta)_\delta$ is a candidate for a solution to the original problem. This program has been performed successfully in several settings, as recent examples we mention [7] and [11].

A radiation condition for truncated domains was introduced in [4], the article contains a well-posedness result for positive absorption $\delta > 0$. Here, we study the limit $\delta \to 0$ of that system.

The method of [4] is taylored for bounded domains, the aim is to implement radiation conditions. We continue this analysis. We emphasise that our method, which uses Gårding's inequality and a compact operator K, is restricted to bounded domains.

Abstract Setting In the first part of this article, we derive the limiting absorption principle in an abstract setting, and consider sesquilinear forms on Hilbert spaces. We recall that the system of [4] is formulated with sesquilinear forms, hence our abstract results can be applied to this "radiation condition on truncated domains".

Given a (real or complex) Hilbert space $\mathcal{H}$, a sesquilinear form b on $\mathcal{H} \times \mathcal{H}$, and $\ell \in \mathcal{H}^*$, we look for an element $u \in \mathcal{H}$ that satisfies the equation

$$b(u, \cdot) = \ell(\cdot)\,. \tag{2}$$

If b is bounded and coercive, then the existence of such an element u follows from the Lax-Milgram lemma. But how can we establish the existence of a solution $u \in \mathcal{H}$ to (2) if b is not coercive? We perform the following approach: instead of b, we consider a family $(b_\delta)_\delta$ of sesquilinear forms such that there is a unique solution $u_\delta \in \mathcal{H}$ to

$$b_\delta\big(u_\delta, \cdot\big) = \ell(\cdot) \tag{3}$$

for each $\delta > 0$. Under suitable assumptions on the family $(b_\delta)_\delta$, we prove that the sequence of solutions $(u_\delta)_\delta$ converges to a solution u to (2) as $\delta \to 0$.

Our result is based on a Fredholm alternative for sesquilinear forms; see Proposition 2.1. We note that the Fredholm alternative is not new, but rather a variant of known results, see [8, 9].

Application to the Helmholtz Equation in a Periodic Waveguide Let us sketch the setting of the truncated problem of [4]. The geometry is described by the numbers $\varepsilon, R > 0$ and $M \in \mathbb{N}$, the truncated domain is $\Omega_R := (-\varepsilon R, \varepsilon R) \times (0, \varepsilon M)$. A parameter $L > 0$ defines the width of the two radiation boxes $(\varepsilon R, \varepsilon(R + L)) \times (0, \varepsilon M)$ and $(-\varepsilon(R + L), -\varepsilon R) \times (0, \varepsilon M)$. The coefficient $a \colon \mathbb{R}^2 \to \mathbb{R}$ is assumed to be periodic in the left half-space and in the right half-space: $a(x) = a_-(x)$ for $x \in (-\infty, 0) \times (0, \varepsilon M)$ and $a(x) = a_+(x)$ for $x \in [0, \infty) \times (0, \varepsilon M)$, where a_- and a_+ are ε-periodic in $\mathbb{R}^2$,

$$a_+\big(x + \varepsilon e_j\big) = a_+(x) \quad \text{and} \quad a_-\big(x + \varepsilon e_j\big) = a_-(x)$$

for every $x \in \mathbb{R}^2$ and $j \in \{1, 2\}$.

The problem is to find a function u_δ which solves (1) in Ω_R and which is radiating in radiation boxes. This problem is recast in [4] with a sesquilinear form β_δ and is written as $\beta_\delta(u_\delta, \cdot) = \langle f, \cdot \rangle$.

The problem for $\delta > 0$ was analysed in [4]. Our abstract results allow to study the limit $\delta \to 0$: Except for a countable set of singular frequencies ω, the solutions u_δ converge to the unique solution u of the original problem $\beta(u, \cdot) = \langle f, \cdot \rangle$.

2 Fredholm Alternative and Limiting Absorption Principle

Our aim in this section is to present a limiting absorption principle for a sesquilinear form defined on a Hilbert space over $\mathbb{R}$ or $\mathbb{C}$. The limiting absorption principle is based on a Fredholm alternative, which we present first.

2.1 *Fredholm Alternative for Sesquilinear Forms*

The Fredholm alternative for Fredholm operators of index 0 is a classical result in functional analysis. As we focus on problems involving sesquilinear forms rather than operators, we present a Fredholm alternative for sesquilinear forms.

We recall that such an alternative is not new. Indeed, Kress [9] proved a Fredholm alternative for bilinear forms on reflexive Banach spaces and characterised the bilinear forms for which such a Fredholm alternative holds. As our version of the Fredholm alternative is slightly different from the one presented by Kress and for convenience of the reader, we provide a proof.

Let $\mathcal{H}$ be a Hilbert space over the field $\mathbb{K} = \mathbb{R}$ or $\mathbb{K} = \mathbb{C}$ and let $b \colon \mathcal{H} \times \mathcal{H} \to \mathbb{K}$ be a sesquilinear form; that is, b is anti-linear in its first argument and linear in the second. Note that b is bilinear in case $\mathbb{K} = \mathbb{R}$. The norm on $\mathcal{H}$ is denoted by $\|\cdot\|$. The *kernel* of b is defined as

$$\ker(b) := \big\{u \in \mathcal{H} : \ b(u, v) = 0 \text{ for all } v \in \mathcal{H}\big\} .$$

The *adjoint* $b^*: \mathcal{H} \times \mathcal{H} \to \mathbb{K}$ of b is given by $b^*(u, v) := \overline{b(v, u)}$. In case $\mathbb{K} = \mathbb{R}$ this reduces to $b^*(u, v) := b(v, u)$. We say that a sesquilinear form $b: \mathcal{H} \times \mathcal{H} \to \mathbb{K}$ satisfies a *Gårding inequality* if there are a constant $c > 0$ and a compact operator $K: \mathcal{H} \to \mathcal{H}^*$ such that

$$\left| b(u, u) + \langle Ku, u \rangle_{\mathcal{H}^*, \mathcal{H}} \right| \geq c \, \|u\|^2 \tag{4}$$

for all $u \in \mathcal{H}$. The space of bounded linear functionals is denoted by $\mathcal{H}^*$. Thus, for $\mathbb{K} = \mathbb{C}$ the compact operator $K: \mathcal{H} \to \mathcal{H}^*$ in the Gårding inequality is required to be an anti-linear map.

Proposition 2.1 (Fredholm Alternative for Sesquilinear Forms) *Let $\mathcal{H}$ be a Hilbert space, and let $b: \mathcal{H} \times \mathcal{H} \to \mathbb{K}$ be a bounded sesquilinear form that satisfies a Gårding inequality. Then the following statements hold:*

(1) Let $\ell \in \mathcal{H}^*$. There exists $u \in \mathcal{H}$ with

$$b(u, \cdot) = \ell(\cdot) \tag{5}$$

if and only if

$$\ell(v) = 0 \quad \textit{for all } v \in \ker(b^*)\,.$$

(2) The dimension of $\ker(b)$ *is finite and equals the dimension of* $\ker(b^*)$.
(3) If $\ker(b) = \{0\}$, *then for every $\ell \in \mathcal{H}^*$ there is a unique $u \in \mathcal{H}$ that solves* (5).

Let us recall that a bounded linear operator $F: \mathcal{H} \to \mathcal{K}$ between two Hilbert spaces $\mathcal{H}$ and $\mathcal{K}$ is called a *Fredholm operator* if both its kernel $\ker F$ and its cokernel $\operatorname{coker} F := \mathcal{K}/\operatorname{im} F$ are finite dimensional. The *index* of F is defined as

$$\operatorname{ind} F := \dim\ker F - \dim\operatorname{coker} F \in \mathbb{Z}\,.$$

The Fredholm index is stable under compact perturbations; that is, if F is a Fredholm operator and $K: \mathcal{H} \to \mathcal{K}$ is compact, then $F + K$ is a Fredholm operator with $\operatorname{ind}(F + K) = \operatorname{ind} F$. Denoting the adjoint of F by $F^*: \mathcal{K} \to \mathcal{H}$, one readily checks that $\ker F^* \cong \operatorname{coker} F$ and $\operatorname{coker} F^* \cong \ker F$; in particular, F^* is a Fredholm operator if and only if F is, and we have $\operatorname{ind} F = -\operatorname{ind} F^*$. For a discussion of Fredholm operators we refer to [3].

We call an anti-linear operator $F: \mathcal{H} \to \mathcal{K}$ an *anti-linear Fredholm operator* if $\mathcal{H} \to \mathcal{K}$, $u \mapsto \overline{Fu}$ is a (linear) Fredholm operator. The following result will be useful in the proof of Proposition 2.1.

Lemma 2.1 *Let $\mathcal{H}$ be a Hilbert space and let $b: \mathcal{H} \times \mathcal{H} \to \mathbb{K}$ be a bounded sesquilinear form that satisfies a Gårding inequality. Then the operator $B: \mathcal{H} \to \mathcal{H}^*$ given by $u \mapsto b(u, \cdot)$ is an anti-linear Fredholm operator of index* 0.

Proof The sesquilinear form b is anti-linear in its first argument and bounded, which implies anti-linearity and boundedness of B. By assumption, b satisfies a Gårding inequality; hence there is a compact anti-linear operator $K\colon \mathcal{H} \to \mathcal{H}^*$ such that (4) holds. The operator $(B + K)\colon \mathcal{H} \to \mathcal{H}^*$ induces a sesquilinear form $a\colon \mathcal{H} \times \mathcal{H} \to \mathbb{K}$,

$$a(u, v) := \big\langle (B + K)u, v\big\rangle_{\mathcal{H}^*,\mathcal{H}} = b(u, v) + \big\langle Ku, v\big\rangle_{\mathcal{H}^*,\mathcal{H}}.$$

Thanks to the boundedness of b and K, the sesquilinear form a is bounded as well. Moreover, a is coercive since b satisfies a Gårding inequality. We can apply the Lax-Milgram lemma to conclude that the operator $(B + K)\colon \mathcal{H} \to \mathcal{H}^*$ is bijective. Consequently, $B + K$ is an anti-linear Fredholm operator of index 0. As the Fredholm index is stable under compact perturbations, we deduce that $B = (B + K) - K$ is also an anti-linear Fredholm operator of index 0. This proves the claim. □

Having Lemma 2.1 at hand, we can perform the proof of Proposition 2.1.

Proof (of Proposition 2.1) By Lemma 2.1 we know that the operator $B\colon \mathcal{H} \to \mathcal{H}^*$, $u \mapsto b(u, \cdot)$ is an anti-linear Fredholm operator of index 0. It is straightforward to show the identity $(\operatorname{im} B)^\perp = \ker B^*$, from which we deduce that $\operatorname{im} B = (\ker B^*)^\perp$, since the range of a Fredholm operator is always closed.

(1) Problem (5) is equivalent to $Bu = \ell$ in $\mathcal{H}^*$. Consequently, $Bu = \ell$ has a solution $u \in \mathcal{H}$ if and only if $\ell \in (\ker B^*)^\perp$. Using the definition of B^*, we find the equation $\langle B^*v, u\rangle = b^*(v, u)$ for all $u, v \in \mathcal{H}$. Hence, by identifying $\mathcal{H}$ with $\mathcal{H}^*$, there holds $B^*v = b^*(v, \cdot)$ for all $v \in \mathcal{H}$. Consequently, $\ker B^* = \ker(b^*)$ and claim *(1)* is proved.
(2) Using the definition of the cokernel, $\operatorname{coker} B \cong (\operatorname{im} B)^\perp = \ker B^*$. As $\operatorname{ind} B = 0$, we find that $\dim\ker B = \dim\operatorname{coker} B = \dim\ker B^*$. Thus $\dim\ker(b) = \dim\ker B = \dim\ker B^* = \dim\ker(b^*)$. Finiteness of the dimensions follows since B is Fredholm operator.
(3) This statement is an immediate consequence of *(1)* and *(2)*. □

2.2 *Limiting Absorption Principle for Sesquilinear Forms*

Given $\ell \in \mathcal{H}^*$, a standard tool to prove the existence of a solution $u \in \mathcal{H}$ to

$$b(u, \cdot) = \ell(\cdot) \tag{6}$$

is the Lax-Milgram lemma. For this tool to work, b has to be coercive. If b is not coercive, one can replace it by a family of suitable sesquilinear forms $(b_\delta)_\delta$ and prove, for each $\delta > 0$, the existence of a solution $u_\delta \in \mathcal{H}$ to

$$b_\delta\big(u_\delta, \cdot\big) = \ell(\cdot)\,. \tag{7}$$

The idea is then to show that the solutions $(u_\delta)_\delta$ of (7) converge in $\mathcal{H}$ to a solution of (6) as $\delta \to 0$.

We provide a sufficient condition for this method to work.

Assumption 2.1 The sesquilinear form $b\colon \mathcal{H}\times\mathcal{H} \to \mathbb{K}$ satisfies the following two requirements:

(i) *(Uniqueness)* The kernel of b is trivial.
(ii) *(Gårding)* There exist a constant $c > 0$ and a compact anti-linear operator $K\colon \mathcal{H} \to \mathcal{H}^*$ such that

$$\left| b(u,u) + \langle Ku, u\rangle_{\mathcal{H}^*,\mathcal{H}} \right| \geq c\,\|u\|^2 \qquad \text{for all } u \in \mathcal{H}.$$

We observe that every coercive sesquilinear form satisfies Assumption 2.1 with $K = 0$.

Lemma 2.2 *Let $\mathcal{H}$ be a Hilbert space and let $b\colon \mathcal{H}\times\mathcal{H} \to \mathbb{K}$ be a bounded sesquilinear form that satisfies Assumption 2.1. Then, for every $\ell \in \mathcal{H}^*$ there is a unique $u \in \mathcal{H}$ such that*

$$b(u,\cdot) = \ell(\cdot)\,.$$

Proof The claim follows by applying Proposition 2.1*(c)*. □

Theorem 2.1 (Abstract Limiting Absorption Principle) *Let $\mathcal{H}$ be a Hilbert space over the field $\mathbb{K}$ and let $b\colon \mathcal{H}\times\mathcal{H} \to \mathbb{K}$ be a bounded sesquilinear form. Assume that $(b_\delta)_\delta$ is a family of bounded sesquilinear forms that satisfies the following requirements:*

(a) For each $\delta > 0$, the sesquilinear form b_δ satisfies Assumption 2.1. Moreover, the compact operator in Gårdings inequality can be chosen independently of $\delta > 0$.
(b) For $u, v \in \mathcal{H}$ fixed, $\lim_{\delta\to 0} b_\delta(u,v) = b(u,v)$.
(c) For every sequence $(u_\delta)_\delta$ in $\mathcal{H}$ with $u_\delta \to u$ weakly in $\mathcal{H}$ there holds:

$$\lim_{\delta\to 0} b_\delta\big(u_\delta - u, v\big) = 0 \quad \textit{for all } v \in \mathcal{H}.$$

Assume further that

$$\ker(b) = \{0\}\,.$$

Then the following holds: For every $\ell \in \mathcal{H}^*$ there exists a sequence $(u_\delta)_\delta$ such that $b_\delta(u_\delta,\cdot) = \ell(\cdot)$, for each $\delta > 0$. The sequence $(u_\delta)_\delta$ weakly converges in $\mathcal{H}$ to an element $u \in \mathcal{H}$ satisfying

$$b(u,\cdot) = \ell(\cdot)\,. \tag{8}$$

Moreover, the limit u is the unique solution to (8).

Proof Fix $\ell \in \mathcal{H}^*$. Lemma 2.2 together with assumption *(a)* imply that, for each $\delta > 0$, there is a unique solution $u_\delta \in \mathcal{H}$ to $b_\delta(u_\delta, \cdot) = \ell(\cdot)$.

Case 1. Assume that $(u_\delta)_\delta$ is bounded in $\mathcal{H}$. We then find a subsequence $(u_\delta)_\delta$ and an element $u \in \mathcal{H}$ such that $u_\delta \to u$ weakly in $\mathcal{H}$. For all $v \in \mathcal{H}$ there holds

$$|b_\delta(u_\delta, v) - b(u, v)| \leq \left|b_\delta\left(u_\delta - u, v\right)\right| + |b_\delta(u, v) - b(u, v)| \, . \tag{9}$$

Thanks to *(b)* and *(c)*, we conclude from (9) that

$$\lim_{\delta\to 0} b_\delta(u_\delta, v) = b(u, v) \qquad \text{for all } v \in \mathcal{H}. \tag{10}$$

On the other hand, for each $\delta > 0$, we have that $b_\delta(u_\delta, \cdot) = \ell(\cdot)$. Combining this with (10) we infer that $b(u, v) = \ell(v)$ for all $v \in \mathcal{H}$ and a solution $u \in \mathcal{H}$ is found. Uniqueness of u follows as $\ker(b) = \{0\}$. As the limit u is unique, we infer that every subsequence of the bounded sequence $(u_\delta)_\delta$ has a subsequence that weakly converges in $\mathcal{H}$ to u, and thus the whole sequence converges weakly.

Case 2. Suppose that $(u_\delta)_\delta$ is unbounded in $\mathcal{H}$. We shall prove that this is impossible. As $(u_\delta)_\delta$ is unbounded, we find a subsequence $(u_\delta)_\delta$ with $\|u_\delta\| \to \infty$ as $\delta \to 0$. Consider the re-scaled sequence $(w_\delta)_\delta$ given by

$$w_\delta := \frac{u_\delta}{\|u_\delta\|} \in \mathcal{H}.$$

Clearly, $(w_\delta)_\delta$ is bounded in $\mathcal{H}$ and thus there is an element $w \in \mathcal{H}$ and a subsequence $(w_\delta)_\delta$ such that $w_\delta \to w$ weakly in $\mathcal{H}$. For each $\delta > 0$, the function $w_\delta \in \mathcal{H}$ satisfies the equation

$$b_\delta(w_\delta, v) = \frac{\ell(v)}{\|u_\delta\|} \qquad \text{for all } v \in \mathcal{H}. \tag{11}$$

As the sequence $(u_\delta)_\delta$ is unbounded in $\mathcal{H}$, we deduce from *(b)*, *(c)*, and Eq. (11) that

$$b(w, v) = \lim_{\delta\to 0} b_\delta(w_\delta, v) = \lim_{\delta\to 0} \frac{\ell(v)}{\|u_\delta\|} = 0 \qquad \text{for all } v \in \mathcal{H}.$$

By assumption $\ker(b) = \{0\}$ and thus $w = 0$.

As ℓ is bounded, we deduce from (11) that $\lim_{\delta\to 0} b_\delta(w_\delta, w_\delta) = 0$. The sesquilinear forms $(b_\delta)_\delta$ satisfy Assumption 2.1; in particular, a Gårding inequality holds. Consequently, there exist a compact operator $K\colon \mathcal{H} \to \mathcal{H}^*$, which is independent of $\delta > 0$, and a constant $c > 0$ such that

$$0 < c = c\,\|w_\delta\|^2 \leq |b_\delta(w_\delta, w_\delta)| + |\langle K w_\delta, w_\delta\rangle| \, . \tag{12}$$

As $(w_\delta)_\delta$ weakly converges to w in $\mathcal{H}$, by compactness of K, we have the strong convergence $\lim_{\delta\to 0} K w_\delta = K w$ in $\mathcal{H}$. Thus, sending $\delta \to 0$ in (12), we find that

$$0 < c \le |\langle K w, w\rangle| \,,$$

This is a contradiction to $w = 0$ and hence the sequence $(u_\delta)_\delta$ cannot be unbounded in $\mathcal{H}$. □

3 Existence Result for a Periodic Bounded Waveguide

In this section we apply Theorem 2.1 to a Helmholtz equation in a bounded waveguide, and extend the analytical results of [4].

3.1 Setting

Let us describe the setting of [4]: Fix a periodicity $\varepsilon > 0$, and define the domain $\Omega_R := (-R\varepsilon, R\varepsilon)\times(0, \varepsilon M)$ with $R > 0$ and $M \in \mathbb{N}$. Assume that $a_+, a_- : \mathbb{R}^2 \to \mathbb{R}$ are two ε-periodic functions for which there exist constants $c_2 > c_1 > 0$ such that $c_1 \le a_+, a_- \le c_2$. Define $a : \mathbb{R}^2 \to \mathbb{R}$ by

$$a(x) := \begin{cases} a_+(x) & \text{for } x \in [0,\infty)\times\mathbb{R}\,, \\ a_-(x) & \text{for } x \in (-\infty, 0)\times\mathbb{R}\,. \end{cases} \tag{13}$$

Fix $\delta, R, M, L > 0$ and $f \in L^2(\Omega_R)$. Below, we will construct functions $u_\delta : \Omega_{R+L} \to \mathbb{C}$ that satisfy the Helmholtz equation with absorption parameter $\delta > 0$,

$$-\nabla\cdot(a\nabla u_\delta) - \omega^2(1+\mathrm{i}\delta)u_\delta = f \quad \text{in } \Omega_R\,, \tag{14}$$

and that are periodic in vertical direction. Furthermore, we construct a vertically periodic function $u : \Omega_{R+L} \to \mathbb{C}$ satisfying

$$-\nabla\cdot\big(a\nabla u\big) - \omega^2 u = f \quad \text{in } \Omega_R\,. \tag{15}$$

To formulate the radiation conditions, we introduce two *radiation boxes*

$$W_{R,L}^- := \big(-\varepsilon(R+L), -\varepsilon R\big)\times(0,\varepsilon M) \quad \text{and} \quad W_{R,L}^+ := (\varepsilon R, \varepsilon(R+L))\times(0,\varepsilon M)\,.$$

The restriction of $u\colon \Omega_{R+L} \to \mathbb{R}$ to $W^{\pm}_{R,L}$ is denoted by $\mathcal{R}^{\pm}_{R,L}(u)$. More precisely, we also shift in the parameter x_1 and set $\mathcal{R}^{-}_{R,L}(u), \mathcal{R}^{+}_{R,L}(u)\colon [0, \varepsilon L)\times[0, \varepsilon M) \to \mathbb{R}$,

$$\mathcal{R}^{-}_{R,L}(u)(x_1, x_2) := u\big(-\varepsilon(R+L)+x_1, x_2\big) \text{ and } \mathcal{R}^{+}_{R,L}(u)(x_1, x_2) := u(\varepsilon R+x_1, x_2)\,.$$

Roughly speaking, the radiation condition at infinity is replaced by the requirement that the solution u of (14) should be a right-going wave in $W^{+}_{R,L}$ and a left-going wave in $W^{-}_{R,L}$. To make this idea precise, we need the Bloch formalism.

Bloch Expansion Set $Y_\varepsilon := \varepsilon(0, 1)^2$. For fixed $L, M \in \mathbb{N}$ we consider the index sets

$$Q_L := \left\{0, \frac{1}{L}, \dots, \frac{L-1}{L}\right\}, \qquad Q_M := \left\{0, \frac{1}{M}, \dots, \frac{M-1}{M}\right\},$$

and

$$I_{L,M} := \{(j, m) :\ j \in Q_L \times Q_M \text{ and } m \in \mathbb{N}_0\}\,.$$

For each $j \in Q_L \times Q_M$ we consider $\mathcal{L}^{+}_j, \mathcal{L}^{-}_j\colon H^1_{\text{per}}(Y_\varepsilon; \mathbb{C}) \to (H^1_{\text{per}}(Y_\varepsilon; \mathbb{C}))^*$ given by

$$\mathcal{L}^{+}_j := -(\nabla + 2\pi \mathrm{i} j/\varepsilon)\cdot\big[a_+(x)\big(\nabla + 2\pi \mathrm{i} j/\varepsilon\big)\big],$$

and

$$\mathcal{L}^{-}_j := -(\nabla + 2\pi \mathrm{i} j/\varepsilon)\cdot\big[a_-(x)\big(\nabla + 2\pi \mathrm{i} j/\varepsilon\big)\big].$$

The definition of these differential operators is motivated by the following observation: If $\psi^{\pm}_j$ is an eigenfunction of $\mathcal{L}^{\pm}_j$ with eigenvalue $\mu^{\pm}_j$, then $x \mapsto \psi^{\pm}_j(x)e^{2\pi \mathrm{i} j\cdot x/\varepsilon}$ is a solution to the Helmholtz equation (15) with $a = a_\pm$, $\omega^2 = \mu^{\pm}_j$, and $f = 0$.

Let $j \in [0, 1]^2$ be a fixed wave vector. We denote by $(\psi^{+}_{j,m})_{m\in\mathbb{N}_0}$ a family of eigenfunctions of the differential operator $\mathcal{L}^{+}_j$. The labelling is in such a way that the corresponding eigenvalues $(\mu^{+}_m(j))_m$ satisfy $\mu^{+}_{m+1}(j) \geq \mu^{+}_m(j)$ for all $m \in \mathbb{N}_0$. Similarly, the eigenfunctions of $\mathcal{L}^{-}_j$ are denoted by $(\psi^{-}_{j,m})_{m\in\mathbb{N}_0}$. We normalise with the condition

$$\fint_{Y_\varepsilon} \left|\psi^{\pm}_{j,m}\right|^2 = 1 \qquad \text{for all } m \in \mathbb{N}_0\,.$$

The eigenfunctions $(\psi^{\pm}_{j,m})_{m\in\mathbb{N}}$ are called *Bloch eigenfunctions*. The following expansion is classical; see, for instance, [1, Lemma 4.9].

Lemma 3.1 (Bloch Expansion) *For $L, M \in \mathbb{N}$ and $\varepsilon > 0$ denote by W the rectangle $(0, \varepsilon L) \times (0, \varepsilon M)$. Let $u \in L^2(W; \mathbb{C})$. Then for both families of eigenfunctions $(\psi^+_{j,m})_m$ and $(\psi^-_{j,m})_m$, the function u admits a unique expansion*

$$u(x) = \sum_{(j,m)\in I_{L,M}} \alpha^{\pm}_{j,m} \psi^{\pm}_{j,m}(x) \mathrm{e}^{2\pi \mathrm{i} j\cdot x/\varepsilon} \quad \text{in } L^2(W, \mathbb{C}), \tag{16}$$

where $\alpha^{\pm}_{j,m} \in \mathbb{C}$ for all $(j, m) \in I_{L,M}$.

To shorten the notation, for $\lambda = (j, m) \in I_{L,M}$ we set

$$U^{\pm}_{\lambda} \colon W \to \mathbb{C}, \quad U^{\pm}_{\lambda}(x) := \psi^{\pm}_{\lambda}(x)\mathrm{e}^{2\pi \mathrm{i} j\cdot x/\varepsilon}.$$

Having the families $(U^+_\lambda)_\lambda$ and $(U^-_\lambda)_\lambda$, the expansion (16) reads

$$u = \sum_{\lambda\in I_{L,M}} \alpha^{\pm}_{\lambda} U^{\pm}_{\lambda} \quad \text{in } L^2(W; \mathbb{C}),$$

where $\alpha^{\pm}_{\lambda} \in \mathbb{C}$ for all $\lambda \in I_{L,M}$.

We recall from [1, Lemma 4.9] the following Plancherel formula

$$\|u\|^2_{L^2(W;\mathbb{C})} = \varepsilon^2 LM \sum_{\lambda\in I_{L,M}} |\alpha^{\pm}_{\lambda}|^2 .$$

Radiation Condition in a Bounded Waveguide The Bloch eigenfunctions $(U^+_\lambda)_\lambda$ and $(U^-_\lambda)_\lambda$ may transport energy in any direction. To indicate in which direction along the x_1-axis energy is transported, we introduce the *Poynting numbers*

$$P^{\pm}_{\lambda} := \operatorname{Im} \fint_{Y_\varepsilon} \bar{U}^{\pm}_{\lambda}(x) e_1 \cdot \left[a_{\pm}(x) \nabla U^{\pm}_{\lambda}(x)\right] \mathrm{d}x . \tag{17}$$

For $L, M \in \mathbb{N}$, we choose two non-empty index sets

$$I^-_{L,M} \subset \left\{\lambda \in I_{L,M} : P^-_\lambda < 0\right\} \quad \text{and} \quad I^+_{L,M} \subset \left\{\lambda \in I_{L,M} : P^+_\lambda > 0\right\} . \tag{18}$$

In order to establish an existence result, we assume $I^{\pm}_{L,M}$ to be finite; see Assumption 3.1. Given the sets $I^-_{L,M}$ and $I^+_{L,M}$, we define the function spaces

$$X^-_{\leftarrow} := \operatorname{span}\left\{U^-_\lambda : \lambda \in I^-_{L,M}\right\} \quad \text{and} \quad X^+_{\rightarrow} := \operatorname{span}\left\{U^+_\lambda : \lambda \in I^+_{L,M}\right\} . \tag{19}$$

The space $X_{\leftarrow}^{-}$ is spanned by all left-going Bloch eigenfunctions U_{λ}^{-} which are defined on the radiation box $W_{R,L}^{-}$. Accordingly, $X_{\rightarrow}^{+}$ is the space spanned by all right-going Bloch eigenfunctions U_{λ}^{+} defined on $W_{R,L}^{+}$.

The function space in which we look for a solution to (14) is

$$V_R := \left\{ u \in H^1(\Omega_{R+L};\mathbb{C}) : \begin{array}{l} u(x_1,\cdot) \text{ is periodic for a.e. } x_1 \in (-R,R)\,, \\ \mathcal{R}_{R,L}^{-}(u) \in X_{\leftarrow}^{-} \text{ and } \mathcal{R}_{R,L}^{+}(u) \in X_{\rightarrow}^{+} \end{array} \right\}, \tag{20}$$

the periodicity is $u(x_1, \varepsilon H) = u(x_1, 0)$ for a.e. $x_1 \in (-R, R)$ in the sense of traces.

The Variational Formulation For a weak formulation of (14) and (15), we introduce the sesquilinear forms $Q_+, Q_-, Q\colon V_R \times V_R \to \mathbb{C}$,

$$Q_+(u,v) := \frac{1}{\varepsilon L}\int_{W_{R,L}^{+}} \bar{u} e_1 \cdot a_+ \nabla v\,, \quad Q_-(u,v) := \frac{1}{\varepsilon L}\int_{W_{R,L}^{-}} \bar{u} e_1 \cdot a_- \nabla v\,, \tag{21}$$

and

$$Q(u,v) := Q_+(u,v) - Q_-(u,v) \tag{22}$$

which we will use to encode the radiation conditions. The signs in the definition of Q are chosen to ensure a positivity property; see the proof of Lemma 3.4. We further define the function $\vartheta : \Omega_{R+L} \to [0,1]$,

$$\vartheta(x_1,x_2) := \begin{cases} 0 & \text{for } |x_1| \geq R+L\,, \\ 1 & \text{for } |x_1| \leq R\,, \\ \frac{1}{L}(R+L-|x_1|) & \text{otherwise}\,. \end{cases} \tag{23}$$

For $R, L, \omega^2 > 0$ and $\delta \geq 0$, we define the sesquilinear form $\beta_\delta\colon V_R \times V_R \to \mathbb{C}$ by

$$\beta_\delta(u,v) := \int_{\Omega_{R+L}} \vartheta a \nabla\bar{u}\cdot\nabla v - \omega^2(1-\mathrm{i}\delta)\int_{\Omega_{R+L}} \bar{u} v \vartheta + Q(u,v)\,, \tag{24}$$

where the coefficient $a\colon \Omega_{R+L} \to \mathbb{R}$ is given in (13).

The following problem was introduced in [4].

Definition 3.1 (Solution Concept) We say that $u\colon \Omega_{R+L} \to \mathbb{C}$ solves the truncated radiation problem with absorption if it is an element of V_R and satisfies

$$\beta_\delta(u,v) = \int_{\Omega_R} \bar{f} v \tag{25}$$

for all $v \in V_R$. If u satisfies (25) with $\delta = 0$, then u is a solution to the truncated radiation problem.

3.2 Existence Result for the Helmholtz Equation with Damping

We need the following assumption to prove the existence of solutions to the truncated radiation problem.

Assumption 3.1 Let $I^-_{L,M}$ and $I^+_{L,M}$ be the index sets of (18), and let $X^-_{\leftarrow}$ and $X^+_{\rightarrow}$ be the function spaces of (19). We demand:

Two different indices, $\lambda = (j, m)$ and $\tilde{\lambda} = (\tilde{j}, \tilde{m})$ in $I^-_{L,M}$ with $\lambda \neq \tilde{\lambda}$, satisfy $j \neq \tilde{j}$. The same holds for $I^+_{L,M}$.

We note that every $\lambda \in I^-_{L,M}$ is a pair (j, m) with $j \in Q_L \times Q_M$ and $m \in \mathbb{N}$, where Q_L and Q_M are finite index sets. Thus, Assumption 3.1 implies that both index sets $I^-_{L,M}$ and $I^+_{L,M}$ are finite. This ensures, in particular, that the function spaces $X^-_{\leftarrow}$ and $X^+_{\rightarrow}$ are finite dimensional and hence closed. $V_R \subset H^1(\Omega_{R+L})$ is thus a Hilbert space.

Remark 3.1 As the spaces $X^-_{\leftarrow}$ and $X^+_{\rightarrow}$ are finite dimensional subspaces of H^1, there exists $C_0 > 0$ such that for all $u \in X^-_{\leftarrow}$ and $v \in X^+_{\rightarrow}$ there holds

$$\|u\|_{H^1(W^-_{R,L})} \leq C_0 \|u\|_{L^2(W^-_{R,L})} \quad \text{and} \quad \|v\|_{H^1(W^+_{R,L})} \leq C_0 \|v\|_{L^2(W^+_{R,L})} \,. \tag{26}$$

A consequence of (26) is a Gårding inequality. We define $K\colon V_R \to V_R^*$, $u \mapsto Ku$ by

$$\langle Ku, v\rangle_{V_R^*, V_R} := 2\omega^2 \langle u, v\rangle_{L^2(\Omega_{R+L})} - Q(u, v)\,, \tag{27}$$

where Q is the sesquilinear form defined in (22). Let us first show that K is indeed compact.

Lemma 3.2 (Compactness of K) *Let R, L, and ω^2 be positive numbers. Let V_R be the function space defined in* (20) *and let Assumption 3.1 hold. Then the operator K defined in* (27) *is compact.*

Proof For brevity we write L^2 instead of $L^2(\Omega_{R+L})$. The compactness of $K\colon V_R \to V_R^*$ is equivalent to the following condition: Every bounded sequence $(u_m)_m$ in V_R admits a subsequence $(u_m)_m$ that weakly converges to some $u \in V_R$ and satisfies

$$\langle Ku_m, v_m\rangle_{V_R^*, V_R} \to \langle Ku, v\rangle_{V_R^*, V_R} \quad \text{as } m \to \infty$$

for all sequences $(v_m)_m$ in V_R with $v_m \to v$ weakly in V_R.

Let $(u_m)_m$ and $(v_m)_m$ be sequences in V_R such that $(u_m)_m$ is bounded and $(v_m)_m$ weakly converges in V_R to $v \in V_R$. As V_R is a Hilbert space, $(u_m)_m$ admits a subsequence $(u_m)_m$ that weakly converges to $u \in V_R$. The function space V_R is a subspace of $H^1(\Omega_{R+L})$ and thus, by the Rellich-Kondrachov theorem, $u_m \to u$ in L^2. Moreover, the weak convergence of $(v_m)_m$ in V_R implies that $v_m \to v$ weakly in L^2 as well as $\nabla v_m \to \nabla v$ weakly in L^2. Using the definition of Q in (21) and (22), the strong L^2-convergence of $(u_m)_m$ and the weak L^2-convergence of $(\nabla v_m)_m$, we deduce that $Q(u_m, v_m) \to Q(u, v)$ as $m \to \infty$. Consequently,

$$\begin{aligned}\lim_{m\to\infty} \langle K u_m, v_m\rangle_{V_R^*, V_R} &= \lim_{m\to\infty} \left(2\omega^2 \langle u_m, v_m\rangle_{L^2} - Q(u_m, v_m)\right) \\ &= 2\omega^2\langle u, v\rangle_{L^2} - Q(u, v) = \langle K u, v\rangle_{V_R^*, V_R} .\end{aligned}$$

This proves the claim. □

Lemma 3.3 (Gårding inequality) *Let R, L, and ω^2 be positive numbers, let V_R be the function space given in* (20)*, and let K be the operator of* (27)*. Suppose that Assumption 3.1 holds. For each $\delta \geq 0$, the sesquilinear form β_δ satisfies a Gårding inequality with the operator K. More precisely, there exists a constant $c > 0$ such that for all $u \in V_R$ there holds*

$$\left|\beta_\delta(u, u) + \langle K u, u\rangle_{V_R^*, V_R}\right| \geq c \, \|u\|^2_{H^1(\Omega_{R+L})} .$$

Proof Fix $u \in V_R$. Using the definition of K, we find that

$$\mathrm{Re}\left\{\beta_\delta(u, u) + \langle K u, u\rangle_{V_R^*, V_R}\right\} = \int_{\Omega_{R+L}} \vartheta a \, |\nabla u|^2 + \omega^2 \int_{\Omega_{R+L}} (2 - \vartheta)\, |u|^2 . \tag{28}$$

From (28) and (26) we deduce that for $u \in V_R$

$$\begin{aligned}\left|\beta_\delta(u, u) + \langle K u, u\rangle_{V_R^*, V_R}\right| &\geq c_1 \, \|\nabla u\|^2_{L^2(\Omega_R)} + \omega^2 \, \|u\|^2_{L^2(\Omega_{R+L})} \\ &\geq \min\left\{c_1, \omega^2\right\} \|u\|^2_{H^1(\Omega_R)} + \omega^2 C_0^{-1} \, \|u\|^2_{H^1(W^-_{R,L} \cup W^+_{R,L})} \\ &\geq C_1 \, \|u\|^2_{H^1(\Omega_{R+L})} .\end{aligned}$$

This proves the claim. □

Lemma 3.4 (Uniqueness for the Homogeneous Problem with $\delta > 0$) *Let R, L, and ω^2 be positive numbers, and let V_R be the function space defined in* (20)*. Fix $\delta > 0$. If $u \in V_R$ satisfies $\beta_\delta(u, \cdot) = 0$, then $u = 0$.*

Proof Let $u \in V_R$ satisfy $\beta_\delta(u, v) = 0$ for all $v \in V_R$. Taking the imaginary part of $\beta_\delta(u, u)$, we find that

$$\omega^2 \delta \int_{\Omega_{R+L}} |u|^2 \vartheta + \operatorname{Im} Q(u, u) = 0 \,. \tag{29}$$

We first show that both terms are non-negative. By definition of the function space V_R, the function u can be expanded in Bloch eigenfunctions in the radiation boxes $W^-_{R,L}$ and $W^+_{R,L}$,

$$\mathcal{R}^-_{R,L}(u) = \sum_{\lambda \in I^-_{L,M}} \alpha^-_\lambda U^-_\lambda \quad \text{and} \quad \mathcal{R}^+_{R,L}(u) = \sum_{\lambda \in I^+_{L,M}} \alpha^+_\lambda U^+_\lambda \,, \tag{30}$$

where $\alpha^\pm_\lambda \in \mathbb{C}$ for all $\lambda \in I^\pm_{L,M}$. The Bloch eigenfunctions $(U^\pm_\lambda)$ are orthogonal with respect to $Q_\pm$, that is,

$$Q_\pm(U^\pm_\lambda, U^\pm_\mu) = 0 \quad \text{for all } \lambda, \mu \in I^\pm_{L,M} \text{ with } \lambda \neq \mu \,. \tag{31}$$

We refer to Lemma 2.4 in [4] for a proof of this orthogonality property. Using the expansion (30) and the orthogonality property (31), we compute that

$$\begin{aligned} Q(u, u) &= Q_+\big(\mathcal{R}^+_{R,L}(u), \mathcal{R}^+_{R,L}(u)\big) - Q_-\big(\mathcal{R}^-_{R,L}(u), \mathcal{R}^-_{R,L}(u)\big) \\ &= \sum_{\lambda \in I^+_{L,M}} |\alpha^+_\lambda|^2 Q_+(U^+_\lambda, U^+_\lambda) - \sum_{\lambda \in I^-_{L,M}} |\alpha^-_\lambda|^2 Q_-(U^-_\lambda, U^-_\lambda) \,. \end{aligned} \tag{32}$$

Using the definition of the Poynting numbers in (17), and the definition of the index sets $I^-_{L,M}$ and $I^+_{L,M}$, we infer from (32) that

$$\operatorname{Im} Q(u, u) = \sum_{\lambda \in I^+_{L,M}} |\alpha^+_\lambda|^2 P^+_\lambda - \sum_{\lambda \in I^-_{L,M}} |\alpha^-_\lambda|^2 P^-_\lambda \geq 0 \,,$$

with equality only if all coefficients α^+_λ and α^-_λ vanish.

Equation (29) has two non-negative terms on the left-hand side and therefore implies $u \equiv 0$ in Ω_R and $\alpha^\pm_\lambda = 0$ for all $\lambda \in I^\pm_{L,M}$. This shows $u \equiv 0$ in Ω_{R+L}. □

Theorem 3.1 (Existence Result for $\delta > 0$) *Let R, L, and ω^2 be positive numbers, $f \in L^2(\Omega_R)$, and let $a \colon \Omega_{R+L} \to \mathbb{R}$ be as in (13). Let V_R be the function space given in (20) and assume that Assumption 3.1 is satisfied. If $\delta > 0$, then there exists a unique solution $u \in V_R$ to the truncated radiation problem with absorption. More precisely, there exists a unique $u \in V_R$ such that for every $v \in V_R$ there holds*

$$\beta_\delta(u, v) = \int_{\Omega_R} \bar{f} v \,,$$

where β_δ is the sesquilinear form defined in (24).

Proof Thanks to Assumption 3.1, V_R is a Hilbert space. Lemma 3.4 implies that the kernel of β_δ is trivial. Lemma 2.2 implies the existence and uniqueness of u. $\square$

3.3 Existence Result for the Helmholtz Equation

In this section, we apply the limiting absorption principle to prove the existence of a solution to the truncated radiation problem. We recall that, by Definition 3.1, a function $u \in V_R$ is a solution provided

$$\beta(u, v) = \int_{\Omega_R} \bar{f} v \qquad \text{for all } v \in V_R\,,$$

where the sesquilinear form $\beta\colon V_R \times V_R \to \mathbb{C}$ is defined as

$$\beta(u, v) := \int_{\Omega_{R+L}} \vartheta a \nabla \bar{u} \cdot \nabla v - \omega^2 \int_{\Omega_{R+L}} \bar{u} v \vartheta + Q(u, v)\,. \tag{33}$$

Theorem 3.2 (Limiting Absorption Principle) *Let R, L and ω^2 be positive numbers, and let $a\colon \Omega_{R+L} \to \mathbb{R}$ be as in* (13). *Let V_R be the function space given in* (20) *and let Assumption 3.1 hold. Assume further that* $\ker(\beta) = \{0\}$. *Then for every $f \in L^2(\Omega_R)$ there exists a sequence $(u_\delta)_\delta$ of solutions to the truncated radiation problem with right-hand side f that weakly converges in V_R to a function $u \in V_R$ satisfying*

$$\beta(u, v) = \int_{\Omega_R} \bar{f} v \qquad \textit{for all } v \in V_R\,. \tag{34}$$

Moreover, this limit u is the unique solution to (34).

Proof We use the limiting absorption principle of Theorem 2.1. Thanks to Assumption 3.1, V_R is a Hilbert space. For each $\delta > 0$, the sesquilinear form β_δ defined in (24) is bounded. Regarding premises *(a)-(c)* of Theorem 2.1: One readily checks that $\lim_{\delta\to 0} \beta_\delta(u, v) = \beta(u, v)$ for all $u, v \in V_R$; hence *(b)* is satisfied. Lemmas 3.3 and 3.4 ensure that *(a)* is satisfied. It remains to show *(c)*.

Let $(u_\delta)_\delta$ be a sequence in V_R with $u_\delta \to u$ weakly in V_R. For every $v \in V_R$, we have

$$\lim_{\delta\to 0}\left[\omega^2(1+\mathrm{i}\delta)\int_{\Omega_{R+L}} \vartheta(\bar{u}_\delta - \bar{u})v\right] = 0 \quad \text{and} \quad \lim_{\delta\to 0} Q(u_\delta - u, v) = 0 \tag{35}$$

and

$$\lim_{\delta \to 0} \int_{\Omega_{R+L}} \vartheta a\big(\nabla u_\delta - \nabla u\big) \cdot \nabla v = 0\,. \tag{36}$$

Combining (35) and (36), we infer that $\lim_{\delta \to 0} \beta_\delta(u_\delta - u, v) = 0$ for all $v \in V_R$, which shows premise *(c)*.

We can therefore apply Theorem 2.1 and deduce the claim. □

Remark 3.2 Using standard arguments for symmetric and elliptic differential operators, one can show that there exists an at most countable set $D \subset [0, \infty)$ such that for all $\omega^2 \notin D$ the kernel $\ker(\beta)$ is trivial.

Comments on Theorem 3.2
Theorem 3.2 contains an existence result: for each $f \in L^2(\Omega_R)$ there is a unique solution $u \in V_R$ to (34). We mention that this existence result can also be obtained without the limiting absorption principle of Theorem 2.1. Indeed, by Lemma 3.3, the sesquilinear form $\beta = \beta_0$ satisfies a Gårding inequality. Thus, by Proposition 2.1, for every $\ell \in V_R^*$ there is a unique solution $u \in V_R$ to $\beta(u, \cdot) = \ell(\cdot)$ provided $\ker(\beta) = \{0\}$.

The existence result of Theorem 3.2 is restricted to bounded domains. This is a consequence of the method of proof: We do not see how Gårding's inequality could be exploited in an unbounded domain. An existence result for unbounded domains with energy methods was recently derived in [14]. This new result actually uses solutions on bounded domains (very much like in the contribution at hand) in order to find solutions on unbounded domains in the limit $R \to \infty$.

The construction of solutions in [14] is done in such a way that $\delta \to 0$ is performed for fixed R, then the limit $R \to \infty$ is performed. In particular, no limiting absorption principle in the unbounded domain is obtained in [14]. We regard it as an additional piece of information that Theorem 3.2 does not only provide an existence result for $\delta = 0$, but that it also provides the limiting absorption principle.

The theorem remains valid for matrix-valued elliptic coefficient fields a. The proof for such a general a requires only minor notational changes.

References

1. Allaire, G., Conca, C.: Bloch wave homogenization and spectral asymptotic analysis. Journal de Mathématiques Pures et Appliquées **77**(2), 153–208 (1998). https://doi.org/10.1016/s0021-7824(98)80068-8
2. Colton, D., Kress, R.: Inverse acoustic and electromagnetic scattering theory. Applied Mathematical Sciences, vol. 93, 2nd edn. Springer-Verlag, Berlin (1998). https://doi.org/10.1007/978-3-662-03537-5
3. Conway, J.B.: A course in functional analysis. Graduate Texts in Mathematics, vol. 96, 2nd edn. Springer, New York (1990)

4. Dohnal, T., Schweizer, B.: A Bloch wave numerical scheme for scattering problems in periodic wave-guides. SIAM J. Numer. Anal. **56**(3), 1848–1870 (2018). https://doi.org/10.1137/17M1141643
5. Fliss, S.: A Dirichlet-to-Neumann approach for the exact computation of guided modes in photonic crystal waveguides. SIAM J. Sci. Comput. **35**(2), B438–B461 (2013). https://doi.org/10.1137/12086697X
6. Fliss, S., Joly, P.: Solutions of the time-harmonic wave equation in periodic waveguides: asymptotic behaviour and radiation condition. Arch. Ration. Mech. Anal. **219**(1), 349–386 (2016). https://doi.org/10.1007/s00205-015-0897-3
7. Hoang, V.: The limiting absorption principle for a periodic semi-infinite waveguide. SIAM J. Appl. Math. **71**(3), 791–810 (2011). https://doi.org/10.1137/100791798
8. Hsiao, G.C., Wendland, W.L.: Boundary integral equations. Applied Mathematical Sciences, vol. 164. Springer, Berlin (2008). https://doi.org/10.1007/978-3-540-68545-6
9. Kress, R.: Fredholm's alternative for compact bilinear forms in reflexive Banach spaces. J. Diff. Equ. **25**(2), 216–226 (1977). https://doi.org/10.1016/0022-0396(77)90201-7
10. Lamacz, A., Schweizer, B.: Outgoing wave conditions in photonic crystals and transmission properties at interfaces. ESAIM Math. Model. Numer. Anal. **52**(5), 1913–1945 (2018). https://doi.org/10.1051/m2an/2018026
11. Radosz, M.: New limiting absorption and limit amplitude principles for periodic operators. Z. Angew. Math. Phys. **66**(2), 253–275 (2015). https://doi.org/10.1007/s00033-014-0399-4
12. Rellich, F.: Über das asymptotische Verhalten der Lösungen von $\Delta u + \lambda u = 0$ in unendlichen Gebieten. Jber. Deutsch. Math. Verein. **53**, 57–65 (1943)
13. Schot, S.H.: Eighty years of Sommerfeld's radiation condition. Historia Math. **19**(4), 385–401 (1992). https://doi.org/10.1016/0315-0860(92)90004-U
14. Schweizer, B.: Existence results for the Helmholtz equation in periodic wave-guides with energy methods. Technical Report, TU Dortmund (2019)
15. Sommerfeld, A.: Die Greensche Funktion der Schwingungsgleichung. Jahresber. Deutsch. Math.-Verein. **21**, 309–353 (1912)

Some Inverse Scattering Problems for Perturbations of the Biharmonic Operator

Valery Serov

Abstract Some inverse scattering problems for the three-dimensional biharmonic operator are considered. The operator is perturbed by first and zero order perturbations, which may be complex-valued and singular. We show the existence of the scattering solutions in the Sobolev space $W^1_\infty(R^3)$. One of the main result of this paper is the proof of analogue of Saito's formula (in different form as known before), which can be used to prove a uniqueness theorem for the inverse scattering problem. Another main result is to obtain the estimates for the kernel of the resolvent of the direct operator in W^1_∞ and to prove the reconstruction formula for the unknown coefficients of this perturbation.

1 Introduction

We consider the following three-dimensional biharmonic operator

$$H_4 u(x) := \Delta^2 u(x) + \mathbf{W}(x) \cdot \nabla u(x) + V(x)u(x), \tag{1}$$

where Δ is the Laplacian and $\cdot$ denotes the dot-product in R^3 for complex-valued vectors in C^3. The bi-Laplacian is perturbed by the first and zero order perturbations, vector-valued function $\mathbf{W}$ and a scalar function V, that may be complex-valued and very singular. More precisely, we assume that $\mathbf{W}$ belongs to $L^\infty(R^3)$ and V belongs to the Kato space K_3, i.e.

$$\sup_{x \in R^3} \int_{|x-y| \le 1} \frac{|V(y)|}{|x-y|} dy < \infty, \tag{2}$$

V. Serov (✉)
University of Oulu, Oulu, Finland
e-mail: vserov@cc.oulu.fi

W. Dörfler et al. (eds.), *Mathematics of Wave Phenomena*, Trends in Mathematics,
https://doi.org/10.1007/978-3-030-47174-3_19

and both have special behaviour at the infinity

$$|\mathbf{W}(x)|, \quad |V(x)| \leq \frac{C}{|x|^{\mu}}, \quad |x| \geq R, \quad \mu > 3, \tag{3}$$

where $C > 0$, and $R > 0$ is big enough.

The motivation to study operators of order four appears for example in the study of elasticity and the theory of vibration of beams. As a concrete example, the linear beam equation [2]

$$\partial_t^2 U(x,t) + \Delta^2 U(x,t) + m(x)U(x,t) = 0,$$

under time-harmonic assumptions $U(x,t) = u(x)e^{-i\omega t}$ leads to the equation

$$\Delta^2 u(x) + m(x)u(x) = \omega^2 u(x).$$

The wave parameter ω is fixed (in general) here, nevertheless we can consider it fixed but big enough. This allows to consider some scattering problems with high frequency for this potential equation. In particular, we can use some numerical methods in that case. For the scattering problems (including linear or nonlinear equations), see for example [3, 9] and references therein. In terms of inverse problems for bi- and poly-harmonic operators it might be mentioned some solutions to inverse boundary value problems, see for example [6]. One can refer also to [14], where the fundamental result concerning the global uniqueness for an inverse boundary value problem was proved. For the operators with vector potential one can mention [13].

The present work is concerned with the following scattering problem for operator H_4 given by

$$H_4 u(x) = k^4 u(x), \quad u(x) = u_0(x) + u_{sc}(x), \quad u_0(x) = e^{ikx\cdot\theta}, \quad \theta \in S^2, \tag{4}$$

where scattered wave u_{sc} and its Laplacian Δu_{sc} are required to satisfy Sommerfeld radiation condition at the infinity

$$\lim_{r\to\infty} r^{\frac{n-1}{2}} \left(\frac{\partial f(x)}{\partial r} - ikf(x) \right) = 0, \quad r = |x|, \quad f = u_{sc} \quad or \quad f = \Delta u_{sc}. \tag{5}$$

The author was originally motivated to start studying scattering for fourth order operators by the article [1] (see also [15]), where the time-evolution of several scattering coefficients for the one-dimensional biharmonic operator were studied. In terms of inverse scattering problems for fourth order operator might be mentioned Iwasaki's results [4, 5]. In these works Iwasaki studied the scattering problem in one-dimensional case and considered the inverse problem as a Riemann–Hilbert boundary value problem with respect to the wave number k in the complex cone $\arg([0, \frac{\pi}{4}]) \setminus \{0\}$.

The main differences of present work is that all considered scattering problems are studying here in the usual Sobolev spaces (compared with the weighted Sobolev spaces in the previous publications) and that the possible local singularities of the unknown coefficients $\mathbf{W}$ and V are stronger than it was considered before. Another important difference is concerned to Theorem 2, where the inverse scattering problem in terms of the Green's functions is considered.

We are looking for the scattering solutions u_{sc} to the Eq. (4) in the Sobolev space $W^1_\infty(R^3)$. Under the Sommerfeld radiation conditions (5) the scattering solutions to Eq. (4) are the unique solutions of the integral Lippmann–Schwinger equation (see [15] for details)

$$u(x) = u_0(x) - \int_{R^3} G_k^+(|x-y|)(\mathbf{W}(y) \cdot \nabla u(y) + V(y)u(y))\, dy, \tag{6}$$

where G_k^+ is the outgoing fundamental solution of the operator $(\Delta^2 - k^4)$ in R^3, i.e., the kernel of the integral operator $(\Delta^2 - k^4 - i0)^{-1}$. This function G_k^+ in R^3 has the following form

$$G_k^+(|x|) = \frac{e^{ik|x|} - e^{-k|x|}}{8\pi k^2 |x|}, \quad k > 0. \tag{7}$$

Since u_0 is just a bounded function with the norm $\|u_0\|_{L^\infty(R^3)} = 1$ it is more convenient to study (in stead of (6)) the equivalent integral equation for the scattered wavc, namely

$$u_{sc}(x) = \tilde{u}_0(x) - \int_{R^3} G_k^+(|x-y|)(\mathbf{W}(y) \cdot \nabla u_{sc}(y) + V(y)u_{sc}(y))\, dy =: \tilde{u}_0 + L_k(u_{sc}), \tag{8}$$

where $\tilde{u}_0 = L_k(u_0)$. As it is shown (see [15]) a solution to the scattering problem (4), (5) also satisfies Eq. (6). This translates the study of the scattering problem to the study of integral equation (6) ((8)). It will be shown that this solution admits for fixed $k > 0$ asymptotic representation

$$u(x, k, \theta) = e^{ikx\cdot\theta} + C\frac{e^{ik|x|}}{k^2|x|}A(k, \theta', \theta) + o\left(\frac{1}{|x|}\right), \quad |x| \to \infty,$$

where $\theta, \theta' = \frac{x}{|x|} \in S^2$, C is known constant, and function $A(k, \theta', \theta)$ is called a scattering amplitude and defined by

$$A(k, \theta', \theta) := \int_{R^3} e^{-ik\theta'\cdot y}(\mathbf{W}(y) \cdot \nabla u(y, k, \theta) + V(y)u(y, k, \theta))\, dy. \tag{9}$$

From the point of view of inverse problems one regards this scattering amplitude as one possible scattering data. For these purposes one requires the scattering amplitude to be known for all possible angles θ and θ' and all arbitrarily high frequencies ($k > 0$ large). Then Saito's formula is given by the following theorem.

Theorem 1 (Saito's Formula) *Assume that* $\mathbf{W}$ *belongs to* $L^\infty(R^3)$, V *belongs to the Kato space* K_3 *and both satisfy conditions* (3). *Then the limit*

$$\lim_{k\to+\infty} k^2 \int\limits_{S^2\times S^2} e^{-ik(\theta-\theta')\cdot x} A(k,\theta',\theta)\, d\theta\, d\theta' =$$

$$= 4\pi^2 \nabla_x \int\limits_{R^3} \frac{\mathbf{W}(y)}{|x-y|^2}\, dy + 8\pi^2 \int\limits_{R^3} \frac{V(y)}{|x-y|^2}\, dy \tag{10}$$

holds in the sense of distributions.

The significance of Saito's formula for inverse problems is apparent from its corollary.

Corollary 1 (Uniqueness) *Let* $\mathbf{W}_1$, V_1 *and* $\mathbf{W}_2$, V_2 *be as in Theorem 1. Let, in addition, Fourier transform of* $\mathbf{W}_1$ *and* $\mathbf{W}_2$ *behaves as* $o(|\xi|^{-1})$ *at the infinity. If the corresponding scattering amplitudes for these coefficients coincide for some sequence* $k_j \to +\infty$ *then*

$$V_1(x) - \frac{1}{2}\nabla\delta \star \mathbf{W}_1(x) = V_2(x) - \frac{1}{2}\nabla\delta \star \mathbf{W}_2(x)$$

in the sense of tempered distributions, where δ *is Dirac delta function and* $\star$ *denotes the convolution.*

The limiting absorption principle can be applied for the operator H_4 to obtain the existence of the integral operator (see [15])

$$\hat{G}_p := \lim_{\epsilon\to+0} (H_4 - k^4 - i\epsilon)^{-1}$$

such that the kernel $G_p(x, y, k)$ for $k > 0$ large enough is the unique solution of the integral equation

$$G_p(x,y,k) = G_k^+(x,y,k) - \int\limits_{R^3} G_k^+(|x-z|)(\mathbf{W}(z)\cdot\nabla G_p(z,y,k) + V(z)G_p(y,z,k))\, dz, \tag{11}$$

From the point of view of inverse problems this kernel G_p can be considered as another possible scattering data. More precisely the following theorem holds.

Theorem 2 (Reconstruction) *Assume that* $\mathbf{W}$ *belongs to* $L^{\infty}(R^3)$*,* V *belongs to the Kato space* K_3 *and* V *satisfies conditions* (3) *and, in addition,* $\mathbf{W}$ *satisfies this condition with* $\mu > 4$*. Then for each fixed* $\xi \in R^3$

$$\mathcal{F}^{-1}(V)(\xi) - \frac{i\xi}{2} \cdot \mathcal{F}^{-1}(\mathbf{W})(\xi) =$$

$$= \lim_{k,|x|,|y| \to +\infty} 32\sqrt{2\pi} k^4 |x||y| e^{-ik(|x|+|y|)} (G_k^+(|x-y|) - G_p(x, y, k)) \tag{12}$$

such that $\xi = k\left(\frac{x}{|x|} + \frac{y}{|y|}\right)$.

Corollary 2 (Uniqueness-II) *Let* $\mathbf{W}_1$, V_1 *and* $\mathbf{W}_2$, V_2 *be as in Theorem 2. If the corresponding kernels* $G_p^{(1)}(x, y, k)$ *and* $G_p^{(2)}(x, y, k)$ *for these coefficients coincide for all* x, y *large enough and some sequence* $k_j \to +\infty$ *then*

$$V_1(x) - \frac{1}{2} \nabla\delta \star \mathbf{W}_1(x) = V_2(x) - \frac{1}{2} \nabla\delta \star \mathbf{W}_2(x)$$

in the sense of tempered distributions, where δ *is Dirac delta function and* $\star$ *denotes the convolution.*

This paper is organised as follows. In Sect. 2 some notations and estimates for G_k^+ are recalled. Then it will be proved the existence and the uniqueness of the solutions to (6) and (11) together with asymptotic behaviour of the scattering solution u and the kernel G_p. Several estimates for u and G_p are also given. Finally in Sect. 3 it will be given the proof of Theorem 1 and 2.

2 Solvability of Direct Scattering Problems

We use the usual definitions of the Sobolev space W^1_∞ and the Lebesgue spaces $L^p, 1 \leq p \leq \infty$. We use also the following definitions for three-dimensional Fourier transform $\mathcal{F}$ and inverse Fourier transform $\mathcal{F}^{-1}$:

$$\mathcal{F}(f)(\xi) = \frac{1}{\sqrt{(2\pi)^3}} \int_{R^3} f(x) e^{ix\cdot\xi}\, dx, \quad \mathcal{F}^{-1}(f)(x) = \frac{1}{\sqrt{(2\pi)^3}} \int_{R^3} f(x) e^{-ix\cdot\xi}\, d\xi.$$

Next, taking into account the definition (7) of $G_k^+(|x|)$ in R^3 we obtain

$$|G_k^+(|x|)| \leq \frac{1}{4\pi k^2 |x|}, \quad |\nabla G_k^+(|x|)| \leq \frac{1}{2\pi k |x|}, \quad k > 0, \quad x \in R^3. \tag{13}$$

We now proceed to prove some estimates for the operator L_k.

Proposition 1 *Let* $\mathbf{W}$ *belongs* $L^{\infty}(R^3)$, V *belongs to the Kato space* K_3 *and both satisfy conditions* (3). *Then the following properties are satisfied.*

1. The function $\tilde{u}_0$ *belongs to* $W^1_{\infty}(R^3)$ *with the estimates*

$$\|\tilde{u}_0\|_{L^{\infty}(R^3)} \le \frac{c_0}{k}, \quad \|\nabla \tilde{u}_0\|_{L^{\infty}(R^3)} \le c_0, \quad k \ge 1,$$

where constant c_0 *is equal to*

$$c_0 = \frac{1}{2\pi}\left(\|\mathbf{W}\|_{L^{\infty}} C_R + C_V + 2CC_{\mu}\right), \quad C_R := \sup_{x\in R^3} \int_{|y|\le R} \frac{1}{|x-y|}\, dy,$$

$$C_V := \sup_{x\in R^3} \int_{|y|\le R} \frac{|V(y)|}{|x-y|}\, dy, \quad C_{\mu} := \sup_{x\in R^3} \int_{|y|\ge R} \frac{1}{|x-y||y|^{\mu}}\, dy \tag{14}$$

with C, R, μ *are as in* (3).

2. *The operator* $L_k : W^1_{\infty}(R^3) \to W^1_{\infty}(R^3)$ *is bounded and satisfies for* $k \ge 1$ *the norm estimates*

$$\|L_k f\|_{L^{\infty}(R^3)} \le \frac{c_0}{k^2}\|f\|_{W^1_{\infty}(R^3)}, \quad \|\nabla L_k f\|_{L^{\infty}(R^3)} \le \frac{c_0}{k}\|f\|_{W^1_{\infty}(R^3)}. \tag{15}$$

Proof Applying (13) one can obtain (see (14))

$$|\tilde{u}_0(x)| \le \frac{1}{4\pi k^2}\left(\int_{|y|\le R} \frac{k|\mathbf{W}(y)| + |V(y)|}{|x-y|}\, dy + C \int_{|y|\ge R} \frac{k+1}{|x-y||y|^{\mu}}\, dy\right) \le$$

$$\le \frac{1}{4\pi k}\left(\|\mathbf{W}\|_{L^{\infty}} C_R + C_V + 2CC_{\mu}\right) \le \frac{c_0}{k}, \quad k \ge 1.$$

By similar method it can be proved that

$$\|\nabla \tilde{u}_0\|_{L^{\infty}(R^3)} \le c_0, \quad k \ge 1.$$

This proves 1. Suppose now that $f \in W^1_\infty(R^3)$. Then

$$|L_k f(x)| \le \frac{1}{4\pi k^2}\left(\int\limits_{|y|\le R} \frac{|\nabla f(y)||\mathbf{W}(y)| + |V(y)||f(y)|}{|x-y|}\, dy + CC_\mu \|f\|_{W^1_\infty} \right) \le$$

$$\le \frac{1}{4\pi k^2}\left(\|\nabla f\|_{L^\infty} \int\limits_{|y|\le R} \frac{|\mathbf{W}(y)|}{|x-y|}\, dy + \|f\|_{L^\infty} \int\limits_{|y|\le R} \frac{|V(y)|}{|x-y|}\, dy + CC_\mu \|f\|_{W^1_\infty} \right) \le$$

$$\le \frac{1}{4\pi k^2}\left(\|\nabla f\|_{L^\infty} \|\mathbf{W}\|_{L^\infty} C_R + \|f\|_{L^\infty} C_V + CC_\mu \|f\|_{W^1_\infty} \right).$$

This proves first inequality from (15). The second inequality from (15) can be proved using (13) similarly. Thus, Proposition 1 is completely proved. □

Proposition 2 *Under the same assumptions for* **W** *and* V *as in Proposition 1 there is a constant* $k_0 > 1$ *such that the function* $u_{sc}(x, k, \theta)$ *defined by the series*

$$u_{sc}(x, k, \theta) = \sum_{j=0}^{\infty} L_k^j(\tilde{u}_0)(x, k, \theta) \tag{16}$$

solves integral equation (8) *((6)) uniquely in* $W^1_\infty(R^3)$, *when* $k \ge k_0$. *Moreover,*

$$\|u_{sc}\|_{L^\infty(R^3)} \le \frac{2c_0}{k}, \quad \|\nabla u_{sc}\|_{L^\infty(R^3)} \le 2c_0 \tag{17}$$

uniformly in $\theta \in S^2$, *when* $k \ge k_0$.

Proof The estimates (15) imply that the norm estimate for operator L_k for $k \ge 1$ is

$$\|L_k\|_{W^1_\infty \to W^1_\infty} \le \frac{2c_0}{k}.$$

Since $\tilde{u}_0$ belongs to $W^1_\infty(R^3)$ this estimate in turn implies that

$$\|u_{sc}\|_{W^1_\infty} \le \sum_{j=0}^{\infty} \left(\frac{2c_0}{k}\right)^j \|\tilde{u}_0\|_{W^1_\infty}.$$

We may choose any $k_0 > max\{1, 2c_0\}$ to conclude that the series (16) converges in $W^1_\infty(R^3)$. Because the operator L_k is linear and maps continuously in $W^1_\infty(R^3)$ the series (16) solves (8). Choosing now $k_0 > max\{1, 4c_0\}$ one can easily obtain (17). Uniqueness of solution follows from the contraction condition of L_k. Proposition 2 is completely proved. □

Concerning the kernel G_p (see integral equation (11)) of the integral operator $(H_4 - k^4 - i0)^{-1}$ one can prove the following result.

Proposition 3 *Under the same assumptions for* $\mathbf{W}$ *and* V *as in Proposition 1 there is a constant* $k_0 > 1$ *such that the function* $G_p(x, y, k)$ *can be defined by the series*

$$G_p(x, y, k) = \sum_{j=0}^{\infty} G^{(j)}(x, y, k), \quad G^{(0)} = G_k^+,$$

$$G^{(j)}(x, y, k) := -\int_{R^3} G_k^+(|x-z|)(\mathbf{W}(z) \cdot \nabla G^{(j-1)}(z, y, k) + V(z)G^{(j-1)}(y, z, k))\, dz \tag{18}$$

which solves integral equation (11) *uniquely, when* $k \geq k_0$. *Moreover,*

$$|G_p(x, y, k) - G_k^+(x, y, k)| \leq \frac{\tilde{c}_0}{4\pi^2 k^3 |x-y|},$$

$$|\nabla G_p(x, y, k) - \nabla G_k^+(x, y, k)| \leq \frac{\tilde{c}_0}{2\pi^2 k^2 |x-y|}, \tag{19}$$

where $\tilde{c}_0 = 2\|\mathbf{W}\|_{L^\infty} C_R + C_V + 3CC_\mu$ *with constants* C_R, C_V, C *and* C_μ *are as in* (14).

Proof To prove (18)–(19) it is needed first to estimate $G^{(1)}$. Indeed, using (13) one can obtain ($k \geq 1$)

$$|G^{(1)}(x, y, k)| \leq \frac{1}{8\pi^2 k^3} \left(\int_{|z| \leq R} \frac{|\mathbf{W}(z)| + \frac{1}{2k}|V(z)|}{|x-z||z-y|} dz + \int_{|z| \geq R} \frac{C + \frac{C}{2k}}{|x-z||z-y||z|^\mu} dz \right)$$

with constant C from the condition (3). Considering now two cases: $|x-z| \leq |z-y|$ and $|x-z| \geq |z-y|$ and taking into account conditions (2), one can obtain ($k \geq 1$)

$$|G^{(1)}(x, y, k)| \leq \frac{1}{8\pi^2 k^3} \left(\frac{2\|\mathbf{W}\|_{L^\infty} C_R}{|x-y|} + \frac{C_V}{|x-y|} + \frac{3CC_\mu}{|x-y|} \right) \leq$$

$$\leq \frac{\tilde{c}_0}{8\pi^2 k^3 |x-y|}, \tag{20}$$

where the constant $\tilde{c}_0$ is as above. Similarly one can obtain

$$|\nabla_x G^{(1)}(x, y, k)| \leq \frac{\tilde{c}_0}{4\pi^2 k^2 |x-y|}. \tag{21}$$

We show now that for $k \geq 1$

$$|G^{(j)}(x, y, k)| \leq \frac{\tilde{c}_0^j}{2k(2\pi k)^{j+1}} \frac{1}{|x-y|}, \quad j = 1, 2, \ldots, \tag{22}$$

and

$$|\nabla_x G^{(j)}(x, y, k)| \leq \frac{\tilde{c}_0^j}{(2\pi k)^{j+1}} \frac{1}{|x-y|}, \quad j = 1, 2, \ldots. \tag{23}$$

By (20) and (21) the claim holds for $j = 1$. Suppose that the claim is proved for $j \geq 1$. The induction hypothesis leads to

$$|G^{(j+1)}(x, y, k)| \leq \frac{\tilde{c}_0^j}{4\pi k^2(2\pi k)^{j+1}} \left(\int\limits_{|z|\leq R} \frac{|\mathbf{W}(z)| + \frac{1}{2k}|V(z)|}{|x-z|} \frac{1}{|z-y|} dz + \right.$$

$$\left. + \int\limits_{|z|\geq R} \frac{C + \frac{C}{2k}}{|x-z||z|^{\mu}} \frac{1}{|z-y|} dz \right) \leq \frac{\tilde{c}_0^j}{4\pi k^2(2\pi k)^{j+1}} \left(\frac{2\|\mathbf{W}\|_{L^\infty} C_R}{|x-y|} + \right.$$

$$\left. + \frac{C_V}{|x-y|} + \frac{3CC_{\mu}}{|x-y|} \right) = \frac{\tilde{c}_0^j}{4\pi k^2(2\pi k)^{j+1}} \frac{\tilde{c}_0}{|x-y|}.$$

This finishes the proof of (22) by induction. The estimate (23) can be obtained similarly (by induction). Choosing now $k_0 > max\{1, \frac{\tilde{c}_0}{\pi}\}$ we obtain the estimates (19). Thus, Proposition 3 is proved. □

Next we study the asymptotic behaviour of the scattering solutions u that provides the scattering data for the inverse scattering problems in the foregoing section.

Proposition 4 *Assume that* $\mathbf{W}$ *is bounded,* V *belongs to the Kato space* K_3 *and both satisfy condition* (3). *Then for fixed* $k \geq k_0$ *the solution* $u(x, k, \theta)$ *to* (6) ((8)) *admits the representation*

$$u(x, k, \theta) = e^{ikx\cdot\theta} - \frac{1}{8\pi} \frac{e^{ik|x|}}{k^2|x|} A(k, \theta', \theta) + o\left(\frac{1}{|x|}\right), \quad |x| \to \infty,$$

The function $A(k, \theta, \theta')$ *is called the scattering amplitude and is defined by Eq.* (9).

Proof Since

$$u_{sc}(y,k,\theta) = -\int_{R^3} G_k^+(|x-y|)(\mathbf{W}(y)\cdot\nabla u(y) + V(y)u(y))\,dy =$$

$$= -\int_{|y|\le|x|^a} G_k^+(|x-y|)(\mathbf{W}(y)\cdot\nabla u(y) + V(y)u(y))\,dy -$$

$$- \int_{|y|\ge|x|^a} G_k^+(|x-y|)(\mathbf{W}(y)\cdot\nabla u(y) + V(y)u(y))\,dy =: I_1 + I_2,$$

where parameter a is chosen such that $0 < a < \frac{1}{2}$. For the integral I_1 we use the following asymptotic

$$G_k^+(|x-y|) = \frac{1}{8\pi k^2|x|}\left(e^{ik|x|}e^{-ik\theta'\cdot y} - e^{-k|x|}e^{k\theta'\cdot y}\right) + O(|x|^{2a-2}).$$

This implies that

$$I_1 = -\frac{1}{8\pi k^2|x|}\int_{|y|\le|x|^a}\left(e^{ik|x|}e^{-ik\theta'\cdot y} - e^{-k|x|}e^{k\theta'\cdot y}\right)(\mathbf{W}\cdot\nabla u + Vu)\;dy + O(|x|^{2a-2}).$$

And this in turn leads as $|x| \to +\infty$ to

$$I_1 = -\frac{e^{ik|x|}}{8\pi k^2|x|}\int_{R^3} e^{-ik\theta'\cdot y}\,(\mathbf{W}\cdot\nabla u + Vu)\;dy + o(|x|^{-1}). \tag{24}$$

Next we consider the integral I_2 and split the region of integration as

$$|I_2| \le \int_{|y|\ge|x|^a} |G_k^+(|x-y|)||\mathbf{W}(y)\cdot\nabla u(y) + V(y)u(y)|\,dy =$$

$$= \int_{|x|^a\le|y|\le\frac{|x|}{2}} |G_k^+(|x-y|)||\mathbf{W}(y)\cdot\nabla u(y) + V(y)u(y)|\,dy +$$

$$+ \int_{|y|\ge\frac{|x|}{2}} |G_k^+(|x-y|)||\mathbf{W}(y)\cdot\nabla u(y) + V(y)u(y)|\,dy =: J_1 + J_2.$$

This finishes the proof of Corollary 1. □

Proof We prove now Theorem 2 (Reconstruction). Based on Proposition 3, one can represent (see (11) and (18))

$$G_k^+(|x-y|)-G_p(x,y,k)=\int_{R^3} G_k^+(|x-z|)[\mathbf{W}(z)\cdot\nabla_z G_p(z,y,k)+V(z)G_p(z,y,k)]\,dz=$$

$$=\int_{R^3} G_k^+(|x-z|)[\mathbf{W}(z)\cdot\nabla_z G_k^+(|z-y|)+V(z)G_k^+(|z-y|)]\,dz+$$

$$+\int_{R^3} G_k^+(|x-z|)[\mathbf{W}(z)\cdot\nabla_z\sum_{j=1}^{\infty}G^{(j)}(z,y,k)+V(z)\sum_{j=1}^{\infty}G^{(j)}(z,y,k)]\,dz:=J_1+J_2,$$

where $k\geq k_0$. We first consider J_1 and divide it into to parts J_1' and J_1'' w.r.t. $|z|<k$ and $|z|>k$, respectively for fixed k big enough. Using conditions (3) and estimates (13) the value J_1'' can be estimated as

$$|J_1''|\leq\frac{c}{k^3}\int_{|z|>k}\frac{1}{|x-z||z-y||z|^{\mu}}\,dz, \tag{26}$$

where constant $c>0$ is independent on x,y,k. Since $\mu>4$ then the latter integral is $o\left(\frac{1}{k^4|x||y|}\right)$ uniformly in k,x,y big enough. For estimation of J_1' we assume that $|x|>k^{4+s}$, $|y|>k^{4+s}$ with $s>0$ and with k is big enough. In this case

$$ik(z-y)=-|y|ik\theta'+ikz,\quad ik|x-z|=ik|x|-ik\theta\cdot z+O(k^{-1-s}),$$

where $\theta=\frac{x}{|x|},\theta'=\frac{y}{|y|}$. These representations lead to the following asymptotic behaviour

$$G_k^+(|x-z|)=\frac{e^{ik|x|}e^{-ik\theta\cdot z}}{8\pi k^2|x|}\left(1+\frac{O(1)}{k^{1+s}}\right),\quad |z|<k,$$

$$\nabla_z G_k^+(|y-z|)=\frac{e^{ik|y|}e^{-ik\theta'\cdot z}}{8\pi k^2|y|}\left(1+\frac{O(1)}{k^{1+s}}\right)\left(-ik\theta'+\frac{ikz}{|y|}\right),\quad |z|<k.$$

This allows to obtain that

$$J_1'=-\frac{e^{ik(|x|+|y|)}}{64\pi^2k^4|x||y|}\int_{|z|<k}(ik\mathbf{W}(z)\cdot\theta'-V(z))e^{-ik(\theta+\theta')\cdot z}\,dz+O\left(\frac{1}{k^{4+s}|x||y|}\right). \tag{27}$$

The next step is: symmetric (w.r.t. θ and θ') form of the latter integral

$$\int\limits_{|z|<k} ik\mathbf{W}(z)\cdot\theta' e^{-ik(\theta+\theta')\cdot z}\,dz$$

leads to the fact that

$$\int\limits_{|z|<k} ik\mathbf{W}(z)\cdot\theta' e^{-ik(\theta+\theta')\cdot z}\,dz = \frac{1}{2}\int\limits_{|z|<k} ik\mathbf{W}(z)\cdot(\theta'+\theta')e^{-ik(\theta+\theta')\cdot z}\,dz.$$

Hence, the first term of the sum (27) can be represented as

$$-\frac{e^{ik(|x|+|y|)}}{64\pi^2k^4|x||y|}\int\limits_{|z|<k}\left(\frac{ik}{2}\mathbf{W}(z)\cdot(\theta+\theta')-V(z)\right)e^{-ik(\theta+\theta')\cdot z}\,dz. \tag{28}$$

For estimation of the term J_2 one can use the estimates (19), (22), (23) and conditions (2) and (3). Then using quite similar technique (with more careful considerations of $G^{(2)}$) as for the proof of the estimates (27) and (28) one can obtain that

$$J_2 = o\left(\frac{1}{k^4|x||y|}\right),\quad k, |x|, |y| \to +\infty. \tag{29}$$

Letting now $k, |x|, |y| \to +\infty$ such that fixed $\xi = k\left(\frac{x}{|x|}+\frac{y}{|y|}\right)$ and combining now (26) - (29) we obtain formula (12). Thus, Theorem 2 is completely proved. □

Corollary 2 can be proved similarly as the proof of Corollary 1.

4 Conclusions

The direct scattering problems for a first order perturbation of the three-dimensional biharmonic operator with singular coefficients ($\mathbf{W}$ belongs to L^∞_{loc} and V belongs to the Kato space $K_{3,loc}$) was studied. It was shown that a solution to scattering problem with certain radiation conditions satisfies the integral Lippmann-Schwinger equation. The same was shown for the Green's function of the perturbed biharmonic operator with these coefficients. These integral equations have the unique solutions in the usual Sobolev spaces $W^1_\infty(R^3)$. The asymptotic behaviour of the scattering solutions for fixed $k > 0$ as $|x| \to +\infty$ was studied and a formula for the scattering amplitude was obtained. Similar asymptotic is obtained for the Green's function when $k, |x|, |y| \to +\infty$.

The main results of this paper, Saito's formula and formula for the Green's functions of the bi-Laplacian and the perturbed bi-Laplacian, were proved under very general assumptions on the coefficients. The proof of these formulas itself was based on explicit calculations starting from the formula for scattering amplitude and from the integral equation for the Green's functions. Some consequences of these results were discussed w.r.t. inverse scattering problems. Namely, the scattering amplitude uniquely determines a combination of the coefficients for the direct problem and in turn gives a uniqueness result for the inverse problem. And the behaviour of the Green's function of the perturbed bi-Laplacian for x, y, k large enough uniquely determines the same combination of the unknown coefficients.

Acknowledgments This work was supported by the Academy of Finland (application no 312123, Finnish Programme for Centres of Excellence in Inverse Modelling and Imaging 2018-2025).

References

1. Aktosun, T., Papanicolaou, V.: Time-evolution of the scattering data for a fourth-order linear differential operator. Inverse Prob. **24**, 055013 (2008)
2. Gazola, F., Grumau, H.C., Sweers, G.: Polyharmonic Boundary Value Problems. Springer, Berlin (2010)
3. Harju, M.: On the direct and inverse scattering problems for a nonlinear three-dimensional schrödinger equation. Ph.D. Thesis, University of Oulu (2010)
4. Iwasaki, K.: Scattering theory for the 4th order differential operators: I. Jpn. J. Math. **14**, 1–57 (1988)
5. Iwasaki, K.: Scattering theory for the 4th order differential operators: II. Jpn. J. Math. **14**, 59–96 (1988)
6. Krupchyk, K., Lassas, M., Uhlmann, G.: Inverse boundary value problems for the perturbed poly-harmonic operator. Trans. Am. Math. Soc. **36**, 95–112 (2014)
7. Lebedev, N.: Special Functions and Their Applications. Dover Publications, New York (1972)
8. Päivärinta, L., Serov, V.: Recovery of singularities of a multidimensional scattering potential. SIAM J. Math. Anal. **29**, 697–711 (1998)
9. Pausader, B.: Scattering for the defocusing beam equation in low dimensions. Indiana Univ. Math. J. **58**, 791–822 (2010)
10. Razborov, A., Sagyndykov, M., Serov, V.: Some inverse problems for the schrödinger operator with kato potential. Inverse Ill-posed Prob. **10**, 395–411 (2002)
11. Saito, Y.: Some properties of the scattering amplitude and the inverse scattering problem. Osaka J. Math. **19**, 527–547 (1982)
12. Serov, V.: Fourier Series, Fourier Transform and Their Applications to Mathematical Physics. Springer, New York (2017)
13. Sun, Z.: An inverse boundary value problem for Schrödinger operators with vector potentials. Trans. AMS **338**, 953–969 (1993)
14. Sylvester, J., Uhlmann, G.: A global uniqueness theorem for an inverse boundary value problem. Ann. Math. **125**, 153–169 (1987)
15. Tyni, T., Serov, V.: Scattering problems for perturbations of the multidimensional biharmonic operator. Inverse Prob. Imag. **12**, 205–227 (2018)

Correction to: Local Well-Posedness for the Nonlinear Schrödinger Equation in the Intersection of Modulation Spaces $M^s_{p,q}(\mathbb{R}^d) \cap M_{\infty,1}(\mathbb{R}^d)$

Leonid Chaichenets, Dirk Hundertmark, Peer Christian Kunstmann, and Nikolaos Pattakos

Correction to:
Chapter 6 in: W. Dörfler et al. (eds.), *Mathematics of Wave Phenomena*, Trends in Mathematics https://doi.org/10.1007/978-3-030-47174-3_6

The original version of the chapter was inadvertently published with an error in the abstract. On line 6 of the abstract, the name 'Okoudjou' was misspelled as 'Oboudjou'. The chapter has now been corrected and approved by the author.

The updated online version of this chapter can be found at
https://doi.org/10.1007/978-3-030-47174-3_6

W. Dörfler et al. (eds.), *Mathematics of Wave Phenomena*, Trends in Mathematics,
https://doi.org/10.1007/978-3-030-47174-3_20

The manufacturer's authorised representative in the EU is Springer Nature Customer Service Centre GmbH, Europaplatz 3, 69115 Heidelberg, Germany. If you have any concerns regarding our products, please contact ProductSafety@springernature.com

Printed and bound by CPI Group (UK) Ltd, Croydon, CR0 4YY
30/09/2026
02225492-0017